T0321139

LIQUID METAL MAGNETOHYDRODYNAMICS

MECHANICS OF FLUIDS AND TRANSPORT PROCESSES

Editors: R. J. Moreau and G. Æ. Oravas

Volume 10

For a list of the volumes in this series see final page

Liquid Metal
Magnetohydrodynamics

edited by

J. LIELPETERIS

Institute of Physics, Latvian S.S.R. Academy of Sciences,
Riga, Salaspils, U.S.S.R.

and

R. MOREAU

Laboratoire MADYLAM, Institut Nationale Polytechnique,
Grenoble, France

KLUWER ACADEMIC PUBLISHERS
DORDRECHT / BOSTON / LONDON

Library of Congress Cataloging in Publication Data

Liquid metal magnetohydrodynamics / edited by J. Lielpeteris and R.
  Moreau.
      p.   cm. -- (Mechanics of fluids and transport processes ; v.
  10)
    ISBN 0-7923-0344-X
    1. Magnetohydrodynamic generators--Congresses. 2. Liquid metals-
  -Congresses. 3. Magnetohydrodynamics--Congresses.   I. Lielpeteris,
  J. (Janis) II. Moreau, René J. III. Series.
  TK2970.L57 1989
  538'.6--dc20   89-15309

ISBN 0-7923-0344-X

Published by Kluwer Academic Publishers,
P.O. Box 17, 3300 AA Dordrecht, The Netherlands.

Kluwer Academic Publishers incorporates
the publishing programmes of
D. Reidel, Martinus Nijhoff, Dr W. Junk and MTP Press.

Sold and distributed in the U.S.A. and Canada
by Kluwer Academic Publishers,
101 Philip Drive, Norwell, MA 02061, U.S.A.

In all other countries, sold and distributed
by Kluwer Academic Publishers Group,
P.O. Box 322, 3300 AH Dordrecht, The Netherlands.

printed on acid free paper

PREFACE

Liquid metal MHD is within the scope of two series of international conferences. One is the International Congress on "MHD Power Generation", held every four years, which includes technical and economical aspects as well as scientific questions. The other if the Beer-Sheva Seminar on "MHD Flows and Turbulence", held every three years in Israël. In addition to these well established meetings, an IUTAM Symposium was previously organized in Cambridge (UK) in 1982 on "Metallurgical Applications of MHD" by the late Arthur Shercliff. It was focussed on a very specific subject developing radiply from the middle of the 1970's. The magnetic field was generally AC, including frequencies high enough for the skin-depth to be much smaller than the typical length scale of the liquide pool. And the development of new technologies, or the improvement of existing ones, was the main justification of most of the researches presented and discussed. Only two participants from Eastern countries attended this Symposium.

By the middle of the 1980's we felt that on this very same topic ideas had reached much more maturity than in 1982. We also realized that a line of research on MHD flows related to fusion reactors (tokamaks) was developing significantly, with particular emphasis on flows at large interaction parameter. Also new ideas were arising on liquid metal dynamos, including geodynamos, self-excitation of magnetic field in laboratory experiments and phenomena observed or expected in fast breeders cooled with liquid sodium, as well as in apparently classical domains like stability and turbulence under the influence of electromagnetic effects. However it was obvious that this research effort was divided between two communities (Eastern and Western) of about the same size (more than one hundred researchers in each), which did not communicate or exchange sufficiently. We were convinced that a meeting of these two, approximately equally well represented, which could not have been foreseen before the time of "perestroïka" and "glasnost", would be particularly stimulating. These were the basic motivations of our proposal to IUTAM to hold a Symposium on "Liquid Metal MHD" in Riga and we were particularly happy when it was accepted.

As one can see on the participants list the goal of having the two communities almost equally well represented was well achieved, taking into account the fact that the host country always participates more than the orthers : 55 from Eastern countries, 29 from Western (including India, Israël and Japan). It was nevertheless surprising that only 3 participants came from North of America. Clearly this reflects some decline in the research on liquid metal MHD in these countries, which does not extend to Western Europe, Japan or the Soviet Union, and which has absolutely no scientific justification. On the contrary, liquid metal MHD appeared, during this five day meeting in Riga, as an active and lively subject, able to renew its main themes and to open new fields of application every decade. The following table of contents, with 10 different oral sessions, all very topical, and 2 large poster sessions, reflects this vitality.

Janis LIELPETERIS

René MOREAU

Table of contents

SCIENTIFIC COMMITEE

Professor **J. LIELPETERIS** *(Co-chairman)*
Institute of Physics, Latvian SSR Academy of Sciences, Riga, USSR

Professor **R. MOREAU** *(Co-chairman)*
MADYLAM, ENSHM de Grenoble, FRANCE

Professor **S. ASAI**
Faculty of Engineering, Nagoya University, JAPAN

Professor **F.R. BLOCK**
Institut für Eisenhüttenkunde, RWTH Aachen, FRG

Professor **V. A. GLUKHIKH**
Efremov Research Institute of Electro-Physical Apparatus, Leningrad

Professor **A. JU. ISHLINSKIJ** *(ex officio)*
National Committee of IUTAM, Moscow, USSR

Professor **P.S. LYKOUDIS**
School of Nuclear Engineering, Purdue University, USA

Professor **P. MINCHEV**
Institute of Metal Science and Technology, Sofia, BULGARIA

Professor **H.K MOFFATT**
Department of Applied Mathematics and Theoretical Physics, University of Cambridge, UK

LOCAL ORGANIZING COMMITTEE

J. BIRZVALKS
Institute of Physics,
Latvian SSR Acad. of Sciences, Riga, USSR

Ju. B. KOLESNIKOV *(Secretary)*
Institute of Physics,
Latvian SSR Acad. of Sciences, Riga, USSR

O. LIELAUSIS *(Chairman)*
Institute of Physics,
Latvian SSR Acad. of Sciences, Riga, USSR

J. LIELPETERIS
Institute of Physics,
Latvian SSR Acad. of Sciences, Riga, USSR

G. K. MIKHAJLOV
National IUTAM Committee,
Moscow, USSR

V.P. POLISHCHUK
Institute of Casting Problems,
Ukrainian SSR Acad. of Sciences, Kiev, USSR

G. ROTBERGS
Institute of Physics,
Latvian SSR Acad. of Sciences, Riga, USSR

Ed. V. SHCHERBININ
Institute of Physics,
Latvian SSR Acad. of Sciences, Riga, USSR

A. V. TANANAEV
Polytechnical Institute, Leningrad, USSR

V. G. ZHILIN
Institute of High Temperatures,
USSR Acad. of Sciences, Moscow

LIST OF PARTICIPANTS

Bulgaria

M.A. DIMITROV

Institute of Metal Science and Technology, Chapaeva 53, Sofia 74

P. MINCHEV

Institute of Metal Science and Technology, Chapaeva 53, Sofia 74

Canada

G.J. BENDZSAK

*Hatch Associates Ltd.
Toronto*

India

A.S. GUPTA

*Department of Mathematics
Indian Institute of Technology
Kharagpur 721302*

P. SATYAMURTHY

*Plasma Physics Division, Bhabha
Atomic Research Center,
Bombay, 400085*

Israël

A. EL-BOHER

*Center for MHD studies, Ben Gurion
University, POB 653, Beer-Sheva*

S. SUKORIANSKY

*Center for MHD studies, Ben Gurion
University, POB 653, Beer-Sheva*

France

A. ALEMANY

*Institute de Mécanique de Grenoble,
BP 68, 38402 St Martin d'Hères*

Ph. CAPERAN

*Institute de Mécanique de Grenoble,
BP 68, 38402 St Martin d'Hères*

J. ETAY

*Laboratoire MADYLAM, ENSHM de Grenoble,
BP 95, 38402 St Martin d'Hères*

Y. FAUTRELLE

*Laboratoire MADYLAM, ENSHM de Grenoble,
BP 95, 38402 St Martin d'Hères*

Ph. MARTY

Institute de Mécanique de Grenoble,
BP 68, 38402 St Martin d'Hères

R. MOREAU

Laboratoire MADYLAM, ENSHM de Grenoble,
BP 95, 38402 St Martin d'Hères

J. SOMMERIA

Laboratoire MADYLAM, ENSHM de Grenoble,
BP 95, 38402 St Martin d'Hères

P. TABELING

Groupe de Physique des Solides, Ecole
Normale Supérieure, 24, rue Lhomond,
75231 Paris

Ch. VIVES

Lab. de MHD, Université d'Avignon,
33, rue Pasteur,
84000 Avignon

F. WERKOFF

Centre d'Etudes Nuclèaires de Grenoble,
85 X, Avenue des Martyrs,
38401 Grenoble

FRG

L. BARLEON

Kernforschungszentrum,
GMBH Postfach 3640, 7500 Karlsuhe

F.R. BLOCK

Institut für Eisenhüttenkunde RWTH
Aachen, Intzestrasse 1, D-5100, Aachen

A. MUEHLBAUER

Institute of Electroheat, University of
Hannover D-3000 Hannover 1

GDR

G. GERBETH

Zentralinstitut für Kernforschung
8051 Dresden

D. HAMANN

Zentralinstitut für Kernforschung
8051 Dresden

A. THESS

Zentralinstitut für Kernforschung
8051 Dresden

Hungary

T. BARTHA

Light Metal Works, H-8005
Szekesfehervar, POB 1,
Verfeci Str. 1-15

M. STEIN

Light Metal Works, H-8005
Szekesfehervar, POB 1,
Verfeci Str. 1-15

Japan

S. ASAI   *Department of Iron and Steel Eng.,*
                        Chikusaku, Nagoya

K. MIYAZAKI   *Department of Nuclear Eng.*
                        University, 2–1 Yamadaoka Suita,
                        Osaka 565

S. OSHIMA   *Tokyo Institute of Technology, Okayama,*
                        2-12-1, Meguro-Ku, Tokyo

E. TAKEUCHI   *Nippon Steel Corp., R&D Lab., Edamitsu*
                        Yawata, Higashi-Ku,
                        Kitakushu- City 805

Switzerland

M. V. ROMERIO   *Département de Mathématiques*
                        Ecole Polyt. Féd., Lausanne

United Kingdom

A.D. GILBERT   *D.A.M.T.P., Cambridge*

R. KENNY   *D.A.M.T.P., Cambridge*

D.C. LILLICARP   *Electricity Council Research Center,*
                        Capenhurst, Chester

A.J. MESTEL   *D.A.M.T.P., Cambridge*

H.K. MOFFATT   *D.A.M.T.P., Cambridge*

USA

C. REED   *Argonne Nat. Lab., Argonne*

M. TILLACK   *Mech., Aerospace & Nucl. Eng. Dept.,*
                        School of Eng. & Appl. Sc.
                        Los Angeles, CA. 90024

USSR

T.N. AITOV   *Polytechn. Inst., Leningrad*

M. Ja. ANTIMIROV   *Polytechn. Inst., Riga*

A. B. BEREZIN   *USSR Acad. Sc., Moscow*

J. BIRZVALKS   *Inst. Phys., Latvian SSR,*
                        Riga, Salaspils

V. BOJAREVICS	*Inst. Phys., Latvian SSR, Riga, Salaspils*
L.L. BULYGIN	*Latv. State Univ., Riga*
Ju. D. CHASHECHKIN	*Inst. Mech. Probl., Moscow*
V.N. DEMJANENKO	*Res. Works Energy, Moscow*
V.N. FOLIFOROV	*Inst. Phys., Latv. SSR Riga, Salaspils*
A. GAILITIS	*Inst. Phys., Latv. SSR Riga, Salaspils*
Ju. M. GELFGAT	*Inst. Phys., Latv. SSR Riga, Salaspils*
L.A. GORBUNOV	*Inst. Phys., Latv. SSR Riga, Salaspils*
Ju. M. GORISLAVETS	*Inst. Electrodyn., Ukr. SSR Kiev*
V.S. GOROVIC	*Inst. Phys., Latv. SSR Riga, Salaspils*
R.K. GORN	*Inst. Cast. Probl., Ukr. SSR Kiev*
A. JAKOVICS	*Inst. Model. Energetics, Kiev*
B.G. KARASEV	*Efremov Inst. Electrophys. Appar. Leningrad*
H. KALIS	*Latv. State Univ., Riga*
I. V. KAZACHKOV	*Inst. Electrodyn., Ukr. SSR Kiev*
I.R. KIRILLOV	*Efremov Inst. Electrophys. Appart., Leningrad*
I.M. KIRKO	*Ural Sc. Center, Perm*
B.A. KOBELEV	*Uralmash, Sverdlovsk*
A.F. KOLESNICHENKO	*Inst. Electrodyn., Ukr. SSR Kiev*
Ju. B. KOLESNIKOV	*Inst. Phys., Latv. SSR Riga, Salaspils*
V.G. KRAVCOV	*USSR Alum. Magn. Inst., Leningrad*

I.V. LAVRENTJEV	*Efremov Inst. Electrophys. Appar., Leningrad*
V.B. LEVIN	*Phys. Techn. Inst., Moscow*
O. LIELAUSIS	*Inst. Phys. Latv. SSR Riga, Salaspils*
J. LIELPETERIS	*Inst. Phys. Latv. SSR Riga, Salaspils*
N.V. LYSAK	*Inst. Electrodyn., Ukr. SSR Kiev*
A. MIKELSONS	*Polytechn. Inst., Riga*
J. MIKELSONS	*Latv. State Univ., Riga*
E.V. MURAVJEV	*Inst. Atomic Energy, Moscow*
V.M. NISOVSKIH	*Uralmash, Sverdlovsk*
I. PLATNIEKS	*Inst. Phys. Latv. SSR Riga, Salaspils*
N.R. POLISHCHUK	*Inst. Cast. Probl., Ukr. SSR Kiev*
V.P. POLISHCHUK	*Inst. Cast. Probl., Ukr. SSR Kiev*
M. PUKIS	*Inst. Phys., Latv. SSR Riga, Salaspils*
K.A. SERGEEV	*Inst. Heat Phys., Novosibirsk*
Ed. V. SHCHERBININ	*Inst. Phys. Latv. SSR Riga, Salaspils*
H.A. TIISMUS	*Polytechn. Inst., Tallin*
L. ULMANIS	*Inst. Phys, Latv. SSR Riga, Salaspils*
V. O. VODJANJUK	*Inst. Electrodyn., Ukr. SSR Kiev*
J.G. ZHILIN	*High Temp. Inst., Moscow*
A.S. ZILBERKLEIT	*Joffe Phys. Techn. Inst., Leningrad*
V. D. ZIMIN	*Inst. Mech. Continua, Perm*
N. Ju. KOLPAKOV	*Inst. Mech. Continua, Perm*

Session A:
Large Interaction Parameter

LIQUID METAL IN A STRONG MAGNETIC FIELD

O.Lielausis
Institute of Physics
Latvian SSR Academy of Sciences
229021 Riga, Salaspils, USSR

ABSTRACT. The study of MHD phenomena in the presence of strong magnetic fields has received a particular attention because of possible
application of liquid metals to projected thermonuclear reactors. A
wide range of phenomena relevant to various types of flows has evoked
a great interest. One should consider both channel flows characteristic
to blanket and flows in the form of films, jets or drops applicable to
devices inside the vacuum chamber for plasma impurity control and first
wall protection. The first experiment of introducing liquid metal into
the discharge chamber of an operating Tokamak has been presented.
Attention is paid to the description of experiments where superconducting magnets are used. Preliminary testing of different LM technologies
under conditions when a dream about high-temperature superconductivity
starts turning into reality is another factor evoking an interest in
MHD phenomena in strong fields.

1. INTRODUCTION

    A new branch in the development of liquid metal (LM) magnetohydro-
dynamics (MHD) characterized by extremely high values of applied magnetic fields is taking shape. The expected applications are related to
two global problems. First of all, there all grounds to ask if it is
not already high time to start preparing for the situation when high-
temperature superconductivity will be turned into technical reality.
Further, there are already very definite proposals for the employment
of LM in the devices with magnetically confined plasma, including
fusion reactors (FR). Just these last applications are used to demonstrate the practical importance of the new branch. The main aim is to
show that the interest in a wide range of MHD phenomena has greatly
increased due to these proposed applications. Attention is paid to the
experiments on superconducting magnets. This new type of MHD devices
is becoming important for effective work in the new direction.

J. Lielpeteris and R. Moreau (eds.), Liquid Metal Magnetohydrodynamics, 3–12.
© 1989 by Kluwer Academic Publishers.

2. LIQUID METAL IN THE SYSTEMS OF PROJECTED FUSION REACTORS

Let us remember the main ideas of LM applications in FR. It should be noted that about 80% of the fusion energy is released in the form of high-energy (14 MeV) neutrons. Blanket surrounding the plasma has to fulfil three essential functions - to absorb the energy of the neutron flux, to breed tritium to maintain a fuel supply (since tritium is not a naturally occuring isotope), and to transfer the heat to the external energy conversion system. All these functions may be provided by a blanket containing LM (Li or Li-Pb eutectics). The major disadvantages (and at the same time the role of MHD) are determined by the need to place the blanket inside the windings, i.e. in a strong (5 ... 12 T) magnetic field. The issue of LM employment in blanket, including MHD effects, are subjects of many articles and reviews.

Much fewer references have been dedicated to another prospect for LM application in FR - in the devices for plasma impurity control and first wall protection. A 3.5 MeV α-particle is also a product of the fusion of deuterium and tritium. After slowing down and heating the plasma, α-particles form a helium "ash" to be removed from the plasma. The removal of these particles, as well as the corresponding α-power, should be accomplished without producing a large impurity content, without damaging the first wall elements. Coping with all the problems, connected with the plasma first wall interactions, with the introduction of external energy for additional plasma heating, let us consider some resulting energetic relations. First of all, it should be mentioned that the heat/particle fluxes to be removed are considerable. For example, the blanket of the well known INTOR reactor, is characterized by a power release of 650 MWt. The INTOR impurity control system has to remove both - the heat (125 MWt) and particles (2×10^{20} He atoms/s). The neutron power (the main component for the blanket) is released in the volume of the lithium containing absorber. The related rate of mean volumetric heat production is rather low, ca. 3 MWt/m³. On the contrary the heat/particle fluxes, guided by the impurity control system, must be received by flat elements, the so called collector plates or contact devices. The problem is that due to some physical or technical reasons the possibility of developing the areas of these elements is restricted. As a result, R & D for high heat/particle flux materials and components is one of the most important technological tasks together with the development of superconducting coils and methods of additional plasma heating. The attempts in designing an efficient contact device (CD), that is a wall component contacting plasma have led to the idea of LM employment to form, in some way, a continuously renewed working surface. The following consideration is mainly based on [1,18].

Liquid metal in the blanket. The very first proposals for LM applications to the FR blankets already cautioned against difficulties connected with MHD. Starting with the late 60-ies the understanding of MHD allowed to advance a lot of positive ideas - to isolate the walls, to match the shape of the channel with the characteristics of the field, to apply an external current for conduction pumping or for the elimination of induced emf, etc. The possibility of transfering an adequate amount of heat by means of heat tubes or via a layer of boiling LM was confirmed by estimates. Experiments carried out are worth mentioning even today. So, in [2] the capacity of a conduction pumping system to work in the presence of a strong magnetic field (B<4T; Ha<22,000) was demonstrated. The measured values of the pressure looses in a closed loop (Fig.1) roughly agreed with predictions based on certain simplifying assumptions. In Fig.2 from [3] it was shown that the temperature differences in a potassium boiler practically do not depend even on a strong (B < 5T) magnetic field. Conclusion that a strong magnetic field has little effect on nucleate boiling of LM was made in [3].

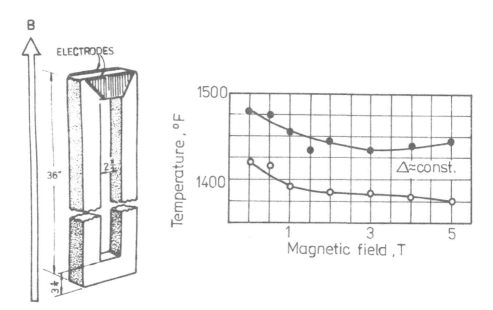

Figure 1. NaK loop designed as a one-sixth scale model of a "racetrack" for a full-scale ORMAK.

Figure 2. Upper and lower limits for the potassium boiler temperature.

It should be mentioned that even nowadays one has to take into account very different concepts by MHD considerations. For example, let us consider the characteristic velocities in different proposals. The mean velocity for a typical self-cooled Li blanket is rather low, ca. 3 cm/s. One can go further and relieve the LM of the coolant function, thus making the LM practically stagnant. At the same time, proposals exist where the velocities of ca. 3 m/s are required – to cool down the fusion zone in a "fast" hibrid blanket [4]. The introduction of heavy lead-type metals makes the range of MHD parameters especially interesting. For a self-cooled Li blanket the values of both the Hartman number and the MHD interaction parameter are very high, i.e. Ha ca. 10^4 and N ca. 10^4. The flow of a heavy metal with a significant velocity, as mentioned above, has a specific feature originating from the fact that in this case, unlike other FR designs, rather high Reynold's numbers are realized. Even in a strong field transient flow conditions with weak or partial laminarization can be expected. Higher than in the laminar case, heat transfer coefficients can be proposed, and this is how to compensate for the worse thermal properties of the coolant as compared with liquid Li. Doubts are expressed [5] that even in the case of a pure Li blanket a fully laminarized flow will be unable to serve for an adequate cooling of the first wall. Here we are touching upon an interesting question - the problem of maintaining considerable velocity disturbances in strong fields. A lot has been done here in the contex of MHD turbulence and heat transfer. Experiments confirm the idea that one has the right to speak about velocity perturbations even in the presence of very strong fields. In Fig.3 from [6] one can see that even at 3T slow convective temperature fluctuations remain in Li. According to [6], from the viewpoint of fusion application, it is interesting that the Nusselt number under a strong field of B = 3T is slightly higher than that under B = 0T.

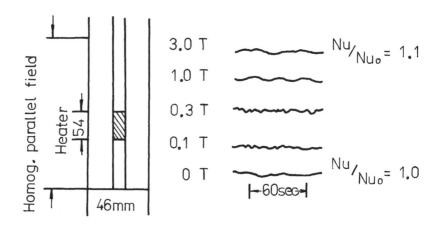

Figure 3. Temperature signals in stagnant liquid lithium heated by a concentrically inserted heater pin.

Nevertheless, the determination of pressure looses in complex LM tubings remains to be the most important task for MHD in the blanket. To be frank, we are rather far from being able to settle this problem. Let us consider (Fig.4) some results from [7] where during the experiment on a Na loop provided with a superconducting magnet values of Ha and N typical for natural conditions were obtained (Ha ca. 10^4; Na ca. 10^4). A simplifying tendency was confirmed. In the presence of a strong field, the coefficient of pressure looses ζ depends practically only on N. Simultaneously a need to improve the accuracy of quantitative estimates were demonstrated. So far the estimates for complex hydraulic elements are based on rather distant analogies. Entrance to or exit out of magnetic gap are mainly used approximations. An example of such estimate, taken from [5], is presented in Fig.4 by line II. The discrepancy between line II and line I (experimental) is considerable. It should be mentioned that values of ζ not for the whole element but for an individual bend are presented.

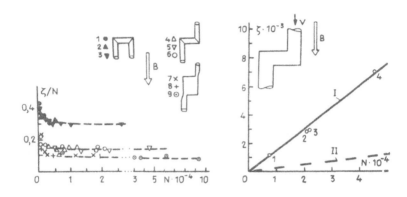

Figure 4. Local pressure losses $\Delta p = \zeta \varrho v^2/2$ in some typical hydraulic elements at high values of Hartman number Ha and MHD interaction parameter N.
a) Ha: 1-5320;2-7950;3-10500;4-8970;5-11900;6-14700;7-9000;
    8-11800;9-14700;
b) Ha: 1-9000;2-15000;3-12000;4-12000.

LM in the system for impurity control and first wall protection.

The use of LM in the control systems enables to design contact devices (CD) with remarkable features - continuously renewed working surface and excellent heat removal capacity. As a basic example the heating of a LM film by a homogeneous flux of heat can be treated [8]. Taking into account that the permissible thermal load for solid CD, limited by erosion and thermal stresses, is usually supposed to be 2 ... 3 MWt/m². one can get that a Li film moving with an acceptable velocity 5 mps is already applicable. At an expected heating length (from 0.5 to 1.0 m) the film with 2 to 3 mm thickness is theoretically capable to carry all the thermal power directed to the divertor. In the case of Ga, the heat accumulation capacity is even higher due to the higher value of $T_{max} - T_{min}$. Above estimates are also valid when the film is substituted by a curtain of jets or droplets. In these cases a problem of providing an opaque curtain arises. This problem can be solved by using multilayer flows.

The idea of employing LM free-surface flows as particle collectors is really exciting. A number of different designs for CD have been presented [1,9,18]. But the problem arises again, that all these flows must be able to work in a strong magnetic field. So, film-type flows have advantages from many viewpoints except the interaction with a magnetic field. It is interesting to remember that a strong influence of a field on a free-surface mercury flow was already described in the Proceedings of the First IUTAM Symposium on MHD [10]. According to the review [11] namely the relation to fusion technology makes the investigation of film-type MHD flows actual today. It is worth noting that new nonstandard MHD interaction may become relevant under these conditions. Thus, a flow of LM (more exactly of Li) in a CD can be looked upon as an example to demonstrate the role of thermocurrents. One has the right to postulate a high drop of temperature along the flow and a low resistance for the circuit shunting the thermo-emf. For both the film and jet-type flow forces, caused by thermocurrents, can even exceed the forces induced by gravitation. It is worth to remember [12] where the general attention was paid to the role of thermocurrents determined by the specific features of Li.

The first suggestions of LMCD [13,14] related to the divertor plates and they based on film or jet-flows. Taking into account their inevitable interaction with the magnetic field the more sophisticated flow patterns, like droplet curtains, were introduced [15,16]. A droplet limiter represents a LM curtain formed by an opaque system of high-velocity drops. By an analogy with an ordinary mechanical limiter it has to scrap off a layer of plasma at its edge. A similar droplet curtain can serve for a CD in the divertor chamber. In principle, the droplet flow can be formed beyond the field coils and then passed through special vacuum ports into the plasma or divertor chamber. Entering the magnetic field the droplets are subjected to electromagnetic forces, resulting in a deceleration, deformation and deviation from the initial trajectory. Nevertheless, the most perspective idea seems to be placing drop generator directly into the chamber, inside the field coils. Hence a number of constraints on the mechanical design as well as the problem of droplet interaction with the nonuniform field can be avoided. But a new problem of R & D for droplet generators capable of working in a strong magnetic field in the direct neighbourhood of plasma arises.

Another important practical question is connected with the design of collecting device for the high-velocity droplet flow. The main problem arises from the spraying of LM. The results of an experimental study [19] devoted to the LM droplet interaction with a LM free surface in the presence of magnetic field are rather interesting here. As it is seen from Fig.5, the magnetic field changes the interaction mode radically and strongly suppresses spraying. Naturally, in the case of film and jet-type flows the above situations remain relevant.

Nevertheless, the main problem still remains whether a contact between the plasma and a LM is permissible at all. An experiment was performed confirming such a possibility, of course, only in the first approximation. Here the author would like to stress that he has the pleasure to present a joint work and the main results are presented in

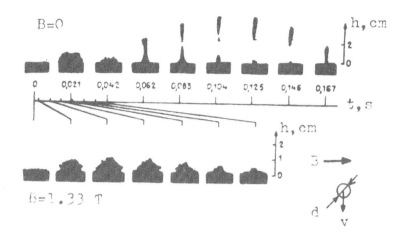

Figure 5. Colision of a InGaSn drop with the surface of the same liquid metal without and in the presence of a magnetic field; v=4.6 m/s; d=3 mm.

[1]. A LM limiter in the form of a jet-drop screen on the Tokamak stand T-3M was installed. The eutectic 67% Ga + 20.5% In + 12.5% Sn with the melting point at 8°C was used. The stand is characterized by the following parameters: the main radius 0.95 m, plasma radius 0.16 m, induction on the axis 1 T, discharge current 40 kA, plasma density $(0.3 \ldots 2) \times 10^{13} cm^{-3}$, discharge time 50 ms, mean heat load on the limiter 1 kWt/cm^2.

Fig.6 shows the lay-out of the LM limiter for T-3M. The loop consists of a supplying main provided with an e.m. pump 1, cooler 2, constant upper and lower level tanks 3 and 4, an e.m. droplet generator 5 [17] and a droplet receiver. The LM screen was formed by 2 ... 4 mm drops, organized in 17 separate jets, located on two lines with a regular distance of 4 mm. The velocity of drops reached 2 ... 5 m/s. In relation to the plasma 6 the droplet screen can be treated as a mechanical limiter, scrapping off a layer of 15 ... 20 mm.

The effect of the LM limiter on the plasma was compared with the influence of an ordinary graphite limiter of the same size (Fig.7).

It was confirmed that in the presence of a contact between the plasma and the LM screen full-scale discharges are quite feasible. During the middle stage of the discharge the plasma parameters for both cases under consideration were similar. The initial stage in the presence of the LM limiter differed in a higher level of emission looses indicating an increased content of impurities. In the presence of the LM limiter there were difficulties with the recurence of the discharges; that shows an additional entry of impurities again.

In principle, a part of these sources could be eliminated. The preliminary purification of the alloy, in particular, from dissolved gases, was insufficient. A presence of undersized droplets exposed to overheating and vaporization was fixed. The applied measures against

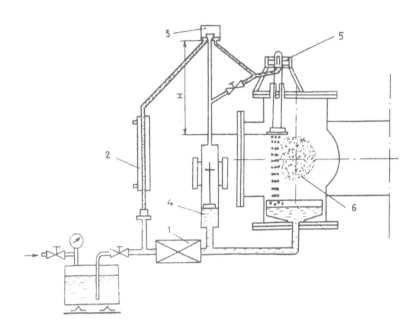

Figure 6. The lay-out of the liquid metal limiter installed on the Tokamak stand T-3M.

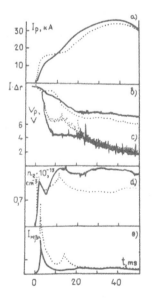

Figure 7. Integral characteristics of the discharge in the presence of a graphite (——) and a LM (———) limiter: a-discharge current; b-readings proportional to horizontal displacement of the plasma ring; c-voltage along the circumference; d-mean density; e-intensity of the spectral line H_β (in relative units).

spraying in the droplet receiver were insufficient as well.

3. DISCUSSION

Employment of LM can have a major impact on the characteristics of such a sophisticated device as fusion reactor. It is clear that the effect will be most marked in the case of a complex application of LM in different systems of the reactor. Blanket represents one of such systems. The above results enable to speak about a real employment of LM in another very important system, i.e. the system for impurity control and first wall protection. It should be mentioned that not only the number of MHD phenomena involved has highly increased due to this application, hence the employment of LM has developed into a real problem of the day, indeed. The blanket belongs to the energy transfer system needed tomorrow. Systems introduced inside the discharge camera should help to solve the general problem - that of plasma confinement.

Nevertheless, the main amount of work, still done, has been devoted to blanket. It would be easier to consider just this element with the aim to give some forecast about the expected development of strong field LM MHD. But one has a possibility to use the excellent review papers, for example [20], here. The importance of moving forward with the development of fusion nuclear technology has been recognized by the main part of the scientific world. The current situation is characterized by a serious shortfall in data and understanding for the most components involved. Complex long-term programs are taking their shape. They include R & D on LM MHD as a very important component. A new generation of experimental facilities is under construction. An effective work on numerics and modelling is to be expected.

There are all grounds to treat the research on MHD phenomena in strong magnetic fields as an interesting, methodically and practically promising branch.

4. REFERENCES

[1] DEM'JANENKO V.N., KARASEV B.G., KOLESNICHENKO A.F. et al., 1988, 'Liquid metal in the magnetic field of a Tokamak reactor', Magnitnaja Gidrodinamika, n.1, 104.

[2] FRAAS A.P., YOUNG F.J., HOLCOMB R.S., 1975, 'Magnetohydrodynamic test of an one-sixth scale model of a CTR recirculating lithium blanket', Ann. Meeting Am.Nucl.Soc. New Orleans, Luisiana, 11 p.

[3] FRAAS A.P., LLOID D.B., MACPHERSON R.E., 1974, 'Effects of a strong magnetic field on boiling of potassium', Oak Ridge Nat.Lab., ORNL TM 4218, 28 p.

[4]MURAV'EV E.V., ORLOV V.V., KHRIPUNOV V.I., 1985, 'Liquid metal cooling of a hybrid Tokamak reactor', Voprosy atomnoj nauki i tekhniki, Ser. 'Termoyadernyi sintez', Iss.4, 24.

[5] ARGONNE NAT.LAB., 1984, 'Blanket comparison and selection study. Final Report', ANL/FPP-84-1.

[6] MIYAZAKI K., YAMASHITA S., YAMAOKA N., 1987, 'Natural convection heat transfer of liquid lithium under transfere and parallel magnetic field', J.Nucl.Sci.Technol., 25(5), 409.

[7] GRINBERG G.K., KAUDZE M.Z., LIELAUSIS O.A., 1985, 'Investigation of local MHD pressure losses on a sodium loop with a superconducting magnet', Magn.Gidrodin., n.1, 121.

[8] LIPOV M.Yu., MURAV'EV E.V., 1980, 'On the development of high-efficient divertor systems for power tokamak reactors', Kurchatov Institute of Atomic Energy, Preprint IAE-3290/8, 31 p,

[9] 'Tokamak concept innovations', 1986, IAEA-TECDOC-373, IAEA, Vienna, 596 p.

[10] ALPHER R.A., HURWITZ H., JOHNSON R.H., 1960, 'Some studies of free-surface mercury magnetohydrodynamics', Rev.Mod.Phys., v.32, n.4, 758.

[11] MURAV'EV E.V., 1988, 'MHD film flows under the conditions of a fusion reactor', Magn.Gidrodin., n.1, 125.

[12] SHERCLIFF J.A., 1979, 'Thermoelectric magnetohydrodynamics', J.Fluid Mech., v.91, n.2, 231.

[13] UWMAC-II, 'A conceptual Tokamak reactor design', Univ. of Wisconsin, 1975, UWFDM-112.

[14] 'Westinghouse compact poloidal divertor reference design', 1977, WEPS-TME-042.

[15] LIPOV M.Yu., MURAV'EV E.V., 1979, 'On the development of high-efficient divertor systems for power Tokamak reactors', Paper presented to the Soviet-American Meeting 'Engineering and Economical Problems of Power Fusion Reactors, Kurchatov IAE, Moscow.

[16] WELLS A.M., 1981, 'A system for handling divertor and energy flux based on lithium droplet cloud', Nucl.Technol./Fusion, n.1,120.

[17] KOLESNICHENKO A.F., 1982, 'Tekhnologicheskie MGD-Ustanovki i Protsessi', Naukova dumka, Kiev, 192 p.

[18] KARASEV B.G., LIELAUSIS O.A., MURAV'EV E.V., TANANAEV A.V., 1987, 'Liquid metals in fusion reactors with magnetic confinement', IAEA-TC-392.3/51, 239.

[19] KAUDZE M.Z.,LIELAUSIS O.A.,1984,'Droplet collision with liquid metal surface in the presence of a magnetic field',Magn.Gidrodin.,n.1,37.

[20] ABDOU M.A., HADID A.H., RAFFRAY A.R. et al., 1988, 'Modelling, analysis and experiments for fusion nuclear technology', Fusion Engineering and Design, 6, 3.

NUMERICAL SOLUTIONS OF THREE-DIMENSTIONAL MHD FLOWS IN STRONG NON-UNIFORM TRANSVERSE MAGNETIC FIELDS*

T. Q. Hua, Argonne National Laboratory,
9700 S. Cass Ave., Argonne, IL 60439 U.S.A.,
J. S. Walker, University of Illinois,
Urbana, IL 61801 U.S.A.

ABSTRACT. Magnetohydrodynamic flows of liquid metals in thin conducting ducts of various geometries in the presence of strong non-uniform transverse magnetic fields are examined. The interaction parameter and Hartmann number are assumed to be large, whereas the magnetic Reynolds number is assumed to be small. Under these assumptions, viscous and inertial effects are confined in very thin boundary layers adjacent to the walls. At walls parallel to the magnetic field lines, as at the side walls of a rectangular duct, the boundary layers (side layers) carry a significant fraction of the volumetric flow rate in the form of high velocity jets. This paper describes the analysis and summarizes the numerical methods for obtaining 3-D solutions (core solutions) for flow parameters outside these layers, without solving explicitly for the layers themselves.

1. Introduction

In a self-cooled liquid-metal blanket of a magnetically confined fusion reactor, the magnetohydrodynamic (MHD) effects are of paramount importance in the design process [1]. The interaction between the circulating liquid metal with the strong magnetic field necessary for plasma confinement results in large electromagnetic body forces which determine the flow distribution of the liquid metal, and produce large MHD pressure gradients. The resulting MHD pressure drop may cause excessive pumping power loss and prohibitively large material stresses. Also, the MHD flow distribution may affect drastically the heat transfer characteristics of the blanket in general and the first wall coolant channels in particular.

In a fusion reactor, the values of the Hartmann number, M, and interaction parameter, N, are typically of the order of $10^3 - 10^5$ [1] so that inertial and viscous effects are confined to thin boundary and shear layers and are negligible outside these layers. Since 1971 [2] and until recently, analysis of 3-D MHD flows in thin-walled conduct-

*Work supported by the U. S. Department of Energy/Office of
Fusion Energy under Contract W-31-109-Eng-38.

13

J. Lielpeteris and R. Moreau (eds.), Liquid Metal Magnetohydrodynamics, 13–19.
© *1989 by Kluwer Academic Publishers.*

ing ducts were based on the thin conducting wall approximation $M^{-1} \ll c \ll 1$ where $c = \sigma_w t/\sigma L$ is the wall conductance ratio. σ_w and σ are the electrical conductivities of the wall and the liquid metal, t is the wall thickness, and L is a transverse characteristic dimension of the duct. Since c is typically 0.01, the thin conducting wall approximation was considered appropriate for fusion reactor blankets. This approximation leads to certain characteristic surfaces in all 3-D MHD flows. The flow requires a large distance, of $0(c^{-1/2})$, to migrate across these surfaces. As a result, the analyses neglect the supposedly small leakage flows across the surfaces by neglecting certain terms containing axial derivative in the core equations.

In 1984 the ALEX (Argonne Liquid Metal Experiment) facility [3] was built at Argonne National Laboratory (ANL) for the purpose of studying liquid-metal MHD flows under fusion relevant conditions in general and providing detailed data for validation of analytical predictions in particular. Values of M and N achieved at ALEX are as high as 6 x 10^3 and 10^5, respectively. Early results pertaining to 3-D MHD flows in a circular thin-walled conducting duct (c = 0.027) in a non-uniform magnetic field revealed that the thin conducting wall approximation provided poor description of the actual flow. Measured details of velocity and electric potential [4] indicated that the error was in the supposedly small leakage flows across the characteristic surfaces neglected in the thin conducting wall approximation. Knowledge gained from the ALEX experiments gives rise to a more realistic approach toward analyzing these flows. The new approach uses the inertialess, inviscid core equations first derived by Kulikovski [5] and simultaneously by Hunt and Ludford [6], coupled with Shercliff's thin-walled boundary conditions [7]. This approach makes no asymptotic assumptions on the value of c and retains the axial derivative terms neglected in the thin conducting wall approximation. A special numerical scheme with a staggered grid along the characteristic surfaces is used and simultaneous integration for pairs of adjacent cells to obtain stable and efficient numerical solutions is employed. This approach has been applied to thin-walled conducting circular duct [8], and rectangular duct [9] of arbitrary conductance ratio. Agreement with the ALEX data is excellent as reported in a companion paper [10]. This paper summarizes the analysis, and presents representative results obtained through the joint analytical efforts at ANL and the University of Illinois.

2. Analysis for Thin-Walled Conducting Ducts

Consider the steady flow of an incompressible liquid metal driven by a pressure gradient along a conducting duct with thin metal walls and with an imposed transverse magnetic field whose strength varies along the duct. The weaker axial magnetic field is neglected in this model because the major electromagnetic body force in the liquid metal arises from the interaction between the fluid flow with the transverse field.

The inertialess, inviscid, dimensionless equations governing the flow of the liquid metal in the core of the flow are:

$$\nabla p = \underset{\sim}{j} \times \underset{\sim}{B}, \quad \underset{\sim}{j} = - \nabla \phi + \underset{\sim}{v} \times \underset{\sim}{B}, \quad \nabla \cdot \underset{\sim}{v} = 0, \quad \nabla \cdot \underset{\sim}{j} = 0 . \qquad \text{(1a,b,c,d)}$$

Here p, $\underset{\sim}{j}$, $\underset{\sim}{v}$, and ϕ are the pressure, electric current density, velocity, and electric potential, normalized by $\sigma U_o B_o^2 L$, $\sigma U_o B_o$, U_o, and $U_o B_o L$, respectively. U_o is the average axial velocity of the fluid, and B_o is a characteristic transverse magnetic flux density. For simplicity, the duct is assumed to be geometrically and electrically symmetric about the midplane perpendicular to the planar magnetic field $\underset{\sim}{B} = B_y (x) \underset{\sim}{y}$. The x, y, z core velocity components u_c, v_c, w_c, and core electric current density components j_{xc}, j_{yc}, j_{zc}, which satisfy the equations (1) and the symmetry conditions, $j_{yc} = v_c = 0$ at $y = 0$, are given by:

$$u_c(x,y,z) = \beta \frac{\partial \phi_c}{\partial z} - \beta^2 \frac{\partial p}{\partial x} , \quad w_c(x,y,z) = - \beta \frac{\partial \phi_c}{\partial x} - \beta^2 \frac{\partial p}{\partial z}$$

$$\text{(2a, 2b) (Cf. 1b)}$$

$$v_c(x,y,z) = -y \, \beta'(x) \frac{\partial \phi_t}{\partial z} + y \frac{\partial}{\partial x} \left[\beta^2 \frac{\partial p}{\partial x}\right] + y \, \beta^2 \frac{\partial^2 p}{\partial z^2}$$

$$+ y \frac{\beta'^2}{2} \frac{\partial}{\partial z} \left[(f^2 - \frac{y^2}{3}) \frac{\partial p}{\partial z}\right] \qquad \text{(2c) (Cf. 1c)}$$

$$j_{xc}(x,y,z) = \beta \frac{\partial p}{\partial z} , \quad j_{zc}(x,y,z) = - \beta \frac{\partial p}{\partial x} \qquad \text{(2d, 2e) (Cf. 1a)}$$

$$j_{yc}(x,y,z) = - y\beta' \frac{\partial p}{\partial z} \qquad \text{(2f), (Cf. 1d)}$$

where $p(x,z)$ is the pressure which is constant along magnetic field lines by virtue of Eq. (1a),
$\beta(x) = B_y^{-1}(x)$, $\beta' = d\beta/dx$, and $y = f(x,z)$ is the liquid metal/wall interface. The electric potential in the core varies along the magnetic field lines according to

$$\phi_c(x,y,z) = \phi_t(x,z) - \frac{1}{2}(f^2 - y^2)\beta' \frac{\partial p}{\partial z} , \qquad \text{(2g)}$$

where ϕ_t (x,z) is the electric potential at the intersection of the magnetic field line and the wall, at $y = f(x,z)$. Equation (2g) is obtained by integrating the y- component of equation (1b), and using equation (2f).

The boundary conditions at the inside surface of a wall which has conductance ratio c_t and is not parallel to the magnetic field lines are

$$\underset{\sim}{v}_c \cdot \hat{n}_t = 0, \quad \left(\underset{\sim}{j}_c - \underset{\sim}{j}_w\right) \cdot \hat{n}_t = 0 \quad \text{at } y = f(x,z) \qquad (3a, 3b)$$

where $\hat{n}_t$ is a unit normal to the wall and $\underset{\sim}{j}_w = -c_t \, \nabla^2 \phi_t$. Conditions (3a, b) neglect the $O(M^{-1})$ jumps in $\underset{\sim}{v}_c$, $\underset{\sim}{j}_c$, and ϕ across the Hartmann layer, which has $O(M^{-1})$ thickness and separates the inviscid core region from the wall. If a wall is parallel to the field lines (side wall), and has conductance ratio c_s, then the appropriate boundary condition is

$$\left(\underset{\sim}{j}_c - \underset{\sim}{j}_w\right) \cdot \hat{n}_s = 0 \qquad (3c)$$

here $\hat{n}_s$ is a unit normal to the side wall, and $\underset{\sim}{j}_w = -c_s \, \nabla^2 \phi_s$. Condition (3c) neglects the $O(M^{-1/2})$ jump in current density across the side wall layer which has $O(M^{-1/2})$ thickness. ϕ_s is the electric potential at the side wall.

The three-dimensional problem with eight variables in the core (p, ϕ, u_c, v_c, w_c, j_{xc}, j_{yc}, j_{zc}) is completely solved once the functions p, ϕ_t, and ϕ_s are determined. The coupled partial differential equations governing p, ϕ_t, and ϕ_s are derived by applying the boundary conditions (3a-3c).

3.   Numerical Methods

The partial differential equations p (x,z) and ϕ_t (x,z) have the forms:

$$\frac{\partial}{\partial x}\left[A_1(x,z)\,\frac{\partial P}{\partial x}\right] + \frac{\partial}{\partial z}\left[A_2(x,z)\,\frac{\partial P}{\partial z}\right] = D\phi_t \qquad (4)$$

$$c_t \, \nabla_w^2 \phi_t = DP \qquad (5)$$

and that governing a side wall electric potential, ϕ_s (x,y), has the form:

$$c_s \ \nabla^2_w \ \phi_s = \pm \ \beta(x) \ \frac{\partial P}{\partial x} \Big|_{\text{side wall}} \quad \text{at } z = \mp 1 \tag{6}$$

where ∇^2_w is the 2-dimensional Laplacian operator at the wall, D is a directional derivative along a characteristic surface described by Holroyd and Walker [11]. Namely, in the xz plane the lines for the directional derivative D are defined by

$$\zeta(x,z) = \int_0^{f(x,z)} \frac{dy}{B_y(x)} = \beta(x) \ f(x,z) = \text{constant}$$

The x-dependence in the functions A_1 (x,z) and A_2 (x,z) involves the variation of the magnetic field, $\beta(x)$ and $\beta'(x)$. The coupled equations (4) and (5) exhibit rather different characteristics for different values of the parameter β'. More particularly, for non-zero values of β' and small values of c_t, the equations alternate between elliptic and parabolic forms. If equations (4) and (5) are solved independently by taking the right-hand-side as known, the truncation error would be $(\Delta z)^2/c_t$ and for $c_t \ll 1$, this error would destroy the solution after just a few iterations. Instead, stable simultaneous solution schemes employing staggered-grid have been developed to solve equations (4-6). The boundary conditions for Eqs. (4-6) in the plane of the duct cross section are derived from among the symmetry conditions, continuity of electric potential and current density, conservation of mass and the irregularity condition. At upstream and downstream the conditions of the flow, such as fully-developed flow, are specified. References [8] and [9] present these schemes, algorithms and the boundary conditions for treating a circular and a rectangular duct, respectively.

4. Analysis for a Nonconducting Circular Duct

In a non-conducting duct, the method of solution is different because the core currents can not enter the wall and must close through the Hartmann layer or within the core itself to complete the circuit. As a result, the $O(M^{-1})$ terms which are neglected in the case of a conducting duct because the wall short-circuits the Hartmann layer, must now be included in the analysis. In other words, solution for the flow in the Hartmann layers must be obtained in order to solve for the core flow. Continuity of velocity, current density and electric potential is used to couple the solutions for the core and the Hartmann layers. Space limitation does not allow detailed presentation of this lengthy and complex analysis. Full presentation of the method and details for solution for a circular duct will be given in a different paper, currently under preparation.

18

5. Summary of Numerical Solutions Capabilities

Numerical solutions have been obtained for flows in non-uniform transverse magnetic fields in conducting ducts of various geometries as well as in non-conducting circular duct. Figure 1 illustrates the geometric configurations treated which include a round duct, a rectangular duct, a duct with arbitrary cross section and a duct of rectangular cross sections whose dimension in the direction parallel to the magnetic field lines changes linearly. Numerical results and comparison with experiments have been presented in references 9, 10, 12, and 13. In general, agreement with experimental data is excellent. The velocity distributions predicted by the MHD solutions are used as input to the energy equation and the temperature in the fluid is calculated in heat transfer modules which are part of the codes.

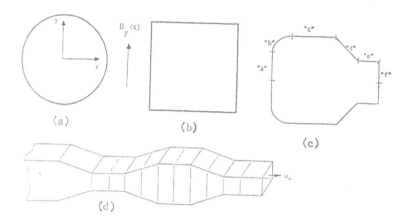

Figure 1. Geometries of ducts (a) circular cross section, conducting and non-conducting ducts; (b) rectangular cross section; (c) arbitrary cross section; and (d) rectangular cross sections with linear expansion or contraction along the direction parallel to the magnetic fields.

References

[1]   SMITH, D. L., et al., 1984, 'Blanket Comparison and Selection Study - Final Report,' Argonne National Laboratory Report, ANL/FPP/84-1.

[2]   HUNT, J. C. R and HANCOX, R., 1971, 'The Use of Liquid Lithium as Coolant in a Toroidal Fusion Reactor, Part 1, Calculation of Pumping Power,' Culham Laboratory, Abingdon, Oxfordshire, England, CLM-R115.

[3]   REED, C. B., PICOLOGLOU, B. F., and DAUZVARDIS, P. V., 1985, 'Experimental Facility for Studying MHD Effects in Liquid-Metal-Cooled Blankets,' Fusion Technology, Vol. 8, Part 2A, 257.

[4]   PICOLOGLOU, B. F., REED, C. B., DAUZVARDIS, P. V. and WALKER, J. S., 1985, 'Experimental and Analytical Investigations of Magnetohydrodynamic Flow near the Entrance to a Strong Magnetic Field,' Fusion Technology, Vol. 10, 860.

[5]   KULIKOVSKI, A. G., 1968, 'On Slow Steady Flows of Conductive Fluid with High Hartmann Number,' Izv.Akad.Nauk SSSR, Mekh.Zhid. i Graza.

[6]   HUNT, J. C. R. and LUDFORD, L. S. S., 1968, 'Three-Dimensional MHD Duct Flows with Strong Transverse Magnetic Fields, Part 1, Obstacles in a Constant Area Channel,' J. Fluid Mech. Vol. 33, Part 4.

[7]   SHERCLIFF, J. A., 1958, 'The Flow of Conducting Fluids in Circular Pipes under Transverse Magnetic Fields,' J. Fluid Mech. Vol. 1.

[8]   TALMAGE, T. and WALKER, J. S., 1987, 'Three-Dimensional Laminer MHD Flow in Ducts with Thin Walls and Strong Magnetic Fields,' Proc. of the 6th Beer-Sheva Seminar.

[9]   HUA, T. Q., WALKER, J. S., PICOLOGLOU, B. F., and REED, C. B., 'Three-Dimensional MHD Flows in Rectangular Ducts of Liquid-Metal-Cooled Blankets,' accepted for publication in Fusion Technology.

[10] PICOLOGLOU, B. F. and REED, C. B., 1988, 'Experimental Investigation of 3-D MHD Flows at High Hartmann Number and Interaction Parameter,' presented at the IUTAM Symp. on Liquid Metal MHD, Riga, USSR.

[11] HOLROYD, J. R. and WALKER, J. S., 1978, 'A Theoretical Study of the Effects of Wall Conductivity, Non-uniform Magnetic Fields and Variable-Area Ducts on Liquid-Metal Flows at High Hartmann Number,' J. Fluid Mech., Vol. 84.

[12] HUA, T. Q., PICOLOGLOU, B. F., REED, C. B., and WALKER, J. S., 1988, 'MHD Thermal Hydraulic Analysis of Three-Dimensional Liquid Metal Flows in Fusion Blanket Ducts,' Int. Symp. on Fusion Nuclear Technology, Tokyo, Japan.

[13] PICOLOGLOU, B. F., REED, C. B., HUA, T. Q., WALKER, J. S., BARLEON, L., and KREUZINGER, H., 1988, 'MHD Flow Tailoring in First Wall Coolant Channels of Self-Cooled Blankets,' Int. Symp. on Fusion Nuclear Technology, Tokyo, Japan.

MHD-FLOWS AT HIGH R_m, N AND Ha

I.V.Lavrent'ev
Efremow Institute of Electrophysical Apparatus
189631 Leningrad
USSR

ABSTRACT.On the basis of MHD flow in a rectangular cross-section channel the influence of N. Ha and R_m, conductivities of the walls and channel geometry on the flow character has been analysed. A transition to the Stokes flow has been considered. An analogy has been traced between the electrodynamic processes at high values of Ha and Rm.

MHD-flows of liquid metals (LM) are described by magnetohydrodynamic equations: the motion of LM by the Navier-Stokes equation and electromagnetic fields by the Maxwell equations and Ohm's law. As seen from the equation written for LM steady flow in dimensionless form

$$N^{-1}(\vec{V}\nabla)\vec{V}=-\nabla(p/N)+Ha^{-2}\nabla^2\vec{V}+\vec{j}\times(\vec{b}_e+R_m\vec{b}_j) \tag{1}$$

MHD flow is determined by the three dimensionless similarity parameters: Hartmann number $Ha=B_0L(\sigma/\eta)^{1/2}$ (B_0 and L are induction and length scales), MHD interaction parameter $N=\sigma B_0^2L/\rho u_0$ (u_0 — characteristic velocity) and the magnetic Reynolds number $R_m=\mu_0\sigma u_0L$, characterizing the effect of magnetic fields induced by liquid motion upon resultant field and thus upon MHD processes in a duct. If N is small, electromagnetic forces do not effect on the flow and hydro- and electrodynamics equations can be solved independently, i.e. in the so called electrodynamic approximation. In practice the Hartmann number and MHD interaction parameter are large enough and electromagnetic force, markedly exceeding inertial and viscous forces in the flow core excluding the thin boundary layers, mainly determines the flow local and integral characteristics weakly depending on the form in which viscous terms are presented. If induced magnetic fields can be neglected ($R_m \ll 1$), the force linearly depends on velocity, since the external field is assumed to be of given value, otherwise (at finite R_m) the dependence becomes nonlinear, since the induced magnetic field depends on velocity. So, in the motion equation one more non-linear term appears. The main difficulty in the calculation of electromagnetic force is that electric currents induced in conducting medium when it moves change magnetic field not only in the region

21

occupied by this medium but also in space surrounding the medium region. Because of extreme complexity of this problem, it should be substantially simplified by averaging the equations and the boundary conditions over one or another space coordinates the change in values is less important for the specific problem in hand.

Based on averaging method for MHD equations suggested in [1] the theory of electrodynamic processes in channels at finite R_m has been developed and experimentally supported [2].

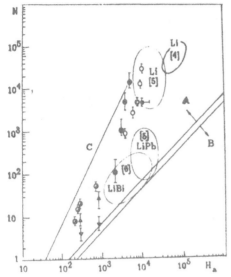

Fig.1. Regions of different flow regimes.

First consider MHD flows at $R_m \ll 1$, i.e. the flows in an external magnetic field. Regions of different flow regimes in the plane Ha-N are shown in fig.1. The region which is above the line "A" (shown by arrow) and is of greatest interest for practical applications and from the standpoint of magnetic hydrodynamics corresponds to the so-called laminarized flow. The turbulent flow region is below the line "B" (arrow down).

The laminarized flow is characterized by a great influence of magnetic field on the flow character (velocity field). This influence shows itself in appearing the Hartmann boundary layers at Ha>>1 on the MHD channel walls parts perpendicular to magnetic field induction vector and in formation of complicated speed structures nonuniform in cross-section and length of channels. As seen from (1) the influence of an inertial term in the hydrodynamic equation decreases with the growth of N and at certain values of $N_*(Ha)$ the flow changes to the so-called Stokes regime described by a linear equation. Fig.1 depicts $N_*(Ha)$ values obtained from experiments [3] for different types of flows, as well as ranges of N and Ha values typical for various projects of LM blankets in magnetic confinement fusion reactors [4-6]. As seen from fig.1, in these devices, especially in the hybrid test thermonuclear reactor (OTR), MHD flow should be inertial and extrapolation of results, obtained from asymptotic solution of the Stokes linear equation, valid for N and Ha lying left and above experimental points in fig.1, to these flows should be treated with extreme care. If we draw a straight line "C" in the plane N − Ha corresponding to a value of Re=2000 which defines a transition from laminar flow to turbulent one, the Stokes flow proves to be a laminar character. In this connection the transition to inertionless Stokes flow regime with the growth of N is of special importance. Nevertheless, results obtained from asymptotic theory at Ha → ∞ and N → ∞ , are of great value, since in some cases analytical solutions can be obtained which provide a better understanding of main physical laws of phenomena considered.

To study main laws of MHD flows at large N and Ha it is most convenient to take flows in rectangular channels at least for two reasons. Firstly, the magnetic field has a pronounced direction – induction vector, which should be taken into account in choosing the shape and orientation of the channel cross–section. For this reason traditional circular cross–section which is optimal from the engineering and technological standpoint is no longer optimal in magnetohydrodynamics. An illustration is conceptual design of LM blanket [7], where in order to reduce hydraulic losses, the heat transfer flow is effected through the slotted channels whose long side is coplanar to toroidal component of tokamak magnetic field.

The second reason is simple theoretical convenience i.e. averaging MHD equations over the channel height or width (depending on the problem), which can be solved easier than three–dimensional equations.

Since the major part of hydraulic channels of any MHD device including the LM blanket comprises constant cross–section channels, considered at first the developed flow in rectangular channels in a uniform magnetic field. Even though these flows have been thoroughly investigated both theoretically and experimentally [8], in particular, by asymptotic methods [9], the flows which are of interest for the blanket considered in [7] and which take place in slotted channels with a long side parallel to the magnetic field with LM-channel wall contact resistance taken into account have not yet been studied. In [10] these flows are analyzed theoretically. It is shown that the channel flow has various structure depending on relative values of Ha, $Ha^{1/2}$, λ, x^{-1}, r, where λ is the ratio of the channel side wall length and transverse one; $x = \sigma_w h_w / \sigma a$ is the wall conductance ratio, σ_w and h_w is their electric conductance and thickness, respectively, a is the transverse wall length,

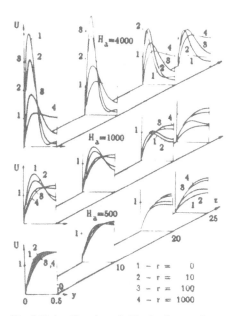

Fig.2.Velocity in slotted channel.

$r = R\sigma/a$ is the relative resistance of electroinsulating film, R is the contact resistance between LM and wall. A common feature for these structures is the formation of the Hartmann boundary layers about Ha^{-1} thick at the transverse walls and shear layers at the side ones where M-shaped of velocity may occur which may have a great effect on heat exchange and mass transfer at side walls. Note, that there may be situations [7] when transverse wall conductance ratio ($x_\perp$) differs from that of side walls ($x_\parallel$), parallel to the magnetic field.

In this case the value of M-shape depends, mainly, on $x_\perp$. In the channel with insulation walls it is nonexistent. Besides it is greatly decreased with the growth of λ at fixed Ha and is increased with the growth of Ha at fixed λ (slot-like parameter).

Fig.2 illustrates design velocity profiles in various cross-sections of the slotted channel with λ = 50, x = 0.01 at various values of r and Ha. It is seen from fig.2 that M-shape increases with a decrease in r and an increase in Ha.

In hydraulic systems of various MHD devices there always are sections, where flow is in a nonuniform magnetic field.

In this case averaging the MHD equation over the channel height (along the magnetic fields) gives a two-dimensional problem which has been solved in [11] numerically, taking into account an inertial term in the motion equation (1). The influence of friction at trans-

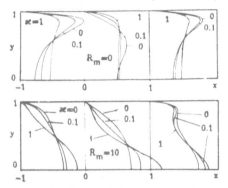

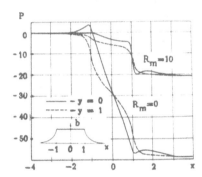

Fig.3. Velocity profiles arising from the magnetic field nonuniformity.

Fig.4.The curves of pressure distribution along the channel.

verse walls (perpendicular to magnetic field) and their electric conductivity as well as the Reynolds magnetic number on local (velocity profile, pressure and electric current density distribution) and integral characteristics of the channel (pressure drop) has been

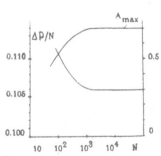

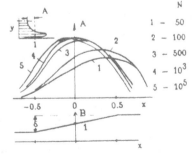

Fig.5.Dependence of normalized pressure drop and maximum extent of profile M-shape on N.

Fig.6.The development of M-shape along the channel.

studied. It has been shown that M-shaped velocity profile arising from the magnetic field nonuniformity decreases at side walls with the increase both in x and R_m as is also seen from fig.3. Simultaneously with increasing x and R_m a tendency is observed to the formation of M-shaped profile, in the flow core middle a velocity hump is formed. Fig.4 shows the curves of pressure distribution along the channel with x = 0 at various R_m and the same N=Ha=100 as in fig.3.

Separately the transition to the Stokes flow with the growth of N has been numerically studied [12]. Fig.5 showing the dependence of normalized pressure drop in the channel ($\Delta p/N$) and maximum extent of profile M-shape A_{max} on N, and fig.6 showing the development of M-shape A(x) along the channel length at various N for Ha=10^3 and x=0, clearly illustrate the transition to the Stokes flow in the channel with λ = 0.1.

The transition occurs at N>10^3. For the same MHD channel the Sto-

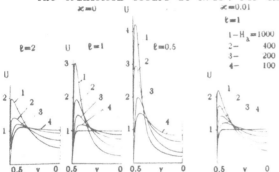

kes flow problem has been solved numerically and asymptotic methods in [13], while in [14] the influence of transverse wall electric conductance on this flow has been studied. Fig.7 shows velocity profiles in the cross-sections of the channel with maximum M-shape depending on Ha at x=0 and x=0.01, and also at various values of magnetic field gra-

Fig.7.Velocity profiles in the cross-section of the channel with maximum M-shape

dients δ/ℓ. Pictures of lines of hydrodynamic and electric currents are given in fig.8 at the same values of Ha and x=0 as in fig.7. It is seen that M-shape increases with the growth of Ha, electric current lines are pressed to the wall and the current value decreases. This is proved by the results of the asymptotic solution (at Ha $\rightarrow$ ∞) of the Stokes problem. In [14] it is shown that the total electric current and, hence, pressure drop $\Delta p/N$ is proportional to $(1/Ha+x)^{1/2}$, i.e. at x=0 and decreases with the growth of Ha as $Ha^{-1/2}$, which is also supported by the results obtained from the numerical solution of the problem (asymptotic dependence was already observed at Ha>500). With the increase in x at x>>1/Ha pressure losses ($\Delta p/N$) are proportional to $x^{1/2}$ which is in a agreement with the result obtained from [15]. From fig.7 it also follows that flow core velocity increases with increasing x.

So, when LM flows in channel with insulating walls in nonuniform magnetic field at Ha>>1, on their side walls (3) velocity boundary humps about $Ha^{-1/2}$ thick are formed, and the electric current in the channel is pressed to the side walls with increasing Ha. In the same channels something like this takes place with electric current also at R_m>>1 [2]. In this case at side walls the current boundary layers

$R_m^{-1/2}$ thick are formed. With the further growth of R_m current layers are also formed at the transverse walls. i.e. the magnetic field is expelled from the flow "current core" and the resultant magnetic field becomes M-shaped in the channel width. Since the main component of the electric field induced in LM is proportional to BV. it also has M-like character in the Stokes flow because of velocity M-shape at the constant magnetic field B. and at $R_m \gg 1$. due to M-shape of field B at constant velocity V. because the thickness of current boundary layers is always larger than that of hydrodynamic layers in real situations, i.e. $Ha \gg R_m$. The similar situation is observed for integral characteristics (channel pressure drop): asymptotic value (at $Ha \gg 1$) of pressure dimensionless drop in the Stokes flow is $(\Delta p/N) \sim Ha^{-1/2}$ and at $R_m \gg 1$ $(\Delta \rho / \rho V^2) \sim R_m^{-1/2}$. Magnetohydrodynamics at high R_m is of great value not only for practical use (creation of MHD pumps, flow coupler and other powerfull devices). but also in physical aspect. which is connected with the appearance of one more nonlinearity in the motion equation (1) due to the dependence of an induced magnetic field on velocity.

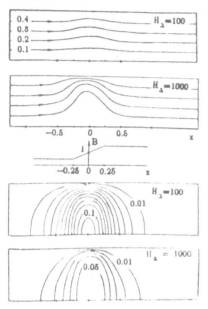

Fig.8.Lines of hydrodynamic and electric currents

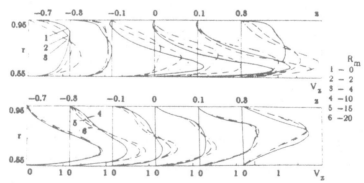

Fig.9.Velocity distributions in some cross-sections of the annular MHD channels.

In conclusion. the problem on the initiation of MHD instability at high R_m should be mentioned. When LM flows in a annular channel (most interesting case for practice) the flow MHD instability can arise at $R_m \gg 1$ not only from azimuthal disturbances but also from a rather high nonuniformity of the magnetic field and velocity in the channel height [16]. Fig.9 gives velocity distributions in some cross-sections of the annular MHD channel (magnetic field is time constant and reverses its sign at z=0). If can be supposed that flow portions with high velocity gradients can influence on flow stability in the channel.

REFERENCES

[1] Lavrent'ev I.V., 1978, "On averaging the electromagnetic field equations in MHD devices with finite magnetic Reynolds number", Magnitnaya Gidrodynamika, N 3. 92.
[2] Lavrent'ev I.V., Shishko A.Ya., 1980, "Electrodynamic processes in MHD channels at high magnetic Reynolds number", Magnitnaya Gidrodynamika, N 3. 81.
[3] Glukhikh V.A., Tananaev A.V., Kirillov I.R., 1987, Magnetohydro-dynamics in Nuclear Energetics Power. M., Energoatomizdat.
[4] Smith D.L. et al., 1985, "Overview of the blanket comparison and selection study", Fusion Technol. (USA), 8, 1(1), 10.
[5] Baker C.C. et al. 1985, "Tokamak power system studies – FY 1985", Argone National Laboratory Report, ANL/FPP–85–2.
[6] Muravjev E.V., Orlov V.V., Khripunov V.I., 1985, "Liquid–metal cooling of hybrid tokamak–reactor. Problems of atomic science and engineering", Thermonuclear Fusion, Issue 4, 24.
[7] Lavrent'ev I.V., 1987, "Liquid–metal blanket of fusion tokamak–reactor", Theses of Rep. of 4–th All–Union Conference on Engineering Problems of Fusion Reactors. Leningrad, January, 1988. M., 326.
[8] Gelfgat Yu.M., Lielausis O.A., Shcherbinin E.V., 1976, "Liquid metal under electromagnetic forces". Riga, 247.
[9] Walker J.S., 1981, "Magnetohydrodynamic flows in rectangular ducts with thin conducting walls", Part 1, J.de Mechan, Vol.20, N 1, 70.
[10] Lavrent'ev I.V., Sidorenkov S.I., Shishko A.Ya., Developed flow in slotted channel in coplanar magnetic field, Magnitnaya Gidrodyna-mika (to be published).
[11] Lavrent'ev I.V., Sidorenkov S.I., 1986, "Influence of MHD channel wall friction on flow at finite R_m",Magnitnaya Gidrodynamika, N 4. 121.
[12] Lavrent'ev I.V., Serebrjakov V.V., "On transition to Stokes flow in MHD channels", Magnitnaya Gidrodinamika (to be published).
[13] Lavrent'ev I.V., Molokov S.Yu., Serebrjakov V.V., Shishko A.Ya., "Stocks flow in rectangular channel with insulating walls in nonuni-form magnetic field", Magnitnaya Gidrodinamika (to be published).
[14] Lavrent'ev I.V., Molokov S.Yu., Serebrjakov V.V., Shishko A.Ya. "Influence of wall electric conductance on stocks flow in rectangular channels", Magnitnaya Gidrodinamika (to be published).
[15] Walker J.S., 1986, "Liquid metal flow in a rectangular duct with a nonuniform magnetic field", J. of Theoretical and Applied Mechanics. Vol.5, N 6, 827.
[16] Kalyutik A.I., Lavrent'ev I.V., Serebrjakov V.V., 1987, "Numeri-cal study of flow in circular MHD channel", 12–th Riga Conference. Vol.1, Salaspils, 171.

MHD PRESSURE DROP OF LIQUID METAL FLOW IN CIRCULAR AND RECTANGULAR DUCTS UNDER TRANSVERSE MAGNETIC FIELD

Keiji MIYAZAKI, Shoji INOUE and Nobuo YAMAOKA
Department of Nuclear Engineering, Faculty of Engineering
Osaka University
2-1 Yamada-oka, Suita, Osaka 565, Japan

Absatract

Simple formulas were derived to estimate the liquid metal MHD pressure drop in a cooling system for fusion use and good agreements were obtained with NaK and Li experiments in use of various SS ducts.

I. INTRODUCTION

The MHD pressure drop is one of the most critical issues for the liquid metal self-cooling of magnetically confinement fusion power reactors. In order to gain better physical understanding and basic data, a series of experimental and theoretical studies have been made on the MHD pressure drop in SS (stainless steel) ducts with simple geometries such as circular, square and rectangular channels by the authors[1-3]. This paper will present a summary of the results, adding some more recent results of NaK experiment with a rectangular SS duct.

II. THEORETICAL ANALYSES

B: Magnetic flux density, u: Flow velocity, U: Mean flow velocity, σ_f: Electric conductivity of fluid, σ_w: Electric conductivity of duct wall.

2.1. Circular Duct[1] (see **Figure 1**)

Voltages V_f in the fluid and V_w in the wall are given as the general solution of Laplace equations:

$V_f = K_1 r \sin\theta$ and $V_w = (K_2 r + K_3/r) \sin\theta$.

Boundary conditions are (1) $\partial V_w/\partial r = 0$ at $r = R_0$, (2) $\sigma_f(\partial V_f/\partial r) = \sigma_w(\partial V_w/\partial r)$

This work was supported by the Grant-in-Aid for Fusion Research from the Ministry of Education, Science and Culture.

29

J. Lielpeteris and R. Moreau (eds.), Liquid Metal Magnetohydrodynamics, 29–36.

at r= R_i and (3) $V_f = V_w$ at r=R_i.

V_f= UBrsinθ /(1+C) where C= $\sigma_w(R_0^2-R_i^2)/\sigma_f(R_0^2+R_1^2)$.

V_w= $R_i^2(r+R_0^2/r)$UBsin θ /[$R_0^2+R_i^2$ +$(R_0^2-R_i^2)\sigma_w/\sigma_f$].

The above derivation is almost the same as the process used by Elrod & Fouse[4] for the analysis of an electromagnetic pump. Using y=r sinθ, dP/dz=-J_yB and J_y=$\sigma_f(-\partial V_f/\partial y +UB)$ for a uniform velocity U, the result is -dP/dz = $K_p\sigma_f UB^2$ where K_p= C/(C+1) and C= $\sigma_w(R_0^2-R_i^2)/\sigma_f(R_0^2+R_1^2)$.

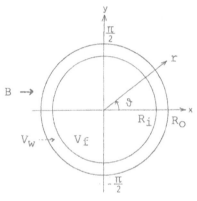

Fig. 1   Model illustration   Fig. 2   Model illustration
          of circular duct   of rectangular duct

2.2. Rectangular Duct[3]   (see **Figure 2**)

The MHD flow analyses have been practiced on the basis of a "core flow" model that both the static electric field strength **E** and the flow velocity **u** have uniform distributions along the applied magnetic field **B** in the core flow region and large jumps of these quantities occur in the secondary boundary layers near the electrodes. According to Walker[5], the coefficient is calculated to be K_p=C(b/a)/(1+a/3b) and C=$\sigma_w t_w/\sigma_f a$, However, this does not agree with the authors' experimental results.

If the uniform distribution is restricted to the resultant electric field (E+uxB), then E and u will be allowed to have nonuniform distributions. An analysis has been made, based on this concept.

(1) Along the electrodes; R_e= a/$\sigma_w t_w$

Current: I(x)= Jx= Ix/a.   Voltage: V(x)= V_0 - $IR_e(x/a)^2/2$.

(2) In the fluid; $R_f = b/\sigma_f a$

Voltage: $V(x) = u(x)Bb - IR_i$. Velocity: $u(x) = u_0 - K_u(x/a)^2$,

where $u_0 = (V_0 + IR_i)/Bb$ and $K_u = IR_e/2Bb$.

Mean velocity: $U = u_0 - K_u/3$. Therefore, $UBb = (V_0 + IR_i - IR_e/6)$.

(3) Along the side walls; $R_w = b/\sigma_w t_w$

$V_0 - IR_e/2 = IR_w$. Therefore, $I = UBb/(R_w + R_e/3 + R_i)$.

(4) Pressure gradient: $-dP/dz = JB = K_p \sigma_f UB^2$,

where $K_p = (R_i/R_w)/[1 + (R_e/3R_w) + (R_i/R_w)]$

or $K_p = C/(1 + a/3b + C)$ and $C = R_i/R_w = \sigma_w t_w/\sigma_f a$.

(5) Electrode voltage: $V(x) = UBb[1 + (1 - x^2/a^2)(a/2b)]/(1 + a/3b + C)$

Side wall voltage: $V(x) = UB(a + b - x)/(1 + a/3b + C)$ for $x = [a, a+b]$.

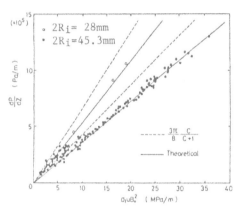

Fig. 3   Pressure drop of NaK
in circular ducts

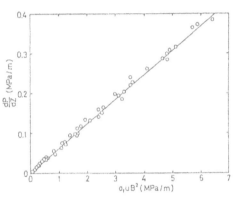

Fig. 4   Pressure drop of Li
in circular duct

III. COMPARISON WITH EXPERIMENTAL RESULTS

Figure 3 shows the plots of pressure drop gradient in the uniform magnetic field in the NaK blowdown experiment with 1-1/2" ($2R_i = 45.3$mm, $t_w = 1.65$mm) and 1" ($2R_i = 28$mm, $t_w = 1.65$mm) 304-SS circular pipe ducts. Good agreements are seen except for a high U and low B ($Ha^2/Re \geq 15$) region. $U = 2$-15m/s ($Re = 8.0\times10^4$-6.2×10^5) & $B = 0.3$-1.75T ($Ha = 710$-4150).

Figure 4 shows the results in the Li loop experiment with a 3/4" ($2R_i = 15.75$mm, $t_w = 1.65$mm) 316-SS circular tube . This time again a good agreement is seen between the theory and experiment. $U = 0.5$-5.0m/s ($Re = 0.9\times10^4$-9×10^4) & $B = 0.1$-1.0T ($Ha = 140$-1380).

The deviation in Carlson's pressure data[6] of Li experiment has been improved by applying the present thick-wall model. The measured voltage distribution on the outer surface agrees well with the theory both for NaK and Li experiments.

Table 1   Key factors of rectangular ducts
of NaK and Lithium experiments

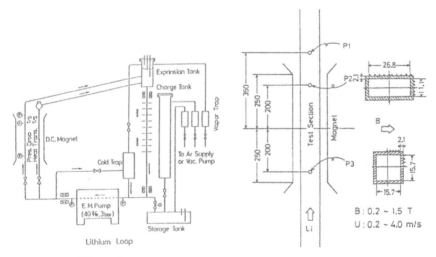

Fluid	$\frac{2a}{2b}$	A.R. ($= \%$)	$C = \frac{a_w t_w}{a_i a}$	Miyazaki parabolic $K_p = \frac{C}{1 + a/3_b \cdot C}$	Walker flat $K_p = \frac{C}{(\%_b \chi_i + \%_{3b})}$
Li	$^{15.7}/_{15.7}$	1	0.0875	0.0616	0.0656
Li	$^{26.8}/_{11.1}$	2.41	0.0517	0.0269	0.0119
NaK	$^{45.5}/_{20.5}$	2.22	0.0462	0.0259	0.0120
NaK	$^{20.5}/_{45.5}$	0.451	0.103	0.0819	0.200

$$\frac{dP}{dz} = K_p \sigma_i U B^2$$

Fig. 5   Setup for Li experiment
with rectangular ducts

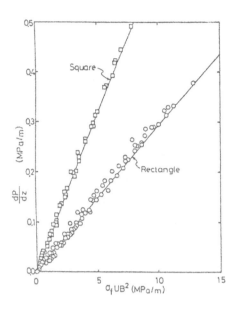

Fig. 6 Pressure gradient of Li flow in rectangular ducts under uniform B

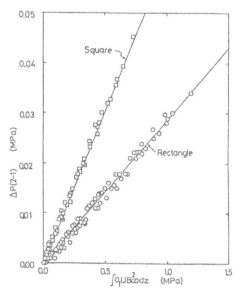

Fig. 7 Pressure drop of Li flow in rectangular ducts under outlet fringing B field

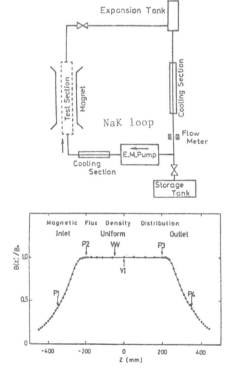

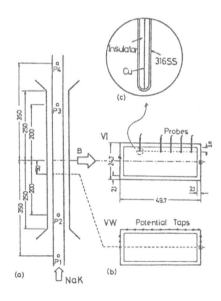

Fig. 8 Setup for NaK flow experiment in rectangular duct and magnetic flux density profile B(z)

Table 1 shows the key factors of rectangular ducts which were used for the experiments. In the FINESSE report $K_p=C$ was adopted. In the case of square ducts, i.e. $a/b=1$, no much difference of K_p is seen between the flat core and parabolic velocity models. However, the difference is outstanding for a/b ratios larger or smaller than unity. The value of K_p is larger nearly by a factor of a/b for the parabolic model than for the flat core model. When the rectangular duct is made a quarter turn or 90 degree rotation, K_p for the parabolic model is increased but this time as much lower than that for the flat core model.

Figure 5 shows a schematic diagram of lithium loop experimenet with the square and rectangular ducts as listed in Table 1.

Figures 6 and 7 show the results on the pressure drop gradient in the uniform field and the pressure drop in the outlet fringing field, respectively. $B=0.2-1.5T$ & $U=0.2-4.0m/s$ (Ha=200-2100 & Re=500-38000).

Figure 8 shows setup diagrams of the NaK experiment. The test duct has a dimension of 45.5mm x20.5mm and $t_w=2.1mm$, made of 304-SS, and fitted to the NaK loop in the two ways by rotating 90 deg., as shown in Table 1. The covered range: $B=0.2-1.5T$ & $U=0.1-0.6m/s$.

Figures 9 and 10 show the pressure gradient in the uniform magnetic region for the two arrangements. The straight lines indicate the theoretical predictions based on the parabolic velocity model. A good agreement is seen for the case of $a/b=2.22$ and a slight deviation to the lower value is seen for the case of $a/b=0.45$. A possible reason of the deviation will be thick-wall effect which is larger for a smaller a/b. The voltages measured on the electrode have a parabolic distribution but are appreciably lower than the present prediction for both cases of Li and NaK experiments, though those on the side wall are linear and have a good agreement. This still remains to be investigated.

Figures 11 and 12 show the plots of pressure drop at the inlet and outlet against $\sigma_f U \int B^2(z)dz$, for the two arrangement. The lines indicate the prediction in use of the same K_p as in the uniform field: $\Delta P= K_p \sigma_f U \int B^2(z)dz$ and $K_p= C/(1+a/3b+C)$. It is noted that no significant difference is discerned between the inlet and outlet.

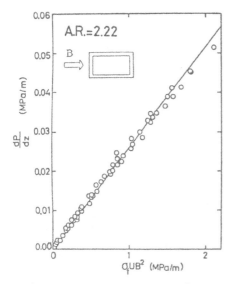

Fig. 9 Pressure gradient of NaK flow in 45.5/20.5 duct under uniform B

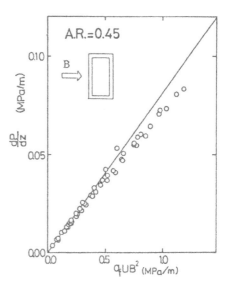

Fig.10 Pressure gradient of NaK flow in 20.5/45.5 duct under uniform B

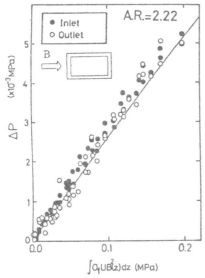

Fig. 11 Inlet and outlet pressure drop of NaK flow in 45.5/20.5 duct

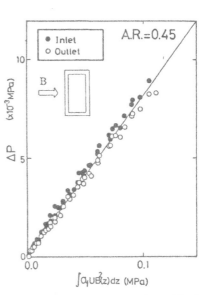

Fig. 12 Inlet and outlet pressure drop of NaK flow in 20.5/45.5 duct

IV. CONCLUSIONS

The theoretical prediction of MHD pressure gradient along the flow can be expressed by $dP/dz = -K_p \sigma_f U B^2$. The coefficient K_p, which is dependent on the shape and size of a duct, can be obtained on an assumption of uniform current density, neglecting fluid-duct friction.

1) For a circular duct, $K_p = C/(C+1)$ and $C = \sigma_w(R_0^2 - R_i^2)/\sigma_f(R_0^2 + R_i^2)$

2) For a rectangular duct with the inner length of 2a along the magnetic field, the inner width of 2b and the thin wall thickness of t_w,

$$K_p = C/(1 + a/3b + C) \quad \text{and} \quad C = \sigma_w t_w/\sigma_f a.$$

3) The MHD pressure drop ΔP in the fringing field B(z) can be approximated by $\Delta P = K_p \sigma_f U \int B^2(z)dz$ for both inlet and outlet.

Validity of the above formulas was verified by the good agreements with the experiments with various test sections. In conclusion, the above formulas will be useful for predicting the MHD pressure drop, in spite of controversy in the parabolic flow velocity profile used in the present analytical model for rectangular ducts. As for the circular duct, the present result will be applicable to a thick wall tube.

REFERENCES

(1) K. MIYAZAKI et al.: MHD Pressure Drop of NaK Flow in Stainless Steel Pipe, Nuclear Technology/Fusion, Vol.4, No.2, pp.447-542 (1983).

(2) K. MIYAZAKI et al.: Flow and Heat Transfer Characteristics in Lithium Loop under Transverse Magnetic Field, ibid., pp.733-738 (1983).

(3) K. MIYAZAKI et al: Magneto-Hydro-Dynamic Pressure Drop of Lithium Flow in Rectangular Ducts, Fusion Technology, Vol.10, pp.830-836 (1986).

(4) H. G. Elrod, Jr. & R. R. Fouse: An Investigation of Electromagnetic Flowmeters, Trans. ASME, (March 1952).

(5) J. S. WALKER: Magnetohydrodynamic Flows in Rectangular Ducts with Thin Conducting Walls, Part I. Constant-area and Variable-area Ducts with Strong Uniform Magnetic Fields, J. de Mechanique, Vol.20, No.1, pp.79-112 (1981).

(6) G. A. KARLSON: Magnetohydrodynamic Pressure Drop of Lithium Flowing in Conductive Wall Pipe in a Transverse Magnetic Field - Theory and Experiment, UCRL-75307, (April 1974).

LINEAR APPROXIMATION APPLICATION LIMITS IN MHD-FLOW THEORY FOR STRONG MAGNETIC FIELDS. EXPERIMENTAL RESULTS

A.V. TANANAYEV, T.N. AITOV, A.V. CHUDOV, V.A. SHMATENKO
Leningrad Polytechnic Institute named after M.I. KALIRIN
195251 Leningrad
USSR

ABSTRACT. The experimental results on the inertialess MHD-flows in a strong magnetic field are presented. Complex shaped duct flows with regular roughless walls and free-surface MHD-flows are investigated. The limits of inertialess Stokes flow existence are defined ; the general asymptotic relations typical for inertialess flows with high Hartmann number are obtained.

Introduction

The advance in development and construction of modern power installation, in general and fusion technology in particular, brings about investigation of MHD-flows with large values of MHD-interaction parameters and Hartmann number : $N \gg 1$, $M \gg 1$ /1/. The majority of theoretical studies of MHD-flows in strong magnetic fields are based on linearized equations of motion (after Stokes) /2, 3/. But this raises the more general question about linearized approximation limit applicability. According to experimental data /4/, the available estimations of linear approximation of the $N \gg M$ type are rather rough and certainly overestimated.

The objective of the paper is to compare the latest experimental results and the linear theory with the view to define the inertialess approach limits. It is also the intention of the authors to attract attention to the theoretical problem of working out the exact evaluation criteria of approximation aproach adquacy for linear MHD-flow equation in strong magnetic fields.

I. Generally accepted principles of the MHD linear theory

The equation of flow motion presented as :

$$N^{-1} (V.\nabla) V = - (p/N) + M^{-2} \nabla^2 V + j \times B \qquad (1)$$

with strong MHD-interactions and with large Hartmann numbers $N \gg M \gg 1$ can be linearized by omitting convective numbers and after that be solved by superposing the solution for the core

$$(p/N) = j \times B, \quad j = - \nabla \varphi + V \times B, \quad j = 0, \quad V = 0 \qquad (2, 3, 4, 5)$$

(whose fluid dynamics is defined by Ohm's law) and for the Hartmann boundary layers (whose inertialess and non-viscous approximations (2) - (5) do not hold true).

By scalar multiplication of (2) by B we shall obtain that the core's pressure is constant along the magnetic field lines. The distribution of the other variables also tends to become constant in the magnetic field direction /1/.

J. Lielpeteris and R. Moreau (eds.), Liquid Metal Magnetohydrodynamics, 37–43.

The important results of the general theory is that (2) - (5) equations are based on the existence in the flow characteristic directions (surfaces) along which the pressure and the potential are constant, while the velocity is tangential. In the non-uniform magnetic field characteristic surfaces are defined by a constant integral value

$$\int_0^{S_o} B^{-1} \, ds = I$$

where integration is carried out along the magnetic field line /3,5/. The existence of the characteristic surfaces is due to the Hartmann boundary layer property to define the value of the currents circulating within the core. In the fully developed flow the current density makes up the value of the order of M^{-1} from the current density in the Hartmann layers. In the non-uniform field region, the current density is

$$j/(\sigma \, V \, B) \sim M^{-1/2} \, . \tag{6}$$

The consequences of (2) and (6) can be the evaluation of the pressure drop in the non uniform region :

$$p^* = \Delta p/(\sigma \, B^2 \, dV) \sim M^{-1/2}$$

or

$$p/(\rho \, V^2) \sim NM^{-1/2} \, . \tag{7}$$

Here, d is the characteristic dimension, σ is the electrical conductivity, ρ is the density, B is the induction and V is the velocity. The value of (7) for the pressure drop is one of the main and easily tested results of the linear approach in free surface boundary flows the evaluation of (7) defines the asymptotics of the height drop levels of the ρgh and using (7) we shall obtain $h^* = h/(V^2/g)$ in inertialess flow which equals $h^* \sim M^{-1/2}N$. Experimental results will be discussed below.

2. The flow through a 180° - Bend

The experimental model was a rectangular cross-section duct with its sides related as $b/a = 10.4$, b = 120 mm. The walls were electrically insulated and the maximum value of flux density in the volume of $600 \times 140 \times 150$ mm^3 was 1.7 T. The liquid metal used was mercury. Figure 1 presents the experimental results /6/. For every curve in Fig. 1 M number is constant.

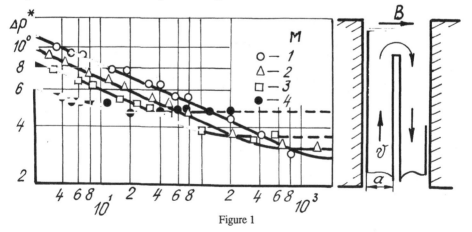

Figure 1

Numbers 1-4 in Fig. 1 corresponds to Hartmann numbers : 450, 353, 294, 185. The asymptotic values of $\Delta p^* = \Delta p_s^*$ correspond to the inertialess flow. The dependance of Δp_s^* (M) in accordance with (7) is easily approximated by the straight line (Fig. 2).

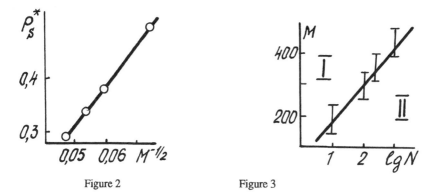

Figure 2   Figure 3

Fig. 3 shows the regions of non-linear I and linear (inertialess) II flow. In region II the flow, at least integrally, can be described by linear equations. In region I the inertia is appreciable to evaluate the hydraulic resistance. In I, the results of electrodynamic approximation approach can be used. However, this evaluation can be overestimated.

3. The flow in the collector model

This flow is an example of a more complicated MHD-flow with 5 sub-ducts of a circular cross-section of the diameter $d_y = 10$ mm and the input and output collectors with $D_y = 26$ mm (Fig. 4). The flow

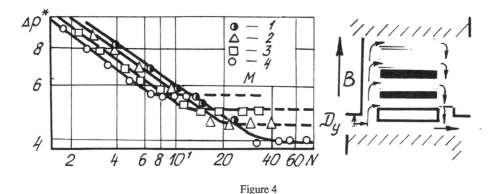

Figure 4

linearization results are mostly the same as for the above flow and can be seen in Figs 5-6. Numbers 1-4 in Fig. 4 correspond to Hartmann numbers 651, 295, 236, 175 respectively. The orientation of the magnetic field along the input and output brings about uniform distribution of flow discharge along the sub-ducts after the general tendency mentioned in section 1. The characteristic time of the flow

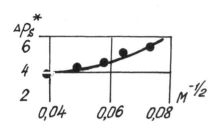

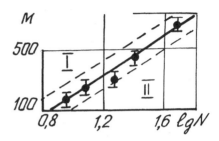

Figure 5   Figure 6

magnetic field interaction is known to be evaluated by [1] :

$$\tau_j \sim \rho /(\sigma B^2)$$

therefore the relation of the convection transfer time $\tau_c \sim d/V$ and gives the value of MHD interaction parameter $N = \tau_c / \tau_j$ characterizing the process of equalization of discharge distribution in the sub-ducts. One can see in Fig. 7 the plot of root-mean-square deviation D is the sub-ducts flow discharge as a function of a uniform flow in a strong magnetic field. One can also see that with the increase of the MHD-interaction parameters, the velocity distribution becomes constant, D decreasing. The most important results of the experiment is that asymptotic value D at N >> 1 increases with the Hartmann number increase (Fig. 7).

4. The flow in rough channel

Such a flow has been throroughly investigated 6 /7/. The attention was paid to the sharp increase of the hydraulic resistance coefficient under the action of the strong magnetic field.

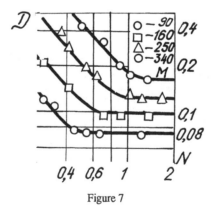

Figure 7

Numerical studies /7/ revealed the transition to the non-separated flow of roughness elements and flow penetration between the roughness elements under the action of the magnetic field, which must be the reason of the resistance increase. Nevertheless, the generally accepted treatment claims that the transition to the non-separated flow is accompanied by the decrease of the hydraulic resistance 6 /8/. As a result, a concept of a "shunt layer" was put forward /9/ to evaluate the rough duct resistance substituting rough walls by electroconducting walls.

Conduction walls parameters were selected on the basis of experimental data approximations.

The abnormal resistance properties of rough ducts can be easily explained if one treats the rough duct as a non-uniform MHD-flow in a strong magnetic field. If the applied field is very strong $h \ll M \ll N$ (where h is the typical roughness size), the distribution of hydrodynamic variables will be such as in section 2 and 3. And the mean resistance value along the longitudinal coordinate will have the same asymptotic $\lambda \sim NM^{-1/2}$. This statement can be proved by the experimental results (Fig. 8).

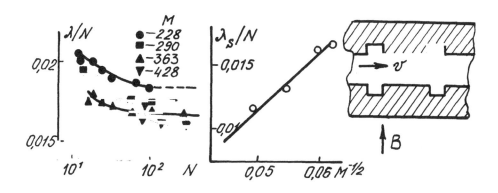

Figure 8

5. The free-surface flow

This flow has been investigated with the help of a model (Fig. 9). The non-uniforming of the free surface profile is caused by the presence of M-shaped velocities before and after the bend /10/ under the action of the vortex electromotive force as well as by the non-uniform distribution of the potential component of the electromotive force with the minimum value in the velocity of the side walls parallel to the field. The behaviour of the non-dimensional depth h^* from N at point 1 is illustrated in Fig. 10. Similar curves were obtained for point 2. In Fig. 10 numbers 1-5 correspond to Hartmann numbers 640, 540, 470, 400, 230. It can be seen that with the increase of N the depth h^* has the tendancy to approach its asymptotic value if M is constant. The asymptotic value of h^*_s corresponds to Stokes flow (Fig. 11). The region of inertialess (I) and inertia (II) flows are shown in Fig. 12.

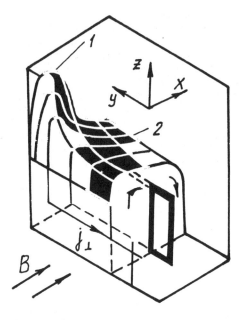

Figure 9

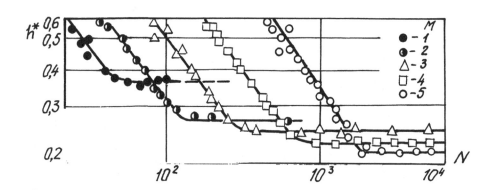

Figure 10

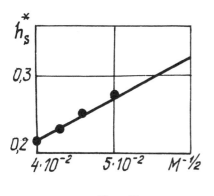
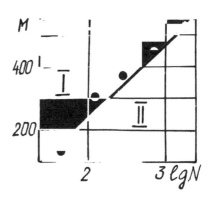

Figure 11   Figure 12

SUMMARY

General regularities of MHD-flow transition to linear flows in a strong magnetic field are revealed by considering characteristic complex MHD-flows. Asymptotic of Stokes approximation are demonstrated. Asymptotic of $NM^{-1/2}$ proved to be observed for : a) typical pressure drop in duct flows, b) typical dimensionless depth $h/(V^2/g)$ in free-surface flows, c) rated hydraulic resistance value λ in rough wall ducts. The limits of linear equation applicability are established.

REFERENCES

1. GLUKHIKH V.A., TANANAYEV A.V., KIRILLOV I.R., 1987, MHD in nuclear power engineering, M. : Energoatomizdat.
2. HUNT J.C.R., SHERCLIFF J.A., 1981, Magnetohydrodynamics at high Hartmann number, Ann. Rev. Fluid Mech. 3, 37.
3. HUNT J.C.R., HOLROYD R.J., 1977, UKAEA Res. Group. Rep., CLM-R169, Adington.
4. AITOV T.N., TANANAYEV A.V., CHUDOV A.V., SHMATENKO V.A., 1987, 1, On Stokes problem in MHD, Riga VII meeting on MHD papers, Riga, 159.
5. KULIKOVSKY A.G., 1973, On conducting non-compressible fluid flows in arbitrary regions with a strong magnetic field, Izv. AN SSSR. MJG. N 3, 144.
6. AITOV T.N., TANANAYEV A.V., YAKOVIEV V.V., 1985, On asymptotic MHD-flows, Izv. AN SSSR. MJG. N 5, 159.
7. BRANOVER H.H., 1970, MHD of incompressible fluids, M. : Nauka.
8. LOYTZYANSKY L.G., 1973, Fluid and Gas Mechanics, M. Nauka.
9. TANANAYEV T.N., 1975, MHD flows in rough tubes, MHD, N 2, 29.
10. AITOV T.N., TANANAYEV A.V., 1984, On the problem of M-shaped velocity developement, Riga Meeting on MHD, Papers, Riga, 1, 171.

Session B:
Fusion Related Flows

APPLICATION OF THE CORE FLOW APPROACH TO MHD FLUID FLOW IN GEOMETRIC ELEMENTS OF A FUSION REACTOR BLANKET

M. S. TILLACK
Mechanical, Aerospace, and Nuclear Engineering Department
University of California, Los Angeles
Los Angeles, CA 90024
USA

ABSTRACT. Inertialess, inviscid solutions of the MHD equations have been numerically obtained for several cases of interest to fusion. This includes conducting pipes with varying magnetic field strength, varying radius, and variable field direction. The solution method has been formulated in a very general way to allow the solution of problems with 3-dimensional magnetic fields and complex channel shapes. The method separates the components of velocity and current parallel and perpendicular to the field, and expresses them in terms of four 2-dimensional functions of integration along the magnetic field lines. These functions of integration are then solved numerically by applying conservation of mass and current at the walls. The full solution can then be reconstructed from the functions of integration. The transformation of the MHD equations to 2-D is computationally very efficient, but can become very cumbersome when both the structure and magnetic geometries are complicated.

1. Introduction

Under sufficiently high magnetic fields, inertial and viscous forces become small in the Navier-Stokes equation. Neglecting these terms, and assuming small magnetic Reynolds number, a linear set of 8 equations results, including Ohm's law, the simplified momentum equation, and conservation of mass and current.

These equations could be solved directly on a 3-dimensional grid; however, the unique properties of the core equations allows transformation of the system of equations to forms which are computationally much more efficient to solve. A solution method is described here based on direct integration of the MHD equations along magnetic field lines, as first suggested by Kulikovskii [1]. The approach reduces the MHD equations to, at most, four 2-dimensional partial differential equations. This solution method is very general, allowing the solution of problems with 3-dimensional magnetic fields and complex channel shapes. Solutions to example problems are presented to demonstrate the capabilities of the method.

2. Description of the Direct Integration Method

The equations describing liquid metal MHD fluid flow include both Maxwell's equations and the Navier-Stokes equation. In dimensionless form, the Navier-Stokes equations is expressed as:

47

J. Lielpeteris and R. Moreau (eds.), Liquid Metal Magnetohydrodynamics, 47–53.
© 1989 by Kluwer Academic Publishers.

$$\frac{1}{N}(\underline{v} \cdot \nabla)\,\underline{v} = -\nabla p + \underline{J} \times \underline{B} + \frac{1}{Ha^2}\nabla^2 \underline{v} \tag{1}$$

where the MHD body force is given by $\underline{J} \times \underline{B}$, and the two dimensionless groups describing the relative magnitude of the terms are the Hartmann number $(Ha=aB\sqrt{\sigma_f/\mu})$ and the interaction parameter $(N=Ha^2/Re$, where $Re=\rho va/\mu)$. The pipe radius is "a", and the fluid density and viscosity are given by "ρ" and "μ".

When the Hartmann number and interaction parameter are sufficiently high, the majority of the flow is dominated by the balance between the pressure gradient and the electromagnetic force. The system of equations then includes the force balance and Ohm's law:

$$\nabla p = \underline{J} \times \underline{B} \qquad\qquad \underline{J} = \sigma_f\,(-\nabla\phi + \underline{v} \times \underline{B}) \tag{2,3}$$

and the conservation equations:

$$\nabla \cdot \underline{v} = 0 \qquad\qquad \nabla \cdot \underline{J} = 0 \tag{4,5}$$

Here the fluid conductivity is "σ_f" and the electric potential "ϕ". Together with the electrical and fluid boundary conditions, Eqs. 2–5 fully determine the velocity profiles in the core. (In this analysis, the magnetic Reynolds number is assumed small, such that the magnetic field is exactly equal to the externally applied field.)

These equations can be separated into components and integrated along magnetic field lines to obtain expressions for the unknown current, velocity, and electric potential fields in terms of two-dimensional functions of integration which are constant along magnetic field lines. The functions of integration are determined through application of the boundary conditions at the surface of the pipe. In this way, Eqs. 2–5, are transformed to a coupled set of partial differential equations [1].

The results are as follows:

$$\underline{J}_\perp = \frac{\underline{B}}{B^2} \times \nabla p \tag{6}$$

$$J_\| = \int (\underline{B} \times \nabla\frac{1}{B^2}) \cdot \nabla p \; dl + A_1 \tag{7}$$

$$\phi = -\int\int (\underline{B} \times \nabla\frac{1}{B^2}) \cdot \nabla p \; dl\,dl + A_1 l + A_2 \tag{8}$$

$$\underline{v}_\perp = -\frac{1}{B^2}\nabla p + \frac{\underline{B}}{B^2} \times \nabla\phi \tag{9}$$

$$v_\| = \int\left((\underline{B} \times \nabla\frac{1}{B^2}) \cdot \nabla\phi + \nabla p \cdot \nabla\frac{1}{B^2} + \frac{\nabla^2 p}{B^2}\right)dl + A_3 \tag{10}$$

Equation 6 is obtained directly from a cross product of $\underline{B}$ with the pressure balance equation, Eq.2. Similarly, Equation 9 is obtained directly from a cross product of $\underline{B}$ with Ohm's law. The parallel components of $\underline{J}$ and $\underline{v}$ are then found by integrating the conservation equations (4 and 5) along the magnetic field lines using the known perpendicular components. Finally, the potential is found by integrating the parallel current, using $J_{\parallel} = -(\partial\phi/\partial l)$. The variable "l" represents the distance along $\underline{B}$.

In order to express boundary conditions on the functions of integration (p, A_1, A_2, A_3), they must be related to physical variables in the duct. For example, if the magnetic field lines point in the x-direction, then (for the general, non-symmetric case) the four functions can be replaced by $\phi_L(y,z)$, $\phi_R(y,z)$, $p(y,z)$, and $v_{xo}(y,z)$, where ϕ_L and ϕ_R refer to the potential evaluated at the left and right walls, p is the pressure (which does not vary along x), and v_{xo} is the x-component of velocity evaluated at $x=x_0$. These four functions are easily related to the original A_1, A_2, A_3, and p. For symmetric problems, only two functions are needed. These are normally chosen to be $p(y,z)$ and $\phi(y,z)$, where ϕ can be evaluated at either wall or at some arbitrary value of x.

The 4 boundary equations which determine the functions of integration are mass and current conservation expressed at the two walls which are intercepted by the field lines. For conducting ducts:

$$\underline{v} \cdot \hat{n} = 0 \qquad (11)$$

$$\Phi \nabla_w^2 \phi = \underline{J}_f \cdot \hat{n} \qquad (12)$$

where $\hat{n}$ is the normal to the wall, $\Phi = (\sigma_w t)/(\sigma_f a)$ is the wall conductance ratio ("t" is the wall thickness and "σ_w" is the wall conductivity), $\underline{J}_f$ is the current density in the fluid, and the Laplacian, ∇_w^2, is performed in the wall only. Equation 12 follows from Ohm's law in the wall $(\underline{J}_w = -\sigma_w \nabla_w \phi)$ and current conservation $(\nabla \cdot \underline{J}_w = 0)$, where the right-hand side $(\underline{J}_f \cdot \hat{n})$ is an effective source term coming from the fluid.

For insulating ducts, currents return through the Hartmann boundary layers, and current conservation is expressed as (see [1]):

$$\left(\frac{1}{Ha}(\nabla \times \underline{v}) + \underline{J}\right) \cdot \hat{n} \qquad (13)$$

Boundary conditions are needed on the functions of integration at the edges of the computational grid. At the inlet and outlet, fully-developed conditions are applied: at the inlet $p=p_1$ and $\partial/\partial z=0$, and at the outlet $p=p_2$ and $\partial/\partial z=0$.

The equations given above are very general. They apply to any duct shape with any conductivity and any magnetic field intensity and direction. The primary restriction is that each field line must pass through walls at no more than two points, and that the walls are smooth (the normal vector is continuous). If a field line passes through more that two points (such as at a wall parallel to the field) then additional information must be supplied to resolve the electric potential along the wall. These special cases can be treated approximately; however, no further attempt is made to discuss them here.

3. Circular Pipe with Varying Transverse and Parallel Magnetic Field

The method described above has been used to solve for the pressures and velocity profiles in a circular pipe with a varying magnetic field having components in both the transverse and parallel directions with respect to the bulk flow. Solutions were obtained using a finite difference approximation to the derivatives.

The geometry is shown in Fig. 1. The magnetic field is given by:

$$\underline{B} = B_x(z)\,\hat{x} \qquad\qquad (14)$$

The computational grid is placed in the natural coordinate system of the magnetic field. The pipe is positioned at an angle, α, with respect to the magnetic field, resulting in both transverse and parallel components. The magnetic coordinate system is Cartesian (x,y,z) and the circular pipe wall is described by the angle θ and longitudinal distance λ.

When $\alpha=0$, the field is entirely transverse. Extensive data have been taken at the ALEX facility for a conducting circular pipe in the entrance and exit region of a transverse magnetic field [2]. Comparisons have been made between the ALEX data and the direct integration method, and there is excellent agreement.

Fig. 2 shows the experimental and calculated axial pressure gradient near the exit from the magnetic field at the center of the pipe $(y=0)$ and at the wall $(y=1)$. A noticeable "hump" exists just upstream of the exit at $y=1$. This behavior is much more pronounced at $y=0$, indicating the tendency for the flow to be retarded near the pipe center.

While experimental verification is not available, problems with $\alpha \neq 0$ have been examined. As an example, angles of 0, 15, and 30° were solved with a magnetic field strength changing from 1 to 0.5 over 1.5 pipe radii. Results are shown in Fig. 3 for the axial velocity profile near a rapid change in magnetic field strength. The classical "M-shaped" profile is apparent, as well as the fact that the parallel field component alters the velocity profile.

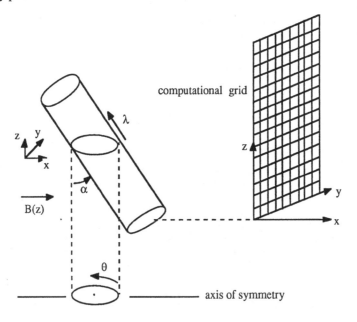

Fig. 1. Geometry for the circular pipe with both parallel and transverse components

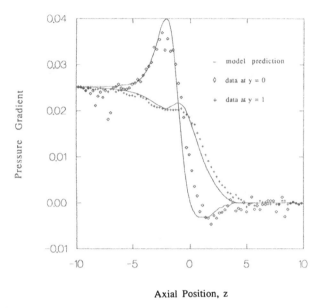

Axial Position, z

Fig. 2. Comparison of pressure gradients with ALEX benchmark data

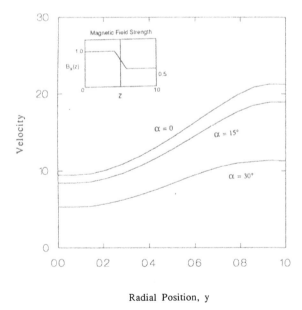

Radial Position, y

Fig. 3. Radial velocity profiles for angled magnetic fields

4. Circular Pipe with Arbitrarily Varying Radius

Another problem solved using the direct integration method consists of a straight circular pipe with a smoothly varying radius located in a uniform transverse field. The geometry is shown in Fig. 4. The radius is given by $R(z)$, where the z coordinate represents the axial distance along the pipe. In this case, symmetry leads to $A_1=0$ and $A_3=0$, and the unknowns are simply p and ϕ – the pressure and electric potential along the center plane of the pipe (x=0).

The normal to the wall, used in Equations 11 and 12, is determined from the slope of the pipe wall. The slope is given by $dR/dz=\gamma$, and the normal is:

$$\hat{n} = \frac{(\hat{x} \sin\theta - \hat{y} \cos\theta)}{\sqrt{1 + \gamma^2}} + \frac{\hat{z}}{\sqrt{1 + 1/\gamma^2}} \tag{15}$$

Numerically, this system of equations is much easier to solve on a rectangular grid. In order to provide this, a coordinate transformation was made to scale the radial coordinate such that it varies between 0 and 1. Rather than the (y-z) plane, the (η-λ) computational plane was used, where:

$$\eta = \frac{y}{R(\lambda)} \qquad \lambda = z \qquad \chi = \frac{x}{R(\lambda)} \tag{16–18}$$

The domain of the solution is now $\chi=0$, $0 \leq \eta \leq 1$, and $0 \leq \lambda \leq L$.

A sample problem shown here is for a simple orifice expansion. The outlet radius is 1.5 times the inlet, with the expansion occurring over 1 pipe radius. The wall conductance ratio is 0.005.

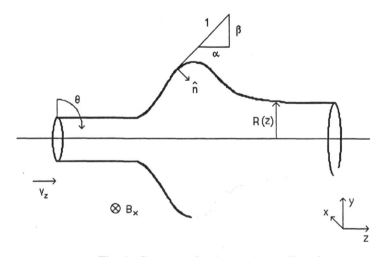

Fig. 4. Geometry for the varying-radius pipe.

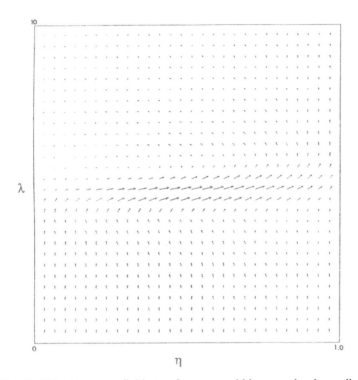

Fig. 5. Velocity vector field at x=0 near a rapid increase in pipe radius

Figure 5 shows the resulting velocity vector field at x=0. The effect of the perturbation is restricted to within a few radii of the region in which the radius is changing.

The pressure varies linearly with λ throughout the bulk of the pipe. Near the perturbation, there is a deep well that is related to the vertical currents. The fluid is accelerated as it enters this region, and decelerated as it climbs up the back side. Near the pipe center, the velocity decreases nearly to zero. Theory predicts for a sufficiently low conductance ratio and rapid expansion that a recirculating eddy may form; however, this is not observed in the present case. (Note: the scaling on the vector field plots tends to exaggerate the vertical component over the axial component. The aspect ratio of the pipe is 10:1.)

References

[1] Kulikovskii A. G., 1968, "Slow Steady Flows of a Conducting Fluid at Large Hartmann Numbers," Fluid Dynamics, vol. 3 (2) 3-10.

[2] Picologlou B. F., Reed C. B., and Dauzvardis P. V., 1986, "Experimental and Analytical Investigations of Magnetohydrodynamic Flows Near the Entrance to a Strong Magnetic Field," Fusion Tech., vol. 10.

EXPERIMENTAL AND THEORETICAL WORK ON MHD AT THE KERNFORSCHUNGSZENTRUM KARLSRUHE THE MEKKA-PROGRAM

L. Barleon, V. Casal, K.J. Mack, H. Kreuzinger,
A. Sterl and K. Thomauske
Kernforschungszentrum Karlsruhe
Institut für Reaktorbauelemente
Postfach 36 40
D-7500 Karlsruhe

ABSTRACT. In order to investigate key features for the development of a selfcooled liquid metal breeder test blanket for the Next European Torus (NET) an extensive experimental program MEKKA (Magnetohydrodynamik Experiment in Natrium-Kalium KArlsruhe) has been initiated at KfK which will be presented.

Experiments with straight ducts of circular or rectangular cross section and with ducts having expansions, contractions or bends will be conducted using a normal conducting 2 Tesla dipole magnet (test volume 0.17 m × 0,5 m × 1,5 m) and eutectic sodium-potassium as liquid metal.

For a further reduction of the MHD pressure drop, the use of Flow Channel Inserts (FCI) within the ducts is foreseen which reduces the effective wall thickness. The basic MHD effects of these FCI's are investigated.

In order to simulate MHD flow in simple geometries computer codes are developed. First theoretical results of MHD flow in a rectangular duct for a Hartmann number of 70 and an Interaction parameter of 1000 are given. The first experimental results of the round duct with and without the FCI confirm the predicted reduction of the MHD pressure drop.

1. INTRODUCTION

Lithium and the eutectic alloy Pb-17Li are attractive breeder materials in fusion reactor blankets. In some concepts liquid metal serves as breeder material only. It circulates slowly for tritium recovery and is cooled by helium or water. A typical example is a concept suggested for NET [1] which uses Pb-17Li as breeder and water as coolant. However, there are concepts in which the same liquid metal serves both breeder material and coolant. Such a self-cooled blanket for NET is under development at Kernforschungszentrum Karlsruhe (KfK). The main advantages are [2]:
- Simple mechanical design with a small number of ducts and welds,
- Reliable afterheat removal by natural convection,
- Higher tritium breeding rate possible than with any other concept.
There are two main issues involved in designing a liquid-metal cooled blanket:
- The corrosion of structural material by the liquid metal limits the maximum interface temperature.
- The strong magnetic field causes a high pressure drop and degrades the heat transport by the liquid metal due to the reduced turbulence.

55

J. Lielpeteris and R. Moreau (eds.), Liquid Metal Magnetohydrodynamics, 55–61.
© *1989 by Kluwer Academic Publishers.*

The proof that this pressure drop can be held low by an appropriate guidance of the flow and a reduction of the wall thickness will decide on the feasibility of such a concept.

The selfcooled liquid metal blanket for NET is characterized by the arrangement of all coolant feed tubes to the blanket at the top of the torus in order to simplify the blanket exchange (fig. 1). This geometry, however, complicates the design of a liquid-metal cooled blanket and increases the magneto-hydrodynamic (MHD) pressure drop because the total blanket cross-section has to be divided into two parts with counter-current flow directions. This results in twice the flow path length and half the flow channel cross-section area compared to a design with access tubes at the top and the bottom.

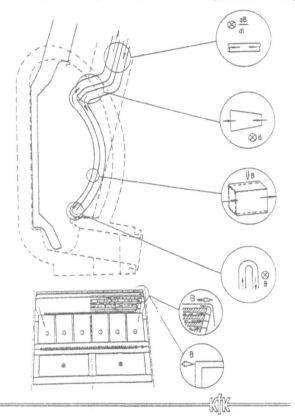

Following this concept key MHD-issues are identified, shown schematically in figure 1, too. There are pressure drop and flow distribution, straight and slightly bent ducts in uniform and highly non-uniform magnetic fields, flow inlet and outlet geometries (expansions and contractions), transitions from pipe to manifolds or vice versa, hairpin shaped bends, abrupt change in flow direction from poloidal to toroidal, and entrance length and flow velocity distribution in toroidal channels.

Fig. 1   The Selfcooled Liquid Metal Blanket Test
           Modul in NET

To overcome the lack of reliable experimental data and to prepare an appropriate analytical design basis the program MEKKA (Magneto-Hydrodynamik-Experiment in Natrium-Kalium KArlsruhe) has been initiated at KfK in 1985. It will be described in this paper.

In an early stage of the blanket design it was recognized that in order to reduce the MHD pressure drop further it is necessary to reduce the wall conductance ratio. For that a new technology called Flow Channel Inserts is developed at KfK. Because the features of this technology were presented in detail in [2] we will present here only the main MHD-features and the experiments done to show the viability of this technique.

2. THE EXPERIMENTAL INSTALLATIONS

For the first step of the experimental program three magnets are available which cover different items of the MHD-program:

A normal conducting dipole magnet (a loan of DESY-Hamburg) is used for the first experiments in straight ducts (see right side of fig. 2). It has a field strenght of 2 Tesla and a test volume of 0,17 m × 0,5 m × 1,5 m. A superconducting Solenoid-Magnet with a field strength of 3.5 Tesla (so-called "CELLO"-magnet), a warm bore of 40 cm diameter, and a length of about 1 m is foreseen to investigate the problem of a poloidal to toroidal bend. A superconducting dipole magnet with 4.5 Tesla, a warm bore of 6 cm diameter and a length of 1 m will be used to investigate the extrapolation to high field strength and high flow velocities in small, presumably insulated flow channels, discussed for liquid metal cooled limiters [3].

The eutectic Sodium-potassium alloy $Na_{22}K_{78}$ is used as liquid metal. NaK is liquid at room temperature thus simplifying the loop design, and most importantly, facilitating the operation of the loop and the conduct of the tests. The loop is designed also to operate at higher temperatures (300 °C) to accelerate and insure the wetting of all MHD-relevant surfaces.

The NaK loop was designed and built at KfK. It consists of a canned motor pump and an additional EM-pump, double tube heat exchanger and an oil cooled cold trap. The flow rate is measured by a gyrostatic flowmeter.

Fig. 2   The Experimental Installation MEKKA

58

3. MHD FEATURES OF THE FLOW CHANNEL INSERTS

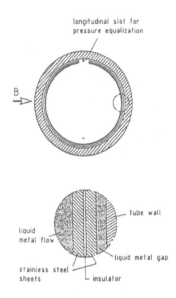

longitudinal slot for
pressure equalization

B

tube wall

liquid
metal flow

liquid metal gap

stainless steel
sheets

insulator

Fig. 3   Principle of a flow channel
          insert

To reduce the MHD pressure drop in the present selfcooled liquid metal blanket design novel flow channel inserts (FCI), as proposed in [2], are employed. With these FCI the load carrying walls and the electric current carrying walls are separated which allows designing the cooling channel structure according to the mechanical needs.

The flow channel insert consists of a thin current carrying inner wall which is insulated by an oxidic layer from the outer wall (fig. 3). The outer wall has a higher wall thickness in order to stiffen the FCI and to support the inner layer. A slot which may be located at any position at the circumference provides the pressure equalization between the duct and the liquid metal filled gap around the insert.

A proof of principle test should show a reduction of the MHD pressure drop according to the reduction of the wall conduction ratio by introducing the flow channel insert. Beyond this test all questions arising from 3-dimensional MHD flow in FCI's like circumferential pressure differences or collapsing forces acting on the thin inner FCI wall have to be investigated.

4. THE EXPERIMENTAL PROGRAM

4.1 The Straight Round Duct without and with Flow Channel Insert

The first experiment in MEKKA is devoted to proof that using Flow Channel Inserts (FCI) reduces the MHD pressure drop .

A straight circular duct with an inner diameter of 130 mm and a wall thickness of 5 mm corresponding to a wall conduction ratio of $C = 0.036$ and a total length of about 6 m is used. Ten axially and 8 circumferentially distributed pressure taps allow far axial and circumferential pressure difference measurements. A flange to attach a traversing mechanism is foreseen to measure velocity profiles at the position of the circumferential pressure taps. A flow straightener at the inlet of the circular ducts consisting of two perforated plates followed by a honeycomb grid flattens the profile a the entrance.

The Flow Channel Insert will be introduced through a flange opening at the downstream end of the test section. The longitudinal slot is aligned with the axial pressure taps on the side to enable axial pressure drop measurements. A simplified version of an FCI is used for this first MHD proof of principle experiment. The 0.5 mm thick inner wall is electrically insulated against the outer wall of the FCI by a 0.5 mm thick ceramic paper.

By introducing the FCI the wall conduction ratio is reduced from 0.036 to 0.0039.

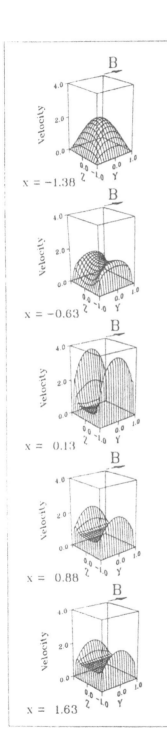

B

Velocity

x = −1.38

B

Velocity

x = −0.63

B

Velocity

x = 0.13

B

Velocity

x = 0.88

B

Velocity

x = 1.63

4.2 The Joint Argonne National Laboratory (ANL)/Kernforschungszentrum Karlsruhe (KfK) Experiment on Flow Tailoring

In a series of two recent papers [4, 5] the idea of improving thermal hydraulic performance by exploiting the electromagnetic forces was advanced. To show the feasibility of such a "MHD flow control" a joint ANL/KfK test was performed. Testing was carried out at ANL's ALEX facility with a test section fabricated at KfK. The test section consists of a series of expansions, contractions and uniform cross section segments. The design of the test section, the MHD analysis carried out by a 3-D computer code and first experimental and theoretical results were presented in [6].

4.3 The Poloidal-toroidal Bend Experiment

The key problem for the design of a selfcooled liquid metal blanket is the pressure drop in a poloidal to toroidal bend.

Up to now two predictions of the pressure drop in such a bend exist which differ by as much as a factor of 10 [3, 7].

Therefore we decided to do a corresponding experiment accompanied by theoretical investigations. A first polidal-toroidal bend experiment will be conducted at the end of 1989 jointly by ANL and KfK in the CELLO magnet in the MEKKA laboratory.

In a knee whose width can be changed stepwise together with dividing the toroidal downstream channels, the pressure drop for different widths, and the velocity distribution will be measured.

5. THEORETICAL WORK ON MHD IN LIQUID METAL

To get a deeper insight into the physics of MHD we try to solve the governing MHD equations numerically. A code has been developed which is capable of simulating the MHD flow in a straight duct of constant rectangular cross-section. Three-dimensional effects are introduced by spatial variation of the magnetic field or the wall conductance ratio, both of which can assume any desired functional form.

The Navier-Stokes-Equations are advanced forward in time by an ADI-method until steady state is reached. At each time step two Poisson-equations for pressure and electric potential are solved by the Fast Poisson Solver H3D of Schumann and Sweet [8]. Spatial discretisation is done on a three-dimensional staggered grid.

With $34 \times 34 \times 34$ grid points the typical cpu-time for a Hartmann-Number M of 50 is 1 hour, rising to 3 1/2 hours for $M = 100$.

As an example of the calculations done, fig. 4 shows the development of the velocity profile in the entrance region of a magnetic field. The field is given by

$$B = (0, B_y, 0) \quad , B_y = 1/(1 + e^{x/x_0}), \quad x_0 = .15,$$

x being the direction of the duct's axis. The plots show the x-component of velocity at different positions around $x = 0$. The parameters used are $M = 100$, N (interaction parameter) $= 10^3$, c (wall conductance ratio) $= .1$.

At the beginning of the field the flow is driven towards the side walls, forming the well-known "M-shaped" profile. Most of the mass-flux is carried by these side-wall jets. As the flow reaches the constant B-field regime, these jets decay, leading to a plug-like profile with only a small overshoot at the side walls. This overshoot is dependent on the wall conductance ratio. With $c = 10^{-3}$, which is a reasonable value for fusion blankets, there is no such overshoot. Nevertheless, the M-shape in the fringing field region still exists. It is caused by induced axial currents which flow in planes perpendicular to B.

6. THE FIRST EXPERIMENTS

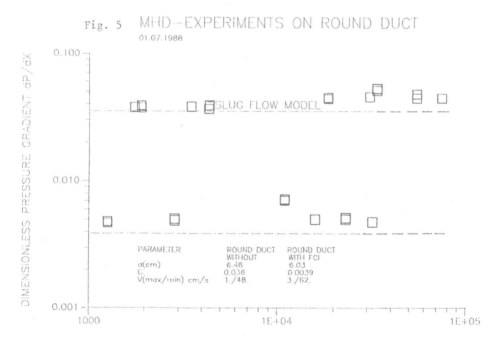

Fig. 5   MHD—EXPERIMENTS ON ROUND DUCT
01.07.1988

The results of the first pressure drop measurements of the thick walled duct are shown in fig. 5, where the dimensionless pressure drop is plotted as a function of the interaction parameter N (upper values) together with the analytical prediction $dP/dx = C/(C + 1) = 0.035$.

The pressure drop measurements were repeated with the FCI moved upstream into the magnet. The results are also plotted in fig. 5 (lower value) together with the analytical prediction for the thin walled (0.5 mm) FCI ($dP/dx = 0.0039$).

The results show the predicted reduction of the pressure drop by a factor of nearly 9.

7. REFERENCES

[1] Dänner, W., Rieger, M. and Verschuur, K.A., 'Progress in Design and Anlysis of the NET Water-Cooled Liquid Breeder Blanekt', Proceedings of the 14th Symposium on Fusion Technology, **Vol. 2**, 1281, Avignon, France, September 7-12, 1986.

[2] Malang, S., Casal, V., Arheidt, K., Fischer, U., Link, W. and Rust, K., 'Liquid-Metal Cooled Blanket Concept for NET", Proceedings of the 14th Symposium on Fusion Technology, **Vol. 2**, 1273, Avignon, France, September 7-12, 1986.

[3] Smith, D.L. et al., 'Blanket Comparison and Selection Study - Final Report', **Vol. 1 through 3**, ANL/FPP-84-1, September 1984.

[4] Walker, J.S. and Picologlou, B.F., 'MHD Flow Control as a Design Approach for Self-Cooled Liquid-Metal Blankets of Magnetic Confinement Fusion Reactors," Fusion Technology, 8, 1, part 2A, 270, 1985 .

[5] Walker, J.S. and Picologlou, B.F., 'Comparison of Three MHD Flow Control Methods for Self-Cooled Liquid-Metal Blankets', Fusion Technology, 10, 3, part 2A, 866, 1986.

[6] Picologlou, B.F., Reed, C.B., Hua, T.Q., Walker, J.S., Barleon, L. and Kreuzinger, H., 'MHD Flow Tailoring in First Wall Coolant Channels of Self-cooled Blankets', Proc. of the ISFNT Meeting, April 10-19, 1988, Tokyo.

[7] Madarame, M., Taghavi, K. and Sillack, M.S., 'The Influence of Leakage Currents on MHD Pressure Drop', Fusion Technology, Vol. 8, 264, July 1985.

[8] Schumann, U., Sweet, R.A., 'Fast Fourier Transforms for Direct Solution of Poisson's Equation with Staggered Boundary Conditions", Journ. Comp. Physics, 75, 123, 1988.

LIQUID METAL TURBULENT FLOW PHENOMENA AND THEIR IMPLICATIONS ON FUSION REACTOR BLANKET DESIGN

S . Sukoriansky[1], H. Branover[1], E. Greenspan[2]
[1]Center for MHD Studies
P.O.B. 653, Beer-Sheva 84105, Israel
[2]Atomic Energy Commission
P.O.B. 7061, Tel-Aviv 61070, Israel

ABSTRACT. The experimental results of heat transfer enhancement in forced liquid metal flows due to the application of a transverse magnetic field are reviewed. It is shown that in non-conducting channels the heat transfer enhancement is due to the creation, by an inverse energy transfer process, of a strong anisotropic turbulence. For this to happen it is necessary to use flow perturbing means (such as grids) which inject to the flow turbulent energy in the form of small scale vortices whose axes are parallel to the field direction. In conducting channels, on the other hand, the magnetic field excites the turbulence without the need for external flow perturbing means. If persisting to the $M > 300$, $Re > 2 \times 10^4$ range arrived at the present experiments, this magnetic field enhanced anisotropic turbulence and heat transfer could significantly improve the design and performance of liquid metal cooled fusion reactors.

1. INTRODUCTION

It is commonly accepted [1-8] that the interaction between the magnetic field and the liquid metal (LM) leads, in the fusion reactor blanket environment, to the suppression of turbulence, i.e., to the establishment of a laminar flow. One consequence of the flow laminarization is a significant reduction in the heat transfer coefficient between the first-wall and bulk LM. When the magnetic field direction is transverse to the LM flow direction, the reduction in the heat transfer coefficient, coupled with an increase in the pressure drop along the LM coolant channel, can significantly complicate the design of self-cooled LM blankets. For example, the poloidal-flow blanket configuration (in which the lithium flows around the axis of the tokamak plasma column) was found [9] to be "geometrically the simplest among all the design options (for a self-cooled liquid metal blanket for tokamaks) considered". However, in designing this blanket it was found [9] that "the average LM velocity required to maintain the maximum interface (i.e., first-wall) temperature at an acceptable level is too high from either the thermal efficiency point-of-view or from the MHD pressure drop (which determines the maximum primary stress level) point-of-view". Consequently, a more complicated toroidal/poloidal flow geometry was selected for the recently designed self-cooled liquid metal blankets [9-11].

The common belief that a strong enough magnetic field suppresses the turbulence in LM flows, thus eliminating the eddy diffusivity and reducing the heat transfer coefficient, is supported by experimental evidence (See review in Refs. 1 and 7). However, more recent experimental results [12-15] indicate that, under certain conditions, the magnetic field can enhance the heat transfer coefficient in LM flows. All of these experiments were conducted with the magnetic field direction perpendicular to the flow direction and, with one exception, using flow channels made of conducting walls (to be referred to as conducting channels). The exception is part of the

63

J. Lielpeteris and R. Moreau (eds.), Liquid Metal Magnetohydrodynamics, 63–69.

64

experiments conducted [13,15] at the Ben-Gurion University (BGU) Center for MHD Studies (CMHDS). Interestingly, each of the above mentioned groups attributed the observed heat transfer enhancement to different phenomena.

In the experiments in Osaka University [12] the heat transfer enhancement is attributed to the interference of the heating element, situated at the center of a cylindrical flow channel, with the LM flow: in the presence of the field the bulk LM flow velocity increases in the channel sections which are perpendicular to the field direction (while the bulk velocity decreases in the channel sections which are parallel to the field). In the experiments at the Tokyo Institute of Technology, [14] on the other hand, the enhancement of heat transfer is attributed to the use of a two-phase medium for the working fluid. In the BGU CMHDS experiments, on the other hand, the heat transfer enhancement is attributed [13,15] to the establishment, by the field, of a strong turbulence and, in the case of conducting channels, also to the magnetic field effect on the velocity profile.

The purpose of the present paper is to review the experimental evidence on the nature of the field enhanced turbulence and heat transfer, and to discuss the implications of these phenomena on the design and performance of LM cooled fusion reactor blankets (including first walls).

The experimental set-up is described in Ref. 13. Fig. 1 shows the flow channel assembly. The heating element provides a uniform heat flux to the upper face of the channel. Most of the measurements pertaining to the non-conducting channel were performed using a grid consisting of 10 mm thick rods spaced 4 mm apart (later referred to as 10/4 mm grid).

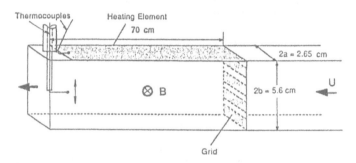

Fig. 1   Schematic diagram of the flow channel used for heat transfer measurements.

2.   NON-CONDUCTING CHANNEL

The effect of the magnetic field on the heat transfer is illustrated in Fig. 2. Shown in the figure is the Nusselt number (Nu) as a function of the interaction parameter ($N \propto B^2/U$) for three Reynolds numbers (Re $\propto$ U). The Nu number arrived at with the maximum magnetic field is significantly higher than the Nu number corresponding to the no field, and even more so, to the field suppressed turbulence flow regime. In the latter regime, Nu $\leq$ Nu($N \approx 0.5$).

Fig. 3 illustrates the effect of the grid orientation on the Nusselt number. When the grid is aligned with its bars parallel to the field direction, the increase in the magnetic field strength can significantly enhance the Nu number. However, when the grid is aligned perpendicular to the field direction, the magnetic field has only a very small effect. In fact, with the perpendicular orientation the grid has practically no effect on the attainable Nu number. The Nu number tendency to increase (in the ⊥B and "no grid" cases) at the high field regime may be due to small turbulence induced at the entrance to the channel.

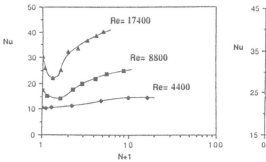

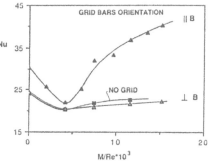

Fig. 2   Effect of the magnetic field on the Nusselt number for liquid metal flow in non-conducting channels. Flow velocity is 15.1 cm/sec (▲), 7.6 cm/sec (■) and 3.8 cm/sec (♦).

Fig. 3   Effect of the turbulence inducing grid orientation on the magnetic field effect on heat transfer in non-conducting channels. Liquid metal flow velocity is 15.1 cm/sec.

Fig. 4 illustrates the magnetic field effect on the LM temperature distribution across the flow channel. It is observed that the application of the field flattens the temperature profile (and reduces the wall-to-bulk temperature difference).

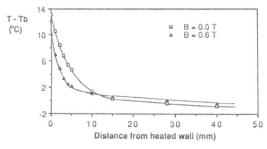

Fig. 4   Effect of the magnetic field on the temperature profile in the non-conducting channel.

The magnetic field effect on the Nu number, illustrated in Figs. 2 and 3 is found to be highly correlated with the field effect on the turbulence structure. The effect of the magnetic field on the turbulence intensity is illustrated in Fig. 5, which represents two sets of measurements using a 5/5 mm grid; in one set the bars are perpendicular to the field direction (Fig. 5a), while in the other set the bars are parallel to the field (Fig. 5b). The honeycomb in these experiments is used to homogenize the velocity profile at the entrance to the test section. It is observed that when the grid bars are oriented perpendicular to the field direction, the magnetic field ($B \propto M/Re$) suppresses the turbulence induced by the grid (Fig. 5a) - as, indeed, is universally expected. However, when the grid bars are oriented parallel to the field (Fig. 5b), the field can significantly augment the turbulence intensity. For this augmentation to take place, the turbulence inducing means (i.e., grid) need be within the influence of the field (compare case IV with cases I to III, Fig. 5b). These and other results (for details see Ref. 16) provide a clear indication that: (a) the magnetic field can significantly enhance the turbulence structure in the plane perpendicular to the field direction, and (b) the decay of this anisotropic turbulence is significantly slower than that of conventional non-MHD shear turbulence.

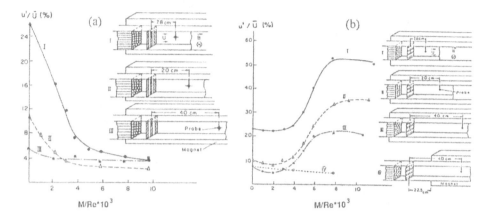

Fig. 5   Effect of the magnetic field on the turbulence intensity in non-conducting channels with the grid bars oriented (a) perpendicular, and (b) parallel to the field direction.

The effect of the magnetic field on the turbulence energy spectrum is illustrated in Figs. 6a and 6b (which correspond to, respectively, Figs. 5a and 5b). The grid orientation is seen to drastically affect the turbulence spectrum evolution in the presence of a magnetic field. When the grid bars are parallel to the field direction (Fig. 6b), the turbulence energy increases with the increase in the magnetic field. The energy density increases most strongly at large scales (small k). However, when the grid bars are perpendicular to the field (Fig. 6a), the turbulence energy decreases with the increase in the field intensity. The above observations, along with additional experimental results brought in Ref. 16, provide a clear indication that turbulence enhancement, which takes place in the plane perpendicular to the magnetic field, results from an inverse energy transfer in which small scale vortices whose axes of rotation are parallel to the field direction grow into large scale vortices. The small scale vortices are created, in the above described experiments, by the grid at the entrance to the channel.

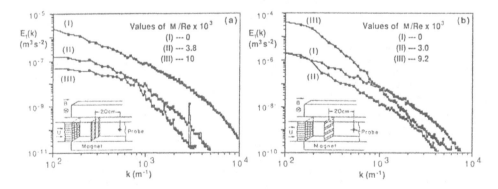

Fig. 6   Effect of the magnetic field on the turbulence spectra in non-conducting channels with the grid bars oriented (a) perpendicular, and (b) parallel to the field direction.

In conclusion, Figs. 3, 5 and 6 show a strong correlation between the heat transfer and turbulence enhancement; an increase in the turbulence intensity (Fig. 5b) is reflected in an increase in the Nu number (Fig. 3; the ‖B case). The enhancement, by a magnetic field, of the heat transfer coefficient in non-conducting channels is due to field created anisotropic turbulence with amplified large scale vortices in the plane perpendicular to the magnetic field.

3.   CONDUCTING CHANNEL

The effect of the magnetic field on the Nu number, temperature, velocity, and turbulence intensity and spectra measured across the conducting channel is illustrated in Figs. 7 to 11. The interaction of the magnetic field with the liquid metal flow in the conducting channel makes it very difficult to accurately calibrate the hot film thermoanemometric probes used for velocity measurements. Also, the adherence of impurities on the probe as well as small temperature drifts introduce experimental errors in velocity measurements. Therefore, the velocity results should be taken as indicating a general trend rather than as providing accurate absolute values.

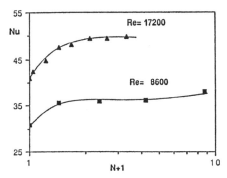 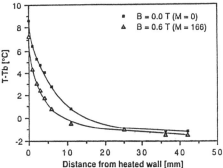

Fig. 7   Effect of the magnetic field on the Nusselt number for liquid metal flow in a conducting channel

Fig. 8   Effect of magnetic field on the temperature distribution across the flow in a conducting channel.

As illustrated in Figs. 7 and 8, the application of the magnetic field improves the heat transfer; the stronger the field the larger the improvement. Surprisingly, this improvement is obtained without the use of any external turbulence inducing means (such as grids).

The improvement in the heat transfer with the application of (and increase in) the magnetic field may be due to the enhancement of turbulence and/or to an increase in the liquid metal velocity near the heated wall. Figs. 9 and 10 indicate that both phenomena take place in our system. The M-shaped velocity profile created by the field (Fig.9) brings about a significant increase in the velocity near the heated wall which could explain, qualitatively, the change of the temperature profile (Fig. 8). Nevertheless, Fig. 10 clearly shows that the application of the magnetic field also significantly enhances the turbulence intensity in the vicinity of the channel wall. The spatial dependence of the turbulence intensity appears to indicate that it is excited by flow instabilities in the vicinity of the peaks of the M-shaped velocity profile (see Fig. 9).

The wide spectrum of fluctuating scales observed in Fig. 11 indicates the existence of turbulence in the vicinity of the channel wall. It therefore appears that the field enhances the heat transfer in conducting channels both by "M-shaping" the velocity profile and by enhancing turbulence. The relative contribution of the two phenomena is yet to be determined.

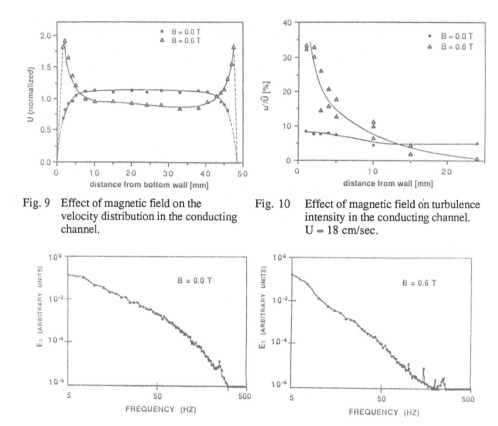

Fig. 9   Effect of magnetic field on the velocity distribution in the conducting channel.

Fig. 10   Effect of magnetic field on turbulence intensity in the conducting channel. U = 18 cm/sec.

Fig. 11   Effect of the magnetic field on the turbulence spectra in the conducting channel. Energy spectra (E_1) are measured 5 mm from the wall parallel to the field.

4.   IMPLICATIONS ON FUSION REACTOR DESIGN

Fusion reactor blanket (and first wall) designs are usually characterized by a higher M and N numbers than accessible with the present experimental setup. Nevertheless, the CMHDS experiments provide no evidence to suggest that the magnetic field enhanced heat transfer will not persist to the MHD flow domain typical of fusion reactors (See Ref. 16 for details). In the following we shall assume that this is really the case.

The anisotropic turbulence which can be established by the magnetic field and the resulting enhancement of the heat-transfer coefficient can improve the design and performance of liquid-metal cooled fusion reactor blankets (including first-wall) in a number of ways, including one or a combination of the following: (a) reducing the coolant flow velocity and, hence, the coolant pumping power requirements; (b) simplifying the blanket design (e.g. by designing a tokamak blanket to have poloidal flow channels rather than a combination of poloidal channels for the bulk of the blanket and toroidal channels for the first-wall [9-11]); (c) reducing the structural material volume fraction (and, hence, improving the blanket tritium breeding ability); (d) increasing the average outlet coolant temperature and, hence, the fusion reactor energy conversion efficiency; (e) increasing the heat flux which can be handled by a liquid metal coolant, thus enabling designing fusion reactors to have a higher first-wall loading and to have liquid-metal (rather than water) cooled high heat-flux components (such as diverter plates).

One illustration of blanket design improvement possibilities is given in Ref. 13. It shows that the magnetic field enhanced turbulence might enable to design a poloidal-flow blanket for a BCSS type tokamak [9,10] to have attractive characteristics: simple geometry, low structure volume fraction, high coolant outlet temperature and relatively low pressure drop.

ACKNOWLEDGMENT

This work was supported by Solmecs (Israel) Ltd. Many thanks are due to Charles Henoch, Dov Klaiman and Ilya Zilberman for their help in the preparation and running of the experiments.

REFERENCES

[1]   BRANOVER H. H., 1967, 'Suppression of Turbulence in Pipes with Transverse and Longitudinal Magnetic Fields,' *Magnetohydrodynamics*, 47, 107.
[2]   HOFFMAN M. A. & CARLSON G. A., 1971, 'Calculation Techniques for Estimating the Pressure Losses for Conducting Fluid Flows in Magnetic Fields,' UCRL-51010, Lawrence Livermore Laboratory,
[3]   HUNT J.C.R. & HANCOX R., 1971, 'The Use of Liquid Lithium as Coolant in a Toroidal Fusion Reactor,' Part I, CLM-R415, Culham Laboratory,
[4]   HANCOX R. & BOOTH J.A., 1971, ibid, Part II, CLM-R116, Culham Laboratory.
[5]   KAMMASH T.,1975, *Fusion Reactor Physics - Principles and Technology*, Ann Arbor Science Publishers Inc., Ann Arbor, Michigan,.
[6]   BRANOVER H., 1978, *Magnetohydrodynamic Flows in Ducts*, John Wiley.
[7]   VISKANTA R., 1985, 'Electric and Magnetic Fields,' *Handbook of Heat Transfer Fundamentals*, W.M. Rohsenow, J. Hartnett and E.N. Ganic, Eds. Chap. 10, McGraw-Hill,
[8]   TAGHAVI K., TILLACK M.S. & MADARAME H., 1987, 'Special Features of First-Wall Heat Transfer in Liquid-Metal Fusion Reactor Blankets,' *Fusion Technology*, 12, 1104,
[9]   ABDOU M. et al., 1983, 'Blanket Comparison and Selection Study,' ANL/FPP-83-1
[10]   CHA Y.S. et al., 1985, 'Design of Self-Cooled Liquid-Metal Blankets for Tokamak and Tandem Mirror Reactors,' *Fusion Technology*, 8, 90.
[11]   MALANG S., 1984, 'Conceptual Design of a Liquid Metal Cooled Blanket for NET,' Internal Report IRB-Nr. 471/84, Karlsruhe Nuclear Research Center.
[12]   MIYAZAKI K. et al., 1986, 'Heat Transfer and Temperature Fluctuation of Lithium Flowing Under Transverse Magnetic Field,' J. of Nucl. Sci. and Technology 23, pp. 582-593.
[13]   SUKORIANSKY S. et al, 1987, 'Heat Transfer Enhancement Possibilities and Implications for Liquid Metal Blanket Design,' Proc. 12th IEEE Symp. on Fusion Engineering, Monterey, CA.
[14]   INOUE A. et al., 1988, 'Studies on MHD Pressure Drop and Heat Transfer of Helium-Lithium Annual - Mist Flow in a Transverse Magnetic Field,' Proc. Int. Symp. on Fusion Nuclear Technology, Tokyo, Japan,   April 1988.
[15]   SUKORIANSKY S. et al., 1988, 'MHD Enhancement of Heat Transfer and its Relevance to Fusion Reactor Blanket Design,' Proc. Int. Symp. on Fusion Nuclear Technology, Tokyo, Japan.
[16]   BRANOVER H. & SUKORIANSKY S., 1987, 'MHD Turbulence with Inverse Energy Cascade and Enhanced Heat Transfer,' in Proceedings of the 25th Symposium on Engineering Aspects of Magnetohydrodynamics, Bethesda, Maryland, 6.4.1.

EXPERIMENTAL INVESTIGATION OF 3-D MHD FLOWS AT HIGH HARTMANN NUMBER AND INTERACTION PARAMETER *

B. F. Picologlou and C. B. Reed
Fusion Power Program
Argonne National Laboratory
9700 South Cass Avenue
Argonne, Illinois 60439   U.S.A.

ABSTRACT. Experimental investigations of 3-D MHD flows in uniform thin conducting wall ducts of circular and square cross section, conducted at Argonne National Laboratory's ALEX facility, are reported. The three-dimensional nature of the flow arises from the spacial variation of the applied transverse magnetic field. Measurements were performed at several Hartmann numbers, M, and interaction parameters, N, with the peak value for M exceeding 6×10^3 and the peak value for N exceeding 10^5. Typical results and their comparison to numerical analysis reported in a companion paper are given, as is a brief description of the ALEX facility and the experimental methods employed. Ongoing activities and plans for future experiments are also discussed.

1. INTRODUCTION

In 1984, the Blanket Comparison and Selection Study (BCSS) [1], an activity that spanned two years and involved the entire fusion technology community of the United States, reached the conclusion that liquid metal cooled blankets represent one of the most attractive design options for tokamak power reactors. In addition, the BCSS identified liquid metal MHD as a feasibility issue for such blankets. This is because, at least for the reactor parameters and the proposed blanket design of the BCSS, uncertainties in the MHD analysis threatened the viability of liquid metal cooled blankets. The uncertainties were brought about by the limited capability for analyzing MHD flows for blanket relevant geometries coupled with a complete lack of detailed experimental data that could be used to support a thermal hydraulic analysis of the blanket.

Given the promise of liquid metal cooled blankets, a liquid metal MHD program was established at Argonne National Laboratory (ANL) for the purpose of developing the analytical tools and providing the experimental data necessary for tokamak blanket design and development. A summary of the analytical work carried out within ANL's MHD program is

---

*Work supported by the U.S. Department of Energy/Office of Fusion Energy under Contract W-31-109-ENG-38.

J. Lielpeteris and R. Moreau (eds.), Liquid Metal Magnetohydrodynamics, 71–77.

given in a companion paper [2]. Detailed description of the ALEX (Argonne Liquid Metal Experiment) facility, specifically designed and built to accomplish the experimental goals of the program, and some experimental results have been presented previously [3], [4], [5], [6].

In the following, the specific goals of our experimental program are reiterated, the capabilities of the ALEX facility and the associated instrumentation system are summarized, the experimental methods used in our investigations are discussed, and representative results from the first two test series are presented. Finally, both ongoing investigations and plans for future experiments are discussed.

2. TEST OBJECTIVES AND THE ALEX FACILITY

The goals of the experimental investigations carried out at ALEX are (a) to provide data of sufficient breadth, resolution, and accuracy for validation of analytical tools and establishing the region of their applicability, and (b) to provide empirical correlations of parameters of engineering importance for these cases which may prove to defy analysis as a result of their geometric complexity or the presence of unusual or unanticipated flow phenomena (e.g., flow instability).

If these goals are to be meaningfully satisifed, both theoretical and experimental investigations should be carried out as close to the prevailing blanket conditions as possible. For ducts in lithium-cooled tokamak blankets, the Hartmann number, M, and interaction parameter, N, are $O(10^3)$-$O(10^5)$ depending on the size of the duct, its orientation with respect to the toroidal magnetic flux density, and the required coolant velocity. It is important to note that, if attention is confined only to first wall coolant channels where the surface heat flux and the high volumetric heat flux deposition make detailed knowledge of the flow structure all the more important, the range of relevant M and N will only extend from $O(10^3)$ to $O(10^4)$. The practical implication of this is that heat transfer experiments involving surface heat fluxes need only cover this smaller range. Single duct isothermal experiments in this range have already been carried out at ALEX.

An additional parameter of paramount importance to the MHD flows under consideration is the wall conductance ratio $c = \sigma_w t_w / \sigma L$, where σ_w, σ are the electrical conductivities of the wall and the liquid metal, and t_w and L are the wall thickness and the characteristic transverse dimension of the duct. For blanket ducts, $c = O(10^{-1})$-$O(10^{-2})$. Also, insulating wall ducts remain a possible option for blanket design and as such need to be investigated. Achieving prototypic values for the wall conductance ratio or using insulating wall ducts in the ALEX experiments is straightforward.

The working fluid for ALEX is the 22Na78K eutectic alloy of sodium and potassium. Not only does it have room temperature properties relevant to MHD thermal hydraulics which are reasonably close to those of lithium at 500°C, but more importantly, it allows testing at room temperature which is essential for detailed local velocity measurements. Such measurements of the flow structure are most important because 3-D

MHD flows at high interaction parameters exhibit unconventional velocity profiles which have a profound effect on heat transfer.

During testing, a variety of flow rates and magnetic flux densities is employed. The NaK working fluid is circulated at flow rates continuously adjustable from 4 to more than 400 liters per minute. The available pressure drop through the test section at 40 ℓ/min is 0.7 MPa. The test articles are located within the highly uniform dipole field of a 2 T conventional electromagnet. The gap between the 0.76 m x 1.83 m pole faces is 20.3 cm. The generous dimension of 0.76 m in the direction transverse to main flow direction allows the use of probe traversing mechanisms and heaters for heat flux experiments simulating first wall heating. The design of the magnet allows the future use of pole face inserts to achieve a variety of magnetic field distributions. The magnet can be moved along the axis of the test article by ±1.22 m from its center position at predetermined constant velocities. This capability not only facilitates servicing of the test section and mounting instrumentation on it but also allows a single instrument to gather data corresponding to any location along the magnet's fringing field.

Although all the tests were conducted at room temperature, the capability of operating the loop at 300°C for a period of several hours to condition the stainless steel test articles after their installation has been proven indispensable. The high temperature operation eliminates the interfacial electrical resistance by reducing the chromium oxide at the inner surfaces. Our experience indicates that errors in excess of 50% can result from lack of such a conditioning.

3. EXPERIMENTAL METHODS

For both the round and square ducts the following measurements were made:

(a) Circumferential distribution of wall potential as function of X/L (X is measured from the edge of the magnet pole face with positive X directed away from the magnet). These measurements are of great diagnostic value because, for the flows under investigation, voltages are essentially constant along magnetic field lines. Hence, wall voltage distributions are translated into voltage distributions in the fluid.

(b) Axial pressure gradient as function of X/L. This is a parameter of engineering importance in that, when integrated along the duct, it provides the overall MHD pressure drop.

(c) Transverse pressure difference as function of X/L. This is the pressure difference developed between the centerline of the duct and the wall in the direction perpendicular to both the magnetic field and the flow. It is the result of axial currents in the fluid driven by 3-D MHD effects. As such, this parameter is a sensitive measure of 3-D effects and its comparison with theoretical predictions is used to ascertain the capability of the latter to model such effects.

(d) Axial velocity at selected transverse locations as a function of X/L. This measurement is made by setting the velocity probe at a fixed transverse location and gathering data while moving the magnet.

(e) Transverse velocity profiles at selected X/L locations. This measurement is made by traversing the velocity measuring probe across the duct while the magnet is stationary.

Measurement of wall voltages is made by stationary electrodes welded on the wall surface or movable spring-loaded electrodes pressed against the wall surface. Measurement of pressures is made with a system of pressure lines, manifolds, and valves using a single pressure transducer. Measurement of velocity is accomplished with a LEVI (Liquid-metal Electromagnetic Velocimeter Instrument) probe. The probe consists of two electrodes whose separation distance of about 1.5 mm is perpendicular to the magnetic field. It is based on a direct application of Ohm's law in moving media and provides accurate measurement of velocity when the current density is much smaller than its short-circuit value. This, of course, is precisely the case for the high Hartmann number flows in thin conducting or insulating wall ducts which are the object of our investigations. The LEVI has proven to be an extremely valuable instrument. It requires no calibration, it is very easy to use, and it has a very fast time response limited only by the response of the voltage measuring instrumentation. Its spacial resolution is determined by the electrode tip separation (1.5 mm in our case). Its output is proportional to the velocity component normal to the distance between the electrode tips averaged over that distance.

4. RESULTS AND DISCUSSION

The computer-driven data acquisition system, and the capability of automated data collection tied to continuous movement of the magnet relative to the instrumented locations of the test article, have been used to collect extensive data at a variety of M and N. Approximately, the range for M extends from 3×10^3 to 6×10^3 for both ducts, whereas the range for N extends from 6×10^2 to 10^4 for the round duct, and 5.5×10^2 to 1.3×10^5 for the square duct. Comprehensive reporting of the results will be given elsewhere. Here, only a small sample of the results is given to provide a demonstration of the variety of data gathered at ALEX and a measure of their quality and resolution.

Figure 1 presents data on the axial pressure gradient along the round duct in the region of the fringing transverse magnetic field. A theoretical prediction for the measured quantity, provided by a fully 3-D numerical solution implemented at ANL [2], is also shown. The numerical solution is based on the equations governing inertialess and inviscid MHD flow (M, N→∞). The only other assumption made is that $t_w \ll L$, so that the electric potential gradient normal to the surface is negligible in the duct wall. The solution contains no adjustable constants. The only input is the distribution of the measured transverse magnetic field, the wall conductance ratio, and the appropriate upstream and downstream boundary conditions (fully developed flow boundary conditions in this case).

Figure 2 presents data on the transverse pressure difference for the square duct. Agreement between analysis and experiment is as exceptional as it was in Fig. 1. It should be pointed out that, in the

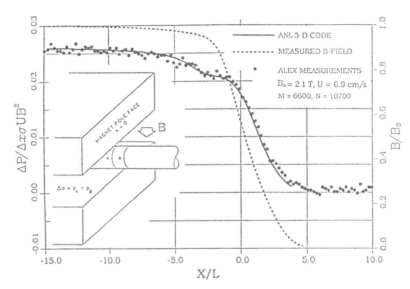

Figure 1. Analysis and experiment for the axial pressure gradient in the
the fringing magnetic field. Round duct with c = 0.027, L = 5.4 cm.

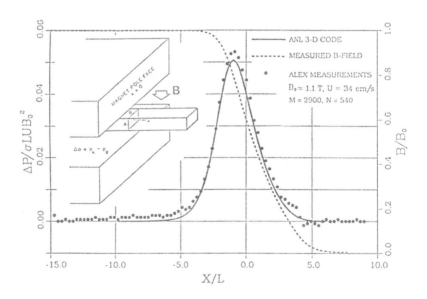

Figure 2. Analysis and experiment for the transverse pressure difference
in the fringing magnetic field. Square duct with c = 0.07, L = 4.8 cm.

fringing field region, axial currents flowing in the walls interact with the transverse magnetic field to cause a pressure difference along the column of liquid metal filling the holes drilled in the wall at the pressure measurement locations. This pressure difference, which depends on the wall thickness, the magnitude of the axial currents, and the location of the holes, modifies the fluid pressure measurement. This effect has been taken into account in the analysis shown in Figs. 1 and 2. Moreover, the theoretical prediction for the pressure gradient accounts for the fact that the experimental pressure gradient is derived as a pressure difference over a finite axial distance of 15.2 cm.

Figure 3 shows a transverse velocity profile for the round duct in the fringing field region. Agreement between analysis and experiment is again outstanding. It is evident that the inertialess inviscid approximation is valid for MHD flows at sufficiently high M and N. The lower limits of M and N for which this approximation is valid will depend on the particular situation, so a general statement cannot be made. Nonetheless, our results indicate that for M and N exceeding 10^3, as they would in most circumstances relevant to fusion blankets, any inertial or viscous effects are likely to be localized.

During our square duct experiments, a laminar instability in the neighborhood of the sidewalls was observed. The presence of the high-velocity jets present in the sidelayers and the associated high velocity gradients are evidently the source of this instability. Full characterization of the instability, and a thorough study of the dependence of its onset on the interaction parameter, Hartmann number, and duct

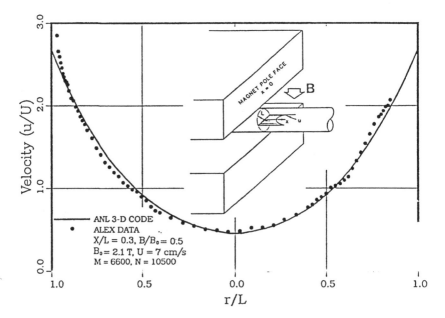

Figure 3. Analysis and experiment for the transverse velocity profile in the fringing magnetic field. Round duct with c = 0.027, L = 5.4 cm.

geometry is an ongoing activity at ALEX. Although the presence of this instability will enhance heat transfer and is therefore desirable, the question about its existence under blanket relevant conditions can only be settled through such thorough study.

Currently, a joint ANL/KfK (Kernforschungszentrum Karlsruhe) activity on proof of principle testing of MHD flow tailoring is being carried out at ALEX. Flow tailoring offers the promise of increased thermal hydraulic performance of the blanket by enhancing desirable features of MHD flows. In the concept under investigation, 3-D MHD flows with high velocities near the sidewalls are created in a uniform magnetic field by variation of the duct geometry (expansions and contractions). Preliminary results support the validity of the concept and are in good agreement with ANL's 3-D MHD code predictions [6].

Following completion of the flow tailoring experiments, extensive investigations of heat transfer in MHD flows in rectangular ducts will be undertaken. In these investigations, uniform heat flux will be applied on a wall parallel to the magnetic field to simulate heating of the first wall in a fusion reactor. ANL's MHD thermal hydraulic code [2] is being used in the detailed design of the test article and its instrumentation. An important objective of the investigations is to gather data on the effect of sidewall instability on heat transfer near fusion reactor relevant conditions.

5. REFERENCES

[1] SMITH, D. L., et al., 1984, 'Blanket Comparison and Selection Study,' Fusion Tech., 8(1), 1.

[2] HUA, T. Q., and WALKER, J. S., 1988, 'Numerical Solutions of Three Dimensional MHD Flows in Strong Uniform Transverse Magnetic Fields,' Proc. IUTAM Symp. on Liquid Metal MHD, Riga, USSR.

[3] REED, C, B., PICOLOGLOU, B. F., and DAUZVARDIS, P. V., 1985, 'Experimental Facility for Studying MHD Effects in Liquid-Metal-Cooled Blankets,' Fusion Tech., 8(1), 257.

[4] PICOLOGLOU, B. F., REED, C. B., DAUZVARDIS, P. V., and WALKER, J. S., 1985, 'Experimental and Analytical Investigations of Magnetohydrodynamic Flow near the Entrance to a Strong Magnetic Field,' Fusion Tech., 10(3), 860.

[5] REED, C. B., PICOLOGLOU, B. F., HUA, T. Q., and WALKER, J. S., 1987, 'ALEX Results - A Comparison of Measurements from a Round and an Rectangular Duct with 3-D Code Predictions,' Proc., IEEE 12th Symp. on Fusion Engr., Monterey, CA.

[6] PICOLOGLOU, B. F., REED, C. B., HUA, T. Q., WALKER, J. S., BARLEON, L., and KREUZINGER, H., 1988, 'MHD Flow Tailoring in First Wall Coolant Channels of Self-Cooled Blankets," Proc. of Intl. Symp. on Fusion Nucl. Tech., Tokyo; to appear in Fusion Eng. & Design.

MAGNETOHYDRODYNAMICS IN NUCLEAR ENERGETICS

V.A.Glukhikh
Efremov Institute of Electrophysical Apparatus
189631 Leningrad
USSR

ABSTRACT. A survey is given on studies being conducted in the USSR on the use of MHD-technology of liquid metals in fast breeders and fusion reactor projects.

1. ELECTROMAGNETIC PUMPS

The industrial application of nuclear power, and primarily, the creation of fast neutron reactors using liquid metal heat transfer (sodium, potassium or their alloys) has in much stimulated the development of "liquid metal" trend in magnetohydrodynamics.

In fast breeders, reactor and steam generator equipment test loops as well as liquid-metal technology various MHD devices are widely used: electromagnetic pumps (EMP), MHD throttles, electromagnetic flow meters and others. The first experiments on using EMP in fast breeders date back to 1955–1963 (EBR-1, DFR, EBR-II reactors).

In the USSR in 1969 first EM pumps were installed in BR-10 reactor. From this time EM pumps are used in auxiliary circuits of all Soviet commercial reactors. They are used in liquid metal charging and cleaning systems, in leak-detection systems of steam generators etc.

A.c. EM induction pumps with cylindrical (CLIP) and helical (HIP) channels are used most extensively. In the Efremov Institute a number of such type pumps has been developed and put into operation (fig.1) [1].The pumps of CLIP type have liquid metal flowrate from 20 to 200 m³/h and rated developed pressure of 0.4 MPa when transferring liquid metal at t° up to 600°. The pumps of HIP-type have flowrate from 2.5 to 20 m³/h at rated pressures of 0.4, 0.8 and 1.6 MPa at the same temperature of sodium. The pumps (HIP up to 350°) and (CLIP up to 450°) have natural cooling. At higher liquid metal temperatures they have indirect water cooling (inductor backs). A number of such pumps has $50-60\times10^3$ trouble-free running hours.

Potential advantages of EM pumps over mechanical pumps make them promising for use in fast breeders main circuits. These are as follows: the possibility of complete pressurization without any seals, the absence of rotating parts, better vibration and noise characteristics,

79

J. Lielpeteris and R. Moreau (eds.), Liquid Metal Magnetohydrodynamics, 79–88.
© *1989 by Kluwer Academic Publishers.*

simple control, convenient and simple maintenance and high reliability.

The Institute has gained a rich experience of manufacture and operation of pumps with flowrates from 150 to 3500 m^3/h in fast breeder main circuits [1]. In the first and second circuits of the BR-10 reactor the four pumps CLIP-3/150 have in total about 120×10^3 running hours. The pumps have linear cylindrical channels and when transferring sodium at 450° they develop a pressure of 0.27 MPa at a flowrate of 150 m^3/h. Their efficiency is 29%. Winding cooling is natural, environment – air, since the reactor is of a loop type.

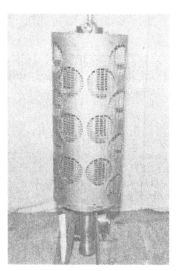 

Fig.1.a.   Fig.1.b.

Figure 1. Electromagnetic pumps: a). CLIP-4/32-2A. b). HIP-4/2.5.

In 1977 in the BOR-60 reactor second circuit the pump CLIP-5/700 was installed which operated without troubles during 34×10^3 hours. Its parameters are: flowrate 625 m^3/h, pressure 0.38 MPa, efficiency 41.8% at sodium temperature of 350°, mass 2800 kg and overall dimensions (diameter × length) 0.6×2.1 m (fig.2). A cylindrical type channel is made with a 180° turn for the flow. The EM pump was located in a mechanical pump tank and, thus, was immersed in liquid sodium. The winding heat was removed by sodium flow in the channel which in a hot state had a reliable heat contact with an inductor.

For the same reactor another immersed design version of pump CLIP-5/850 with forced water cooling of winding was made. But this design was found to be not quite reliable and the pump was not installed in the reactor. During the tests with sodium at 350° the following parameters were obtained: flowrate 830 m^3/h, pressure 0.47 MPa, efficiency 39.5%.

The next important step was the development of the pump CLIP-3/3500 (fig.3) for semicommercial operation in one of the second loops BN-350 fast reactor at sodium temperature of 300°. Test of this pump was carried out in 1986. In nominal regime the following data were obtained : flowrate 3600 m^3/h, pressure 0.3 MPa, consumed power 1 MW, voltage 650 V, current 3000 A at 50 Hz, efficiency 30%. These test confirmed design parameters and showed that the pump met the operation requirements in the reactor second loop. The pump operated at low excessive input pressure and had better vibration and noise characteristics than a mechanical pump.

Fig.2.   Fig.3.

Figure 2. CLIP-5/700.

Figure 3. CLIP-3/3500.

Semicommercial operation of CLIP-3/3500 should confirm the expedience of the replacement of mechanical pumps by EM pumps in fast breeder main circuits. It should be noted that now EM pumps of any power can be built for the loop type reactor with indirect cooling of winding. Immersed pumps (pool type reactor) where a version without external cooling is most promising require commercial production of high temperature (500-550°), high voltage (several kV) winding wires and electroinsulating materials. To our opinion, the prospect of using EM pumps in fast breeder main circuits consists not in a simple replacement of mechanical pumps, but in designing the reactor equipment which most completely uses potential possibilities of EMP. Such an example is given in [2] where EM pump is combined with a heat-exchanger.

Abroad as far as we know the interest to use EM pumps in fast breeder main circuits arises episodically. In 1978 the General Electric Company built an EMP with flowrate of 3300 m³/h and a gaseous nitrogen cooled winding for the FFTF reactor [3]. Unfortunately the

 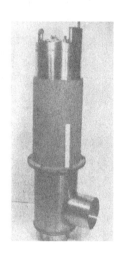

Figure 4. MHD-throttles.

pump was not tested and program financing was stopped. Recently the Company reported on the development of PRISM fast modular reactor using 4 EM pumps in its first circuit. The pump design parameters are: flowrate 2400 m³/h, pressure 0.84 MPa, efficiency 42%, winding temperature 400-450°. External cooling system are nonexistent.

2. MHD THROTTLES

In LM systems of reactors and test loops another type of MHD devices is used, though not so extensively. These are MHD throttles [1]. The throttle provides flowrate control by electromagnetic action on LM flow. Compared to mechanical valves they have the following advantages : smooth precision control of flowrate in pipelines of any diameter, law coefficient of hydraulic resistance in "open" position, high speed of response, reliability and service life. The Institute has developed a number of MHD throttles with constant magnetic fields and cylindrical and helical channels for flowrates from fractions of m³/h to several thousand m³/h and pressures from a few hundredth of MPa to several MPa (fig.4).

3. MHD COUPLERS

When discussing the prospects of induction pump application in auxiliary and main circuits of fast breeder it is interesting to consider the possibility of using for these purposes flow couplers whose principle of operation is described in Patent [4] and consists in hydraulic power transmission from the generator channel to pump channel electrically connected to the former in the presence of external magnetic field.

In the Institute the studies on MHD couplers (flow couplers – FC) are carried out with intervals beginning from 1977. The theory of such

Figure 5. The first MHD-coupler designed. manufactured and tested in the Efremov Institute in 1983.

type devices was developed in [5,6]. The possibilities of using large-scale FC with flowrates of 10^3–10^4 m^3/h and hydraulic power of several MW to pump LM in the fast breeder first circuit were estimated. It was shown that the efficiency of FC can achieve 60–70%. In [7-9] an important question of the influence of end effects and systems demagnetization at the high magnetic Reynolds number on the effectiveness of large-scale FC operation has been studied.

Theoretical investigations show that in spite of large number of design versions. all FC can be divided into flowrate converters when pump sections require higher flowrate than generator sections which can be obtained at much higher pressure in generator sections than in pump sections. and pressure converters, when, on the contrary, in generator sections flowrate is higher and pressure is lower than in pump sections. The difference between them is that magnetic field and / or channel height of their generator and pump sections (size along induction vector of external magnetic field) should be different. This is seen from FC integral characteristics [6] written below

$$\Delta p_g = k \sigma B_g^2 V_g L (\beta_g - S^*) / \alpha.$$

$$\Delta p_p = k \sigma B_g^2 V_g L p_0 (1 - \beta_p S^*) / \alpha,$$

$$\Delta p_p / \Delta p_g = p. \quad Q_p / Q_g = q. \quad Q_{g,p} = V_{g,p} a_{g,p} b_{g,p}.$$

$$p_0 = B_p b_g / B_g b_p, \quad \eta_e = pq, \quad S^* = p_0 q < 1. \tag{1}$$

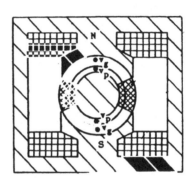

Figure 6. Cross-section of coaxial FC.

G and P refer to generator and pump sections, respectively. Δp – developed pressure, Q – flowrate, V – velocity, B – induction, L – FC effective length, σ – electric conductance of liquid metal, k – form factor, S^* – coupling coefficient, α, β depend on geometry, thickness and electric conductance of channel walls and end effects.

At p>1 we have a pressure converter and at q>1 – flowrate converter.

The first FC designed, manufactured and tested in the Institute in 1983 was a multichannel device (pressure converter) (see fig.5) [15]. FC was tested in a liquid metal circuit with working fluid Na-K (78% K) at temperatures up to 100°. The induction pump installed in the circuit made it possible to study device operation at a pressure up to 0.6 MPa and flowrate up to 50 m³/h in the generator section in a wide range of coupling coefficient S^* at magnetic field induction up to 0.4 T. As tests show, FC of this type has low efficiency because electric current is closed in subchannels but not in a copper busbar, and not only because of poor contact with the copper busbar, but also because of end effect connected with the device finite length.

The negative influence of end effect increases with increasing the flowrate and this design of FC is hardly promising for the first circuit. The ring-shaped FC used in conception of "Double Pool" reactor [10] is also not promising as shown in [8]. Besides, the above mentioned FC refer to pressure convertors, while for liquid metal pumping in the first circuit at the cost of hydraulic power of the second circuit the pressure converter is needed.

Coaxial FC of pipe-in-pipe type where generator and pump channels are located one above another in magnetic field of radial direction are more universal. In the Institute a number of such devices was made and tested in which magnetic field was produced by electromagnets with exciting winding and permanent magnets. The tests confirmed service-ability of such type devices and validity of basic decisions adopted in their designing. The generator and pump channels in models were

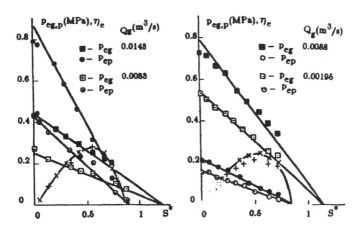

Figure 7. Experimental and theoretical values of pressures and EM ef-ficiency of the device operating as pressuare (a) and flowrate (b) converter.

made by coaxial arrangement of three different diameter pipes fixed by special insertions so that each model had, in principle, two iden-tical FC not connected electrically with each other. Inside the inner pipe there was a ferromagnetic core and the entire device was encom-passed by magnetic system. The difference in height (in magnetic field direction) of generator and pump channels made it possible to study the operation of FC in pressure converter and flowrate converter regimes. Cross-section of a coaxial FC having an electromagnet with core 0.7 m long is shown in fig.6. Generator channels were hydrauli-cally connected in parallel to an external pump and pump channels were (also in parallel) connected to a hydraulic sodium circuit whith another electromagnetic pump and a valve, which made it possible to study FC operation in the whole range of variation of effective coupling coefficient S* [11]. Fig.7 shows experimental and theoretical (curves plotted by formulas (1)) pressure values in generator and pump sections, as well as the value of electromagnetic efficiency of the device operating as pressure (fig.7.a) and flowrate (fig.7.b) converter. As seen from fig.7 the formulas (1) adequately describe FC operation and they can be used for the calculation of such couplers.

It should be noted that the efficiency of the considered coaxial FC is lower than that obtained on special demonstration models [12,13]. But with the increase of FC scale its efficiency should increase. A main advantage of coaxial FC design is its simplicity, ease of manufacture and reliability which makes it possible to consider these FC as modules for the construction of larger couplers, e.g. for the reactor first circuit.

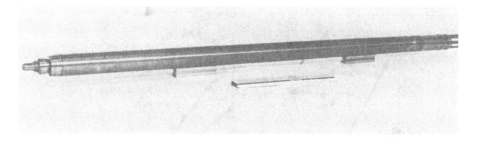

Fig.8. Coaxial FC for BOR-60 reactor.

Besides, the development of fast breeders requires the creation of small-size MHD devices located in the immediate vicinity from the reactor core, and having high reliability at high temperatures and high level of radiation. Devices satisfying these requirements are coaxial flowcouplers whose magnetic fields are produced by high-coercitive permanent magnets.

These FC can be used in independent loops with a circulation circuit (ILC) located in reactor core fuel assembly box. ILC serves to study the behaviour of fuel elements with promising constructional and fissible materials under severe conditions without worsening the reactor safety. Similar devices can be used in failed element detection systems of fast breeders.

Fig.8 shows an MHD device designed and manufactured in the Institute for in - pile investigations on the BOR-60 reactor. The device consists of a coaxial FC using permanent magnets which is controlled by a demagnetization winding, and of a d.c. pump with an annular channel which provides the heat transfer circulation in emergency operation. The device was tested in reactor conditions on the loop with sodium at 450-500°. Achieved design parameters were : $\Delta p \approx 0.2$ MPa, $Q \approx 1.5$ m^3/h at efficiency $\approx$ 12%.

Thus, the possibility was demonstrated to use FC in auxiliary circuits of a failed element detection system, a cleaning system and for in-pile investigations.

Based on operating experience with flow couplers, it may be concluded that they are promising for pumping the LM in the first circuit of the pool type fast breeder at a sacrifice in hydraulic power developed by the second circuit pumps. Quantitative estimates of this scheme effectiveness show that FC are competitive with induction pumps in power characteristics, and have advantages in reliability and in overcoming the problem of emergency cooling.

Now a modular design of FC consisting of separate coaxial couplers located around a reactor core under a heat-exchanger is thought of as suitable.

4. LIQUID METALS IN FUSION REACTORS

Advances in the development of LM technology for fast breeders made it possible to use LM as a universal heat transfer for fusion reactor blankets. LM has perfect thermal properties, low vapour pressure and is radiation-resistant. LM may also be used as raw material for tritium regeneration and medium for neutron breeding, and as a component of protection of magnetic system coils from fast neutrons and gamma rays. In this case Li and Pb and also their alloys are promiising. Main criterion in selecting the heat transfer liquid is reliability and safety of blanket design with selected type of liquid. In this case LM are out of competition because for reactor heat removal with their help it is sufficiently to have low pressures in hydraulic systems (due to high boiling temperature of LM) which simplifies the design and increases its safety.

The realization of such a design, however, encounters serious difficulties connected with pumping the electroconducting LM in strong magnetic fields of the reactor. In order to reduce pressure, efforts were made to find means, providing the reduction of currents flowing in the channel walls. It can be made either by reducing the effective electric conductance of walls (multilayer walls) or electric contact of LM with wall (electroinsulating film coatings). Both methods require serious technological developments whose success is rather problematic. But even in the case of success the design scheme using the LM transported across a strong toroidal magnetic field by pipes to distributing collectors in the blanket region will hardly solve this problem because of rather high pressure losses in such hydraulic system even on the assumption of perfect insulation of walls from LM. Therefore it is necessary to look for other methods to decrease electromagnetic retarding forces, for instance, rational selection of cross-section shape and channel orientation in the magnetic field.

In the Institute a hybrid LM blanket with PbBi heat transfer on the basis of conceptual design [14] is being developed. Design data of resistance coefficient relative to LM flow in slotted channels and first results of experiments on electroinsulating properties of film coatings make it possible to hope for successful realization of the design developed.

In conclusion, another promising field of using LM in fusion reactors should be mentioned: systems of plasma impurity control and first wall protection. Comprehensive information on these systems is given in other reports of the present Symposium.

REFERENCES

[1] Glukhikh V.A., Tananaev A.V., Kirillov I.R., 1987, Magnetohydrodynamics in Nuclear Energetics. M., Energoatomizdat.
[2] Deverge C., Lefrere J.P., Peturaid P., Sauvage M., 1984, 'The Use of Electromagnetic Pumps in Large LMFBR Plants'. In: Liquid Metal Engineering and Technology, Vol.3, 245. Proc. of the 3-d Intern.Conf., Oxford.
[3] Kliman G.B., 1979, 'Large Electromagnetic Pumps', Electrical Machines and Electromechanics. Vol.3, N 3, 129.
[4] Pulley O.O., 1953, 'Improv. in or Relating to Electromagn. Liquid Metal Pump. Syst.', Brit.Pat.745460, cl.35, AIX, 4.11.1954.
[5] Vasil'ev V.F., Kirillov I.R., Lavrent'ev I.V., Semikov G.T., 1978, 'MHD-Generator-Pump System For Fast Breeders', Magnitnaya Gidrodynamica, N 4, 125.
[6] Birzvalk Yu.A., Karasev B.G., Lavrent'ev I.V., Semikov G.T., 1983, 'MHD-Generator-Pump System'. In: Eight International Conference on MHD Energy Conversion. M., v.4, 69.
[7] Birzvalk Yu.A., Lavrent'ev I.V., Shishko A.Ya., 1983, 'Longitudinal End Effect in Flow Coupler'. Magnitnaya Gidrodynamica, N 2, 111.
[8] Lavrent'ev I.V., Shishko A.Ya., 1985, 'Integral Characteristics of Two-Pole Flow Coupler at Finite Magnetic Reynolds Number', Magnitnaya Gidrodynamica. N 2, 103.
[9] Andrianov A.N., Lavrent'ev I.V., Shishko A.Ya., 1987. 'Electrodynamic Processes in Flow Coupler at Large Magnetic Reynolds Number'. 12-th Riga Conf. on MHD. Vol.2, 75.
[10] Hattori S et al., 1987, 'Annular Flow Coupler-Incorporation Double Pool Reactor Approach'. Proc. Int'l Conf. Fast Breeder Systems: Experience Gained and Path to Economical Power Generation. Richland USA (Log N MP2-14).
[11] Andrianov A.N., Karasev B.G., Lavrent'ev I.V., Semikov G.T., Shishko A,Ya., 1987, 'Studies of Operating Characteristics of Coaxial Flow Couplers'. 12-th Riga Conf.on MHD, Vol.2, 87.
[12] Nathenson R.D., Alexion C.C., Xecton A.R., Gray O.E., 1985,'Demonstration of Flow Coupler for the LMFBR'. In: Single and Multiphases Flows in an Electromagnetic Fields, AIAA, New-York, Vol.100, 533.
[13] Miyazaki K. et al., 1987, 'Analysis and NaK Experiment of Electromagnetic Flow Coupler'. Proc.Int'l Conf.Fast Breeder Systems: Experience Gained and Path to Economical Power Generation, Richland USA (Log N TP2-3).
[14] Lavrent'ev I.V., 1987, 'Liquid Metal Blanket of Fusion Tokamak-Reactor'. In: Abstracts of Rep. of 4-th All-Union Conf. on Engineering Problems of Fusion Reactors., Leningrad, January, 1988. M., 326.
[15] Andrianov A.N., Bezgachev E.A., Karasev B.G., Lavrent'ev I.V., 1984, 'Test MHD-Generator-Pump Installation'. 11-th Riga Conf. on MHD, Vol.2, 79.

Poster Session 1:
DC Fields

LIQUID METAL FLOWS WITH POLYDISPERSED SOLID AND GASEOUS INCLUSIONS

V. I. BLINOV, N. Kh. EHRKENOV, Yu. M. GORISLAVETS
A. F. KOLESNICHENKO, A. A. KUCHAEV, A. V. LJUTKEVICH
Inst. Electrodynamiks, Ukr. SSR Acad. Sci., Kiev
252680, Kiev-57, pr. Pobedy, 56
USSR

ABSTRACT. This paper examines the flow of liquid metal with non conducting particles in a cylindrical channel under the influence of radial and azimuthal electromagnetic forces, caused by the interaction of the current, induced in the channel, with internal and external magnetic fields. We obtain solutions of the equations of the electromagnetic field and of the motion of the liquid metal with non-metallic insertions, allowing the study of the behaviour of such particles in the metal fusion flow with regard to its coagulation. The integral characteristics of the flow, including motion velocity, concentration, mass increase of the coagulating particles in time, are defined. Experimental measurements and their comparison with results of the calculation are presented.

Liquid metal flows with non-electroconductive polydispersed inclusions in the magnetic fields are accompanied by separation, their character being mainly defined by the relationship of the force of the electromagnetic and inertial nature, affecting the particles or bubbles of the impurity. The polydispersity of impurity results in a variety of velocities of unequally sized particles (everything which is said further will be true for bubbles as well) and the gas being adhered on their surface, collision of the particles is accompanied by their irreversible coagulation. The density of the conglomerates formed can be less than liquid metal density.

The behaviour of such polydispersed impurity during its rotation around a cylindrical channel axis presents a certain interest. Assume that the movement of liquid metal with particles is possible due to azimuthal electromagnetic forces, which have appeared as a result of the interaction of the channel axis current with intrinsic and external magnetic fields. This case is characteristic of the channel induction furnaces, whose magnetic fields are formed by the inductor and by the auxiliary electromagnetic systems.

First, the electromagnetic field in the liquid metal of the cylindrical channel applying to the channel induction furnace with regard to rotating (azimuthal) movement was defined. The second stage included the calculation of the parameters of the movement of the non-electroconductive coagulating particles in the liquid metal flow.

An idealized calculating model for the definition of the electromagnetic field in the cylindrical channel is provided in Fig. 1. Under the influence of the inductor current I_I the current with the density δ is induced in the channel. Beside the inductor the system contains an auxiliary coil whose current I_c is out of phase with the current I_I. With such a coil available, rotating torque can be achieved causing the azimuthal liquid metal movement in the channel.

Model of the calculation has been constructed with the following assumptions :

1. Inductor, auxiliary coil, cylindrical channel and magnetic core are all axially symmetric around the z axis.

2. Quasi-stationary currents of the inductor and auxiliary coil with frequency ω' are uniformly distributed on their cross section.

J. Lielpeteris and R. Moreau (eds.), Liquid Metal Magnetohydrodynamics, 91–96.
© 1989 by Kluwer Academic Publishers.

3. The magnet is not saturated, its relative magnetic permeability being : μ_F = const.

4. There is an azimuthal speed component U_φ only in the cylindrical channel.

5. Since the size and concentration of the particles are small their influence on the electromagnetic field distribution and metal moving speed in the channel is ignored.

With such assumptions the problem of the electromagnetic field can be solved using the method of integral equations [1, 2]. Because of the accepted idealization, the electromagnetic field of the "inductor-coil-channel" system presented in the cylindrical coordinate system ρ, θ, z is two-dimensional and axially symmetric, so, as it is known, it is enough to make its calculation in one meridian plane θ = const.

The integral equations, describing the distribution of the electric field E in the liquid metal channel and the density of the connected magnetization current σ, at the boundary of the magnetic core regarding the induced density current δ_i, are as follows :

$$\dot{\tilde{E}}(Q) + j\lambda_o\gamma \iint\limits_{Dm} \dot{\tilde{E}}(R)\,T(Q,R)ds_R + j\lambda_o \oint\limits_{c_F} \dot{\tilde{\sigma}}(R)T(Q,R)dl_R +$$

$$j\,\lambda_o \iint\limits_{Dm} \dot{\tilde{\delta}}_i(R)T(Q,R)ds_R = \dot{F}_1\,(Q), \quad Q \in D_m, \tag{1}$$

$$\dot{\tilde{\sigma}}(R) + \mathbf{æ}\!\int\limits_{c_F} \dot{\tilde{\sigma}}(R)S(Q,R)dl_R + \mathbf{æ}\!\iint\limits_{Dm} \dot{\tilde{\delta}}(R)S(Q,R)ds_R = \dot{F}_2(Q),\, Q \in C_F, \tag{2}$$

where $\dot{\tilde{E}}(Q) = \sqrt{\rho_Q}\,\dot{E}(Q)$, $\dot{\tilde{\sigma}}(Q) = \sqrt{\rho_Q}\,\dot{\sigma}(Q)$, $J = \sqrt{-1}$, Q - is the point of observation, R - is the lowing point, $\lambda_o = \mu_o\omega'/2\pi$, $\mathbf{æ} = (\mu_F - 1)\,[\pi(\mu_F + 1)]^{-1}$, D_m - is the zone, which presents by itself the cutting of the liquid metal by the meridian plane, C_F - is the line, marking the magnetic core border line (Fig. 1), γ - is the specific electrical conductivity of the liquid metal, $T(Q,R)$ and $S(Q,R)$ - are the nuclei integral of the equations [1].

Equations (1) and (2) can be written :

$$\dot{F}_1\,(Q) = -j\,\lambda_o \iint\limits_{D_I} \dot{\tilde{\delta}}_I\,(R)T(Q,R)ds_R - j\,\lambda_o \iint\limits_{D_c} \dot{\tilde{\delta}}_c\,(R)T(Q,R)ds_R \tag{3}$$

$$\dot{F}_2\,(Q) = -\mathbf{æ}\!\iint\limits_{D_I} \dot{\tilde{\delta}}_I\,(R)S(Q,R)ds_R - \mathbf{æ}\!\iint\limits_{Dc} \dot{\tilde{\delta}}_c\,(R)S(Q,R)ds_R, \tag{4}$$

$$\dot{\tilde{\delta}}\,(R) = \gamma\dot{\tilde{E}}(R) + \dot{\tilde{\delta}}_i\,(R), \dot{\tilde{\delta}}_i\,(R) = \sqrt{\rho_R}\,[U_\rho\,(R)\,\dot{R}_z\,(R) - U_z\,(R)\,\dot{B}_\rho\,(R)]\,\gamma, \tag{5}$$

where D_I and D_c - are the zones, presenting the meridian plane which cutting the inductor and the coil turns correspondingly, $\tilde{\delta}_i\,(R) = \rho_R\,\dot{I}_I\,W_I/S_I$, $\tilde{\delta}_c\,(R) = \rho_R\,\dot{I}_c\,W_c/S_c$, B - is magnetic field induction, $U_\rho\,(R) = -U_\varphi\,(R)\sin\varphi$, $U_z(R) = U_\rho(R)\cos\varphi$.

Having assigned the velocity U_φ and induction B and having solved the equation system (1)-(2) beforehand, it is possible to define the electrical field E and the current density in the liquid metal. After that the magnetic field induction can be defined by means of vector potential differentiation

$$A(Q) = \frac{\mu_o}{2\pi\sqrt{\rho_Q}} \iint\limits_{D_m} \dot{\tilde{\delta}}(R)\,T(Q,R)\,ds_R + \frac{\mu_o}{2\pi\sqrt{\rho_Q}} \iint\limits_{D_l} \dot{\tilde{\delta}}_l(R)\,T(Q,R)\,ds_R$$

$$+ \frac{\mu_o}{2\pi\sqrt{\rho_Q}} \iint\limits_{D_c} \dot{\tilde{\delta}}_c(R)\,T(Q,R)\,ds_R + \frac{\mu_o}{2\pi\sqrt{\rho_Q}} \oint \dot{\tilde{\sigma}}(R)\,T(Q,R)\,dl_R \,. \tag{6}$$

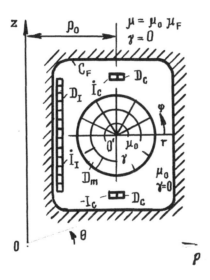

Fig. 1 An idealized calculating model for the definition of the electromagnetic field in the cylindrical channel.

Having calculated the induction $\dot{B}$, it is not difficult to find the azimuthal constituent forces f_φ, causing the liquid metal rotation and the radial forces f_r :

$$f_\varphi = - \mathrm{Re}\,(\dot{\delta}.\, \overset{*}{\dot{B}}_z)\,\sin\varphi + \mathrm{Re}\,(\dot{\delta}.\, \overset{*}{\dot{B}}_\rho)\,\cos\varphi,$$

$$f_r = \mathrm{Re}\,(\dot{\delta}.\, \overset{*}{\dot{B}}_z)\,\cos\varphi + \mathrm{Re}\,(\dot{\delta}.\, \overset{*}{\dot{B}}_\rho)\,\sin\varphi \,. \tag{7}$$

To find the rotating metal velocity, let us consider the channel with the radius R and the length L with the distributed forces f_φ in the cylindrical system r, φ, x (Fig. 1). Let us assume that besides U_φ in general there is axial velocity U_x in this channel, caused by the influence of external forces (the transit flow velocity for instance). Then the equation of the liquid metal movement in the cylindrical channel wil be as follows :

$$U_x \frac{\partial U_\varphi}{\partial x} - \nu \left(\frac{\partial^2 U_\varphi}{\partial r^2} + \frac{1}{r} \frac{\partial U_\varphi}{\partial r} + \frac{\partial^2 U_\varphi}{\partial x^2} - \frac{U_\varphi}{r^2} \right) = \frac{<f_\varphi>}{\rho_m}, \tag{8}$$

where $U_\varphi = U_\varphi (r, x)$, $U_x = U_x (r) = const$, $<f_\varphi>$ - is φ - averaged electromagnetic density in the channel, ν and ρ_m - is kinematic viscosity coefficient and liquid metal density.

The equation (8) has been solved numerically on the basis of the local one-dimensional method [3] under the following limiting constraints :

$$U_\varphi \, (x = 0) \qquad = U_\varphi \, (x = L) \qquad = U_\varphi \, (r = 0) \qquad = U_\varphi \, (r = R) \qquad = 0. \tag{9}$$

The solution for equations (8) and (9) is found in defining the distribution of the azimuthal velocity U_φ along the radius and the direction of the cylindrical channel. Having chosen the part of the channel, where $\partial U_\varphi / \partial x \approx 0$, wa obtain the value of velocity $U_\varphi (r)$, and then U_ρ and U_z (in the coordinate system ρ, θ, z), which can be used for the calculation of the induced currents density δ_i in the equation (5).

Thus, as a result of the common solution of the equations of electromagnetic field (1) and (2) and equation of movement (8), we shall obtain the current density distributions, electromagnetic forces and azimuthal component of the liquid metal velocity in the channel.

The behaviour of the inclusions of impunity depends to a great extent upon the character of its polydispersity.

It is assumed that the initial granulometric function of the polydispersed inclusions corresponds to the logarithmically normal distribution law [4].

$$n(d) = \frac{1}{\sqrt{2\pi} \, d\sigma_z} \, \exp \left[- \frac{(\ln d - \ln d_o)}{2\sigma_z} \right] , \tag{10}$$

where d - is the diameter of the provisionally spherical particle, σ_z - is the standard deviation of the logarithm of the size of the particles.

Let us consider forces, effecting the separate particle

$$m_i \frac{d^2 r}{dt^2} = F_c^r + F_e^r + F_R^r + F_{em}^r + F_K^r ,$$

$$m_i \frac{d}{dt} \left(r \frac{d\varphi}{dt} \right) = F_R^\varphi + F_{em}^\varphi + F_m^\varphi + F_K^\varphi , \tag{11}$$

where m_i - is mass of the particle with regard to joined liquid mass, F_c - is centrifugal force, F_e - is hydrodynamic expulsive force, F_R - is hydrodynamic resistance force, F_{em} - is electromagnetic force, F_K - is Coriolis force, F_M - is Magnus force. These forces are presented in the following way :

$$F_c^r = \frac{\pi}{6} \rho_i d_i^3 r \omega^2, \quad F_e^r = - \frac{\pi}{6} \rho_m d_i^3 r \omega^2, \quad F_R^r = - \frac{\pi}{4} c_d^r d_i^2 \rho_m \left(\frac{dr}{dt} \right)^2 ,$$

$$F_{em}^r = - \frac{\pi}{8} d_i^3 f_r, \quad F_K^r = \frac{\pi}{3} d_i^3 \rho_i \cdot r \frac{d\varphi}{dt}, \quad F_R^\varphi = \frac{\pi}{4} \cdot c_d^\varphi d_i^2 \rho_m \cdot \rho^2 \left(\frac{d\varphi}{dt} \right)^2 ,$$

$$F_{em}^\varphi = - \frac{\pi}{8} d_i^3 f_\varphi , \quad F_K^\varphi = \frac{\pi}{3} d_i \rho_i \omega \frac{dt}{dt}, \quad F_M^\varphi = \frac{\pi}{3} d_i^3 \rho_m \omega_i \frac{dr}{dt} ,$$

where ρ_i - is the apparent particle's density, $\omega = U_\varphi/r$ - is angular velocity of the rotation of the particle in relation to the axis of its own in the non-uniform velocity flow, $c_d = f(Re)$ - is the drag of the spherical particle, Re - Reynolds number.

The speed of the mass increase of the particle of j-fraction on account of the coagulation with the particles of a smaller size of i-fraction is equal to [4].

$$\frac{dm_j}{dt} = \sum_{i=1}^{i} \varepsilon_{j,i} \, k_{j,i} \, m_i \, (m_j - m_i) N_i \,, \tag{12}$$

where $\varepsilon_{j,\,i}$ - is the capture coefficient, taking into consideration the mutual power influence of coagulating particles ($\varepsilon_{j,i} \leqslant 1$), $k_{j,\,i} = \pi/4 \, (d_j + d_i)^2 \, | \, V_j \, V_i|$ is the coagulation constant, V_j, V_i - are j- and i-particles velocities in relation to the liquid metal, N_i - is the specific calculating concentration of the particles of i-fraction.

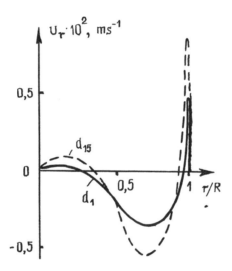

Fig. 2 The particles radial velocity in the channel.

The numerical solution of the equations (11) and (12) allows us to define the parameters of movement of the particles with regard to their coagulation in the rotating liquid metal flow.

The calculation of the liquid metal flow with polydispersed inclusions has been done on the basis of the initial data, which are as follows : channel radius R = 0.07 m, liquid metal - aluminium, initial mass constant of particles $\varepsilon_0 = 0.0025$, their size rang d_{min} - $d_{max} = 10^{-6}$ m - 10^{-3} m is divided into 15 fractions, relative particles density $\rho_i/\rho_m = 0.8$, ampere-turns of the inductor $I_I \, W_I = 5.6$ kA, (I_I , I_c) = 90 degrees. As a result, the distribution of the electromagnetic forces in the channel cross-section is achieved, the velocity of rotating liquid metal movement is defined, as well as the relative velocity of particles movement.

The calculations have illustrated that there are three distinctive zones in the cross-section of the channel and the radial velocity of the particles of each zone has a different direction (Fig. 2). In the by-axial and by-wall zones the particles move to the channel wall, and in the middle zone to the axis. The middle zone expands with the reducing of ρ_i/ρ_m and the increasing of I_cW_c. The heat wall zone supplies the wall by depositing particles, which causes the thickining of the channel wall. Meanwhile the particles from the by-axial zone are coming to the middle zone, where most of the coagulation forming the large particle conglomerates takes place. Thus, the concentration of the particles in the liquid metal is reduced in time because of their depositing on the walls, as well as their being absorbed by large conglomerates, which rise to the surface and move away from the channel.

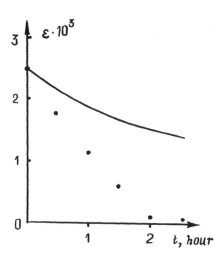

Fig. 3 The variation of the calculating particles mass constant in time.

Fig. 3 illustrates the variation of the calculated particles mass constant ε in time. The experimental data shown by dots in Fig. 3, are obtained on the installation whose parameters correspond to the initial data of the calculation. The differences between calculation and experimental data are caused by ignoring the axial component of the particles movement. Taking into consideration this movement should accelerate the coagulation and therefore the impurity concentration should be reduced more quickly as in the experiment.

References

[1] TOZONY O.V., 1967, "The calculation of electromagnetic fields on the computers" (in Russian), Kiev : Technika, 252.
[2] PETRUSHENKO E.J., TOZONY O.V., 1965, "The calculation of field in the magnetic systems" Science & Technique in the city economy (in Russiant), Kiev : Budivelnik, vol. 3, 3-18.
[3] SAMARSKI A.A., 1961, "Introduction into the computing circuits theory" (in Russian), Moskow : Nauka, 656.
[4] STERNIN L.E., 1974, "The foundations of the gasodynamics of two-phase flowings in nozzles" (in Russian), Moskow : Mashinostroenie, 212.

DISPERSION OF SMALL PARTICLES IN MHD FLOWS

G. GERBETH, D. HAMANN
Central Institute for Nuclear Research, Rossendorf
P.O. Box 19
8051 Dresden
GDR

ABSTRACT. Calculations are presented of the asymptotic dispersion coefficient tensor of small non-conducting particles in a turbulent liquid metal flow exposed to an external homogeneous magnetic field. The equation of motion of a small sphere is used assuming that only the coefficients but not the structure of this equation are changed in a magnetic field. The liquid is assumed to be unbounded. Therefore, the fluid correlation tensor in a magnetic field is mainly determined by the interaction parameter. The dispersion tensor is anisotropic due to the two preferred directions of the system, namely, the directions of the magnetic field and the deterministic particle velocity. Special emphasis is put on the influence of this deterministic velocity on the dispersion tensor due to crossing trajectory effects.

Introduction

The description of a turbulent bubbly flow using a diffusion equation for the local void fraction α (r, t)

$$\left(\frac{\partial}{\partial t} + v_o \frac{\partial}{\partial r}\right) \alpha = \frac{\partial}{\partial r} \left(D \frac{\partial \alpha}{\partial r}\right) \tag{1}$$

is a known method in ordinary two-phase flow. This approach can be extended to a liquid metal gas flow exposed to an external magnetic field /1/. The magnetic field is taken into account by its influence on the diffusion tensor D and the relative velocity of the bubble v_o. Of course, the main problem is the determination of D depending on the homogeneous magnetic field B and the flow parameters. Simplifying assumptions are necessary in order to make an analytical solution possible. Therefore, an unbounded liquid with a constant mean velocity is presumed. In this case, the diffusion tensor is a function of time only. An important characteristic of the particle diffusion process is the constant asymptotic tensor :

$$D = \lim_{t \to \infty} D(t).$$

The aim of the present paper is to determine this asymptotic diffusion tensor depending on the external magnetic field and the deterministic particle velocity.

97

J. Lielpeteris and R. Moreau (eds.), Liquid Metal Magnetohydrodynamics, 97–102.

2. Diffusion Theory

The analysis is carried out in a frame of reference moving with the constant mean velocity of the flow. The fluid is specified by an Eulerian velocity field $u(x, t)$ whereas the particle velocity w is separated into a deterministic and a random part

$$w(t) = v_0 + v(t)$$

The deterministic drift v_0 is caused by gravitational or buoyant forces and is assumed to be constant for convenience. The influence of v_0 on the particle diffusion was first pointed out by Yudine /2/ and is known as the effect of crossing trajectories due to the passing of the particle from one domain of strongly correlated fluid to another. The diffusion tensor is determined by the statistical characteristics of the random part $s(t)$ of the particle trajectory

$$D_{ij}(t) = 1/2 \ dR_{ij}(t) /dt \qquad (2)$$

where

$$R_{ij}(t) = <s_i(t) \ s_j(t)> \qquad (3)$$

is the mean-squared random displacement of a particle. These statistical characteristics are given by a statistical description of the turbulent velocity field and the interaction between fluid and suspended particle. In the absence of a magnetic field isotropic homogeneous and stationary turbulence is assumed characterized by the turbulence intensity u_0 ans the wave number k_0 of the energy containing eddies. The model spectrum of Kraichnan /3/ is used for the Fourier transform $\phi^{(0)}_{ij}(k, \omega)$ of the Eulerian correlation tensor.

The interaction between fluid and suspended particle is given by the equation of motion of the particle. For a small sphere this equation is known as

$$\frac{d}{dt} v(t) + Sv(t) + C \int_0^\infty d\tau dv/d\tau \ (t-\tau)^{-1/2} =$$

$$\left[R \frac{\partial}{\partial t} u(x, t) + Su(x, t) + C \int_0^\infty d\tau \partial u/\partial \tau (t-\tau)^{-1/2} \right]_{x = r(t)} \qquad (4)$$

assuming a small particle Reynolds number, small particle diameter $d << k_0^{-1}$ and $\lambda = d^2 \ u_0 k_0 \ v^{-1} << 1$ (v -kinematic viscosity). The coefficients are constant and given by

$$R = 3\rho f / (2\rho p + \rho f) \quad , \quad S = 36\eta \ d^{-2} / (2 \ \rho p + \rho f) \quad ,$$

$$C = 18 \ (\rho f \eta / \pi)^{1/2 \ d^{-1}} / (2 \ \rho p + \rho f) \qquad (5)$$

where ρp, ρf, η denote particle density, fluid density and dynamic viscosity, respectively. An essential difficulty in calculating D arises from the nonlinearity contained in the response of the particle to the velocity fluctuations : the fluid velocity $u(x, t)$ determining the force on the particle has to be taken at the unknown particle position $x = r(t) = v_0 t + s(t)$. An approximate solution of this problem was

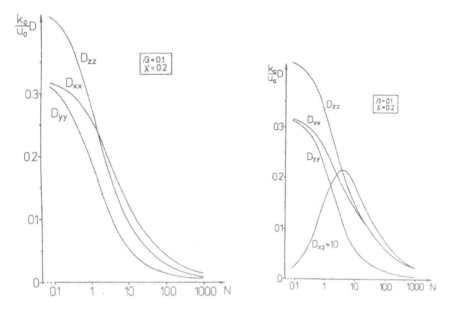

Figure 1 a : Diffusion coefficients in a transverse magnetic field $v_O \sim e_z$, β = const = 0.1, λ' = 0.2.

Figure 1 b : Diffusion coefficients in an inclined magnetic field $v_O \sim e_z$, $B \sim (e_x + e_z)$, β = const = 0.1, λ' = 0.2.

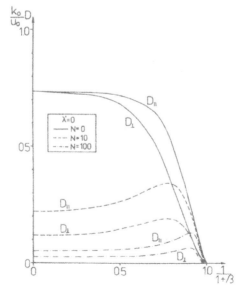

Figure 2 : Diffusion coefficients in a longitudinal magnetic field $v_O \sim e_z$, N = const = 0 ; 10 ; 100, $D_{II} = D_{zz}$, $D_\perp = D_{xx} = D_{yy}$.

given by Reeks /4/ using an iterative procedure and in a complementary way by Pismen, Nir /5/ using the independence approximation. Their analysis was based on Stokes drag only and is extended here to the equation of motion (4).

3. Influence of an External Magnetic Field

Which modifications are introduced by the presence of an external magnetic field ? The magnetic field changes the equation of motion (4), the deterministic particle velocity vo and the fluid correlation tensor ϕ_{ij} (k, ω). The change of v_0 involves a reduction of v_0 due to an increased particle drag and a change of direction if the field is not parallel or perpendicular to v_0. Results on stationary drag coefficients are summarized in /6/. Calculation of the diffusion tensor requires the knowledge not only of the stationary drag but the whole equation of motion (4). Up to now nothing is known about this for MHD flows. For non-conducting particles it is assumed that only the coefficients R, S, C but not the structure of this equation is changed in a magnetic field (R', S', C'). The influence of a magnetic field on the fluid correlation tensor was determined in the second order correlation approximation in /7/ and can be written for Re = u_0 / vk_0 >> 1 in the simple form

$$\phi_{ij}(k, \omega) = \phi_{ij}^{(o)}(k, \omega) \left[1 + (Nk\omega_o/\omega)^2\right]^{-1} \qquad (6)$$

where Nk = N(ekeB)^2 has been introduced. N = σB^2 /ρfu_ok_0 is the ususal interaction parameter and ek, eB denote unit vectors in the direction of k and B, respectively. Thus all suppositions for the calculation of D are given.

4. Results

The final expression for the diffusion tensor is

$$D_{ij} = \int_0^\infty dt \, G_{ij}^s(t) + 2(C'/S')G_{ij}^s(o) \qquad (7)$$

with

$$G_{ij}^s(t) = \int d^3k \, Q_{ij}^s(k, t) \, e^{-1/2 \, k_i \, k_j \, \alpha_{ij}(\tau)} \qquad (8)$$

$$Q_{ij}^s(k, t) = \phi_{ij}^s(k, t) \cos(kv_ot) + \phi_{ij}^a(k, t) \sin(kv_ot) \qquad (9)$$

where ϕ_{ij}s,a are the symmetric and antisymmetric part of the fluid correlation tensor. For a detailed derivation of this result, we refer to /8/. The expression for the tensor α_{ij} in (8) is given in various approximations in /8/, too. The additional term in (7) is due to the long time velocity response imparted by the Basset history force.

What are the general features of these diffusion coefficients ? The diffusion tensor is anisotropic due to the two preferred directions of the problem, namely, the directions of B and v_0. The tensor D depends on the dimensionless parameters N, $\beta = u_0/v_0$ and $\lambda' = 2 (C'/S')^2 u_0 k_0$. This last parameter is unknown and in the following arbitrarily set to $\lambda' = 0$ or $\lambda' = 0.2$ since $\lambda \ll 1$ was a prerequisite of the equation of motion in the case of vanishing magnetic field.

In the following figures the diffusion coefficients are made nondimensional (related to u_0/k_0) and the direction of v_0 is chosen as z-direction. In Fig. 1a, b the diffusion coefficients are presented for $\beta = 0.1$. They show the expected decrease with increasing magnetic field. More interesting is the dependence on β shown in Fig. 2 for a longitudinal magnetic field. In the case of vanishing magnetic field, this dependence is monotonous. This property is destroyed by a magnetic field : the diffusion coefficients increase with decreasing β up to a certain value $\beta = \beta m$.

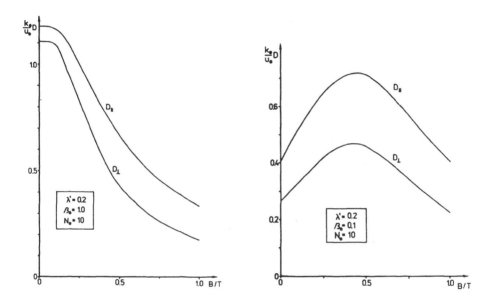

Figure 3 a : Magnetic field dependence of longitudinal diffusion coefficients. $N_0 = 10$, $\lambda' = 0.2$, $\beta_0 = u_0/v_{00} = 1.0$.

Figure 3 b : Magnetic field dependence of longitudinal diffusion coefficients. $N_0 = 10$, $\lambda' = 0.2$, $\beta_0 = u_0/v_{00} = 0.1$.

The two dependences on N and β are combined in Fig. 3 a, b. Here the characteristic parameters $N_0 = N$ (B = 1T) and $\beta_0 = u_0/v_{00}$ are introduced where $v_{00} = v_0$ (B = 0). For the slope of v_0 with increasing magnetic field a realistic curve is used for small bubbles in liquid sodium /1/. The shape of the curves in Fig 3a, b depends strongly on the parameter β_0. This is no unexpected result concerning the strong dependence of D_{ij} on β for $\beta < 1$ in Fig. 2. The interesting result is that the crossing trajectory effect leads to increasing diffusion coefficients in an increasing magnetic field.

5. Conclusions

The diffusion model for a dispersed flow can be extended to an MHD flow. Restricting to the combined effect of an external magnetic field and a deterministic particle velocity on the diffusion coefficient tensor the main result is that increasing D_{ij} are obtained for increasing magnetic field in a certain range of the parameter β due to the crossing trajectory effect. This effect has a significant influence on D_{ij} for $\beta < 1$ with $D_{ij} \rightarrow 0$ for $\beta \rightarrow 0$. In a lot of real situations the order of magnitude of this parameter is $\beta \sim 1$.

The main uncertainly in the presented calculation of D_{ij} is the equation of motion of a sphere in MHD flow. Investigations on this problem are in progress.

References

/1/ GERBETH G., HAMANN D., 1987, Kernenergie, Vol. 30, 149.

/2/ YUDINE M. I., 1958, Adv. Geophys., Vol. 6, 185.

/3/ KRAICHNAN R. H., 1970, Phys. Fluids, Vol. 13, 22.

/4/ REEKS M. W., 1977, J. Fluid Mech., Vol. 83, 529.

/5/ PISMEN L. M., NIR A., 1978, J. Fluid Mech., Vol 84, 193.

/6/ LIELAUSIS O., 1975, Atomic Energy Review, Vol. 13, 527.

/7/ KRAUSE F., RAEDLER K. H., 1980, Mean-Field Magnetohydrodynamics and Dynamo Theory, Akademieverlag Berlin, 133.

/8/ GERBETH G., HAMANN D., 1986, ZfK-601.

TWO-PHASE FLOW STUDIES IN MERCURY-AIR LIQUID METAL MHD GENERATORS

P. Satyamurthy, N.S. Dixit, T.K. Thiyagarajan,
N. Venkatramani, V.K. Rohatgi
Plasma Physics Division
Bhabha Atomic Research Centre
Bombay 400 085, India

ABSTRACT. A two phase liquid metal MHD system consisting of Mercury and Air is built to study MHD Generator behaviour at high void fractions. Average void fraction, pressure, and voltage fluctuations and electrical continuity have been studied. It is found that all the frequency of fluctuations of pressure and voltage are within 100 Hz. There is an increase in the rate of electrical discontinuity between electrodes with increase in void fraction. Gamma ray technique using Cs-137 and Co-60 satisfactorily measures void fraction.

1. INTRODUCTION

Two phase Liquid Metal Magnetohydrodynamic Generator (LMMHD) systems consisting of Liquid Metal as electrodynamic fluid and some suitable vapour or gas for thermodynamic fluid have been proposed for electrical power conversion from various heat sources like solar energy, Fast Breeder Reactor, Industrial waste heat etc. |1| . For solar energy conversion and for Industrial heat sources with source temperatures around 200°-$300^{\circ}C$, the most promising liquid metals are Mercury, Lead-Bismuth alloy etc. |2| . Sodium or NaK is more suitable for Fast Breeder Reactor system.

In LMMHD systems one of the important problem areas is the study and optimisation of the MHD Generator when two phases are present. A large void fraction (around 0.85 at exit) has low frictional losses and reduced inventory of the Liquid Metal |3| and on the other hand, one encounters problems like electrode discontinuity, fluctuations due to slug flow and inefficient energy transfer from thermodynamic fluid to liquid metal (high slip). A detailed study is required to understand the performance of two phase flows at high void fractions.

In view of this a small liquid metal facility consisting of Mercury-Air flow at room temperature has been built and various problems have been studied at different void fractions ($\alpha \simeq 0.2$ to 0.9). Some of the experimental results are presented here.

103

J. Lielpeteris and R. Moreau (eds.), Liquid Metal Magnetohydrodynamics, 103–109.
© *1989 by Kluwer Academic Publishers.*

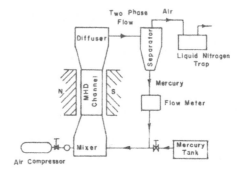

Fig·I Schematic of the LMMHD facility

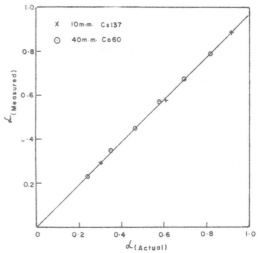

Fig. 2 Measured Vs Actual Void fraction

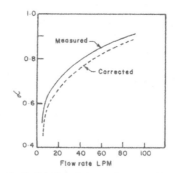

Fig·3· Void fraction as a function of
air flow rate

2. SYSTEM DESCRIPTION

The loop consists of a Mixer, MHD channel ($160 \times 40 \times 5$ mm^3 with
B = 0.1 T) with copper electrodes, diffuser, hydrocyclonic separator,
MHD flow meter as shown in Fig. 1. The diagnostics consists of gamma
ray spectrometer for measuring void fraction |4| , Piezo-electric
transducer for pressure fluctuations, high speed framing mode camera
to study flow pattern. The continuity between electrodes through
mercury has also been measured. In addition to open circuit voltage,
flow rate of air and mercury, static pressure in the channel have been
measured.

3. EXPERIMENTS

3.1. Void Fraction by Gamma Ray Attenuation Technique

Gamma rays from Cs-137 (6 mCi) or Co-60 (5 mCi) has been used to
measure void fraction with NaI(Tl) scintillation detector system. The
void fraction is determined by measuring the intensities I_a of the
unattenuated beam ($\alpha=1$), I_1 the attenuated beam after passing through
mercury only ($\alpha=0$) and I of the beam after passing through two phase
flow. The void fraction is given by |5|

$$\alpha \;=\; \ln \frac{I}{I_1} \bigg/ \ln \frac{I_a}{I_1} \tag{1}$$

The validity of the technique has been proved by simulating an MHD
channel with known void fractions. Fig. 2 gives measured void fractions versus actual void fractions and the agreement is within 5% .
Actual void fraction in the MHD channel has been measured as a
function of flow rate and the results are shown in Fig. 3. Also since
α fluctuates strongly due to slug nature of the two-phase flow, the
dynamic correction has to be applied to get average α |6| . The
corrected average α is also shown in the same figure. As can be seen,
with higher flow rates of air, α tends to flatten. Similar results
have been obtained by Unger et al |7| in mercury-steam system.

3.2. Electrical Continuity

Internal continuity as a function of time has been measured for various
void fractions. A typical recorded curve is shown in Fig. 4. In
Fig. 5, percentage of discontinuity as a function of α is shown. We
see, as expected, the discontinuity increasing with α. It is interesting to see even at relatively smaller α, we see the presence of
discontinuity and this can be attributed to slug nature of flow and
smaller aspect ratio of the present channel.

106

Fig. 4 Recorded Continuity/ OC
discontinuity between electrodes for α = 0.9

SC

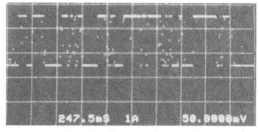

Fig. 5 Percentage of
discontinuity between electrodes as a function of α

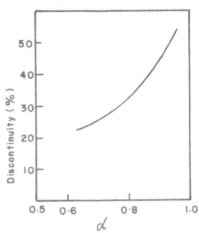

Fig. 6a Open circuit voltage
as a function of time for
α = 0.8 (negative voltage
corresponds to upward flow)

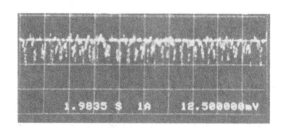

Fig. 6b Frequency spectrum
of voltage fluctuations
(scale : x-axis; 1 div.=3 Hz)

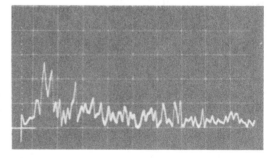

3.3. Voltage Fluctuations

The open circuit voltage fluctuates due to velocity fluctuations as
well as continuity fluctuations. A typical voltage fluctuation and the
frequency distribution of the fluctuation is shown in Fig. 6. There
are no high frequency components and all are within 100 Hz. For an α
of 0.80 the peak voltage was 11 mV and average around 3 mV. Because
of highly fluctuating nature of the flow and strong correlation between
α fluctuations and velocity fluctuations of mercury, we cannot directly
determine mercury velocity from open circuit voltage. However,
assuming homogeneous flow, the approximate average velocity is around
1 m/s.

3.4. Pressure Fluctuations

The pressure fluctuations have been measured using a piezo electric
transducer at the center of the channel and its frequency distribution
is plotted in Fig. 7. Like voltage fluctuations there are no high
frequency components present. Toma et al |8| found high frequency
fluctuations ($\sim$ 1 KHz) near the entrance of the mixer arising due to
velocity slip. It may be concluded here that these waves may have been
completely attenuated at the center of the channel or alternatively at
high void fractions they are not present.

3.5. Flow Structure by High Speed Photography

The flow has been photographed by high speed camera at 500 frames per
second. The nature of flow was slug and visually one could observe
discontinuity between the electrodes some times. In addition some
times there has been a simultaneous back flow near the electrodes. A
typical sequence is shown in Fig. 8.

4. CONCLUSIONS

1) The flow is slug type with large voltage fluctuations. All the
frequency fluctuations at the center of the channel are within 100 Hz
similar to pressure fluctuations.

2) There are no high frequency pressure fluctuations and frequencies
are within 100 Hz in the middle of MHD channel.

3) The Gamma-ray technique satisfactorily measures void fraction of
mercury-air system. Co-60 gives better accuracy when two-phase flow
cross-section increases above 1 cm while Cs-137 can be used for smaller
cross-section.

4) Electrical discontinuity is present for all void fractions (from
0.2 to 0.9). However, the percentage of discontinuity increases sharply
with void fraction. To prevent discontinuity between electrodes the
aspect ratio should be increased.

Fig. 7 Frequency spectrum of
pressure fluctuations
(scale : x-axis; 1 div. = 6 Hz)

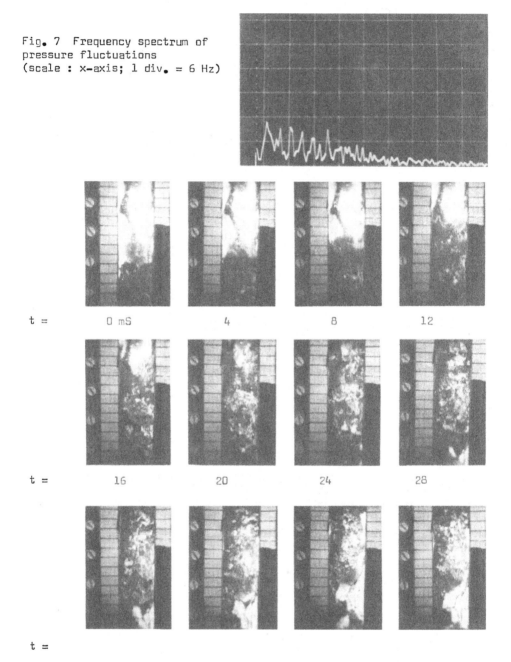

t =   0 mS   4   8   12

t =   16   20   24   28

t =

Fig. 8 High Speed Photographs of the flow (500 frames/second)

REFERENCES

1   1981, Liquid Metal MHD Energy-Conversion System : 'Concept,
    Applications, Status', ANL Report.
2   El-Boher A., Lessin S., Unger Y. and Branover H., 1986,
    'Testing of OMACON Type Liquid Metal MHD Power Facilities',
    9th International Conference on MHD Electrical Power Generation
    Japan, Nov. 17-21, 892.
3   Pierson E. S. and Herman H., 1981, 'Economic Analysis of
    Solar-Powered High-Temperature Liquid Metal MHD Power Systems',
    ANL Report.
4   Petrick M. and Swanson B.S., 1958, 'Radiation Attenuation
    Method of Measuring Density of a Two-Phase Fluid', The Review
    of Scientific Instruments, Vol. 29, No. 12, 1079.
5   Wentz L. B., Neal L. G. and Wright R. W., 1968, 'X-ray
    Measurement of Void Dynamics in Boiling Liquid Metals',
    Nuclear Applications, Vol. 4, 347.
6   Harms A. A. and Forrest C. F., 1971, 'Dynamic Effects in
    Radiation Diagnosis of Fluctuating Voids', Nuclear Science and
    Engineering, Vol. 46, 408.
7   Unger Y., El-Boher A., Lessin S. and Branover H., 1986,
    'Two-Phase Metal-gas Flows in Vertical Pipes', 9th International
    Conference on MHD Electrical Power Generation, Japan,
    Nov. 17-21, 743.
8   Toma T., Yoshino K. and Morioka S., 1986, 'Mechanism of Velocity
    Slip and Associated Turbulence in Accelerating Nozzle Flow of
    Bubbly Liquid', 9th International Conference on MHD Electrical
    Power Generation, Japan, Nov. 17-21, 722.

LIQUID METAL MHD GENERATORS IN TWO-PHASE FLOW SYSTEMS

A. El-Boher, S. Lesin and H. Branover
Center for MHD Studies
P.O.B. 653, Beer-Sheva 84105, Israel

ABSTRACT.   Results of experimental studies of Liquid Metal MHD Faraday type constant cross-section generators are presented in comparison with existing theories. Experiments have been performed on large scale integrated Liquid Metal MHD energy conversion facilities Etgar-3 and ER-4 operating with heavy metals (lead-bismuth alloy, mercury) and steam. It is shown that existing calculation methods are adequate for designing such generators with confidence.

1.   INTRODUCTION

Development of Liquid Metal MHD (LMMHD) energy conversion systems was started in the beginning of the 1960's [1] and since then a number of concepts of such systems have been developed and several systems have been actually built and tested. A relatively complete review of the history of developing LMMHD energy conversion systems can be found in ref. [2,3]. At present the most advanced LMMHD program is the Etgar program elaborated at the Center for MHD Studies of Ben-Gurion University in Israel. The Etgar-3 pilot plant has been operated for about 2000 hours (accumulatively) and the design of the commercial demonstration plant Etgar-5 (1.3 MW_e) is in advanced stages and will be operational in mid-1990. The Etgar program is based on a gravitational heavy liquid metal two-phase flow MHD system concept with a single phase flow constant cross section Faraday type MHD generator. (The operation description of an Etgar type system is given in section 3.)

The present paper concentrates on the performance of a single phase flow LMMHD generator. This relatively simple device has been thoroughly investigated theoretically, beginning from the classical work of Sutton et al. [4].

In this paper experimental results are compared with existing theories, while the role of end effects is specially emphasized. Attention has been given to the problem of electrical contact resistance between the Liquid Metal and the electrodes.

Performance of the generator as a part of the entire two-phase flow LMMHD energy conversion system is also analyzed. Here the phenomena of carry-over of steam into the generator has been specially investigated.

2.   PREVIOUS STUDIES OF LMMHD FARADAY TYPE GENERATORS AND PURPOSES OF THE PRESENT WORK

Early theoretical investigation of Faraday type MHD pumps of the type considered here have been published by Barnes [5] in 1953 and Blake [6] in 1956 which already addressed the problem of end effects and indicated that by changing the shape of the fringing field in the end zones (beyond the electrodes) it is possible to influence the shunting currents and to decrease their negative effect on the efficiency of the device.

A much more expanded and advanced treatment of the same problem, as related also to generators, has been performed in the well known work by Sutton et al. [4]. This work addressed several situations regarding the end effects for loaded generators: abrupt termination of the magnetic field at the ends of the electrodes, overhanging of the magnetic field beyond electrodes and exponential decay of the field in the end zones. One of the most important conclusions of this work

111

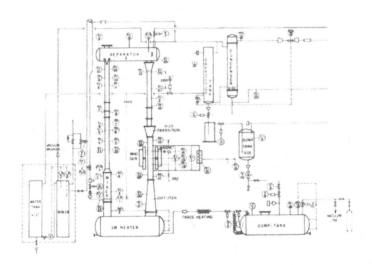

Fig. 1. P&ID of the Etgar-3 facility

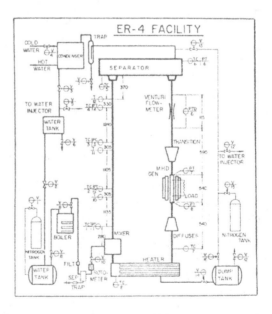

Fig. 2. P&ID of the ER-4 facility

was the desirability of an exponential distribution of the decaying magnetic field in the end zones. Among more recent studies the two-dimensional numerical study by Gherson and Lykoudis [7] should be mentioned. This work considered also the introduction of electrically insulating vanes into the end zones for further decrease of shunting currents and improvement of the generator efficiency. It was shown that an 0.86-0.87 generator electrical efficiency should be practically attainable (not taking in account viscous effects).

A number of LMMHD Faraday type pumps have been studied experimentally [8]. However, single phase constant cross section Faraday type generators remain little investigated experimentally. Some experimental studies have been performed with two-phase generators [9]. However, since those generators are of different geometry (divergent channels) and since the electrical conductivity of a two-phase liquid metal gas flow is not yet precisely known, it is difficult to use these results in relation to single phase generators. Questions of contact electrical resistance between the liquid metal and the generator electrodes, as well as the influence on generator performance of the carry over of steam bubbles into the generator, has not been dealt with so far.

In view of the above the present work was undertaken for: 1) direct experimental verification of Sutton's theory and of the general predictions of numerical studies; 2) evaluation of the performance of a generator under conditions when a small amount of steam bubbles is carried over into the generator; 3) assessment of the contact electrical resistance between the liquid metal flow and the electrodes.

3.   EXPERIMENTAL FACILITIES AND TEST PLAN

The Etgar type LMMHD energy conversion system has been extensively studied at the Center for MHD Studies at Ben-Gurion University.

A schematic of an Etgar system is shown in the Process and Instrumentation Diagram (P&ID) of the Etgar-3 system (Fig. 1). The system consists of two vertical pipes, connected at the bottom and sharing a separator tank at the top. The left pipe is the riser and the right pipe is the downcomer. The system is filled with a hot liquid metal (LM) and circulation of this LM occurs by introducing a thermodynamic fluid, through a mixer, into the riser. The two-phase fluid flows up the riser, driven by the pressure difference between the two columns of fluid. It is separated at the top, and flows down the downcomer as single phase LM. Electric power is generated in a single phase LMMHD generator mounted in the downcomer. Special features of this system and its advantages are analyzed elsewhere [10].

Two liquid metal MHD power conversion facilities: ER-4 facility (Etgar-type research facility which is the fourth in the Etgar series) and Etgar-3 have been built and tested. The LMMHD power conversion systems development program puts special emphasis on extensive supporting studies relted to various aspects of two-phase liquid metal flow phenomena. The results were published elsewhere [11].

3.1   ER-4 Facility

The facility served as an intermediate step in the upscaling from the "table-top" facility to Etgar-3 pilot plant. However, it also provides very convenient conditions for detailed experimental studies of a number of physical phenomena related to liquid metal MHD power facilities in general and the LM natural circulation concept in particular. The operation of the ER-4 facility can be understood from the schematic Fig. 2 presenting the P&ID. The parameters of the ER-4 system are given in Table 1.

The MHD channel was built from four welded stainless steel plates. The inner surface was coated by enamel to give an electrical insulation. The channel has copper electrodes mounted to the channel by screws and a Viton O-ring to prevent leaks. The channel also has a copper compensating bar where the load electrical current flows in the direction opposite to that between the electrodes inside the channel. Thus, the total induced current crossing the magnetic field is zero. The external electrical circuit of the generator (load) is represented by a piece of water-cooled copper pipe. The load resistance can be adjusted by changing the effective length of the pipe and its temperature. The channel does not have insulating vanes for reducing the end-effects. With a load

Fig. 3.   View of Etgar-3 MHD generator

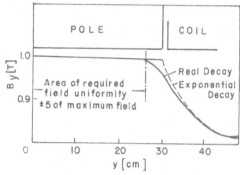

Fig. 4.   Etgar-3 longitudinal magnetic field distribution

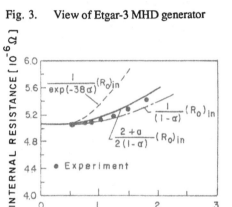

Fig. 5.   Comparison of experimental values with various models for internal resistance $(Ro)_{in}$ vs. LM superficial velocity

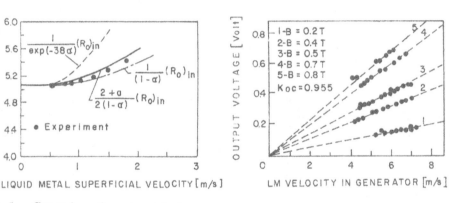

Fig. 6.   Etgar-3 open circuit output voltage vs. generator LM velocity

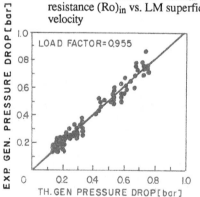

Fig. 7.   Comparison of experimental to theoretical open circuit generator pressure drop in Etgar-3 at different magnetic fields B=0.2-0.8T

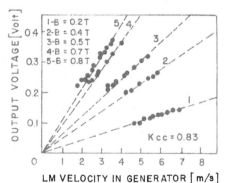

Fig. 8.   Etgar-3 closed circuit output voltage vs. LM velocity in generator at different magnetic fields

resistance of $R_L = 46\mu\Omega$, the average closed circuit load factor calculated according to Sutton [4] is $K_{cc} = 0.725$. The data acquisition system is described elsewhere [10].

The load factor geometry is kept constant in all runs. For a given value of gas flow rate, the magnetic field intensity is varied in steps from 0.0 to 0.8 tesla. This procedure is repeated for different values of gas flow rates, varied in steps from $0.538 \cdot 10^{-3}$ - $4.0 \cdot 10^{-3}$ kg/sec. and mixture quality from $2.0 \cdot 10^{-5}$ to $10.2 \cdot 10^{-5}$, respectively.

3.2 Etgar-3 Facility

Etgar-3 was designed as an integrated power system to both demonstrate the system's unique features and performance potential. It represents the intermediate stage between laboratory scale experiments and an industrial scale demonstration project.

The schematic of the Etgar-3 facility is given in Fig. 1. The parameter of the Etgar-3 system are given in Table 2. The MHD channel, illustrated in Fig. 3, was built similar to ER-4 MHD channel and with an average theoretical closed circuit load factor $K_{cc} = 0.83$.

The operation procedure is similar to that of the ER-4 facility. The procedure is repeated for different gas flow rates, varied in steps from 0.019-0.1 kg/s and mixture quality, varied from $0.5 \cdot 10^{-4}$ to $2.7 \cdot 10^{-4}$. The magnetic field intensity is varied in steps from 0.2 to 0.8 tesla for a given value of gas flow rate.

TABLE 1 - ER-4 system parameters		TABLE 2 - Etgar-3 system parameters	
Liquid metal	Mercury	Liquid metal	Lead Bismuth
Volatile liquid	Steam	Volatile liquid	Steam
High temperature in cycle	431.3 K	High temperature in cycle	423 K
Low temperature in cycle	338.6 K	Low temperature in cycle	338 K
Mixer pressure	5.34 bar	Mixer pressure	4.9 bar
Thermal input	7.0 KW	Thermal input	97.5 KW
Average void fraction in upcomer	0.3	Average void fraction in upcomer	0.4
Effective height of the system	5.0 m	Effective height of the system	7.5 m
Upcomer diameter	0.078 m	Upcomer diameter	0.203 m
Downcomer diameter	0.078 m	Downcomer diameter	0.203 m
MHD generator channel width	0.02 m	MHD generator channel width	0.06 m
MHD generator electrode spacing	0.10 m	MHD generator electrode spacing	0.15 m
MHD generator electrode length	0.35 m	MHD generator electrode length	0.525 m
Magnetic field	0-0.80 T	Magnetic field	0-0.80 T

4. EXPERIMENTAL RESULTS AND DISCUSSION

Before we present and analyze the experimental results it is necessary to clarify some details of the conditions under which the experiments were performed so that the identity of conditions supposed in respected theory and in reality would be established.

First of all we will address the question of the magnetic field distribution in the end zones of the generator. As indicated above, the theory developed by Sutton et al. [4] supposes exponential distribution of the firing fields. Measurements of the magnetic field in the Etgar-3 generator show that between the poles the field is constant within 2% accuracy, while beyond the poles magnetic field distribution is very close to an exponent $B = B_0 \exp\left[-\left(\dfrac{y-0.3}{0.075}\right)\right]$ (see fig. 4).

The next question was related to the contact electrical resistance between the liquid metal in the channel and the electrodes. This question was clarified by measuring the generator internal resistance between the electrodes, using a micro ohm meter, at LM superficial velocities 0-0.5 m/s. At this velocity range no carry over of bubbles into the downcomer was observed, thus it is certain that a single phase resistance value was measured. The experimental values were compared with

116

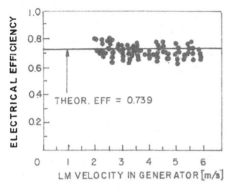

Fig. 9. Etgar-3 electrical generator efficiency compared with the mean theoretical value at different magnetic fields B=0.4-0.8T

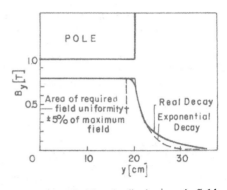

Fig. 10. ER-4 longitudinal magnetic field distribution

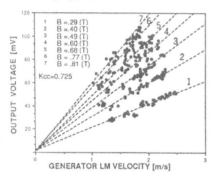

Fig. 11. ER-4 closed circuit output voltage vs. LM velocity in generator at different magnetic fields

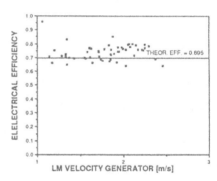

Fig. 12 ER-4 electrical generator efficiency compared with the mean theoretical value at different magnetic fields B=0.29-0.81T

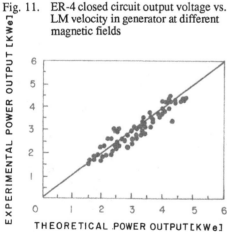

Fig. 13. Comparison of experimental to theoretical power output in Etgar-3 at different magnetic fields B=0.5-0.8T

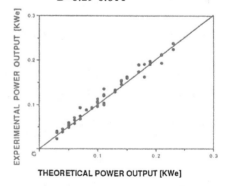

Fig. 14. Comparison of experimental to theoretical power output in ER-4 at different magnetic fields B=0.29-0.81T

calculated values and found to be 4.5% lower than the calculated values (Fig. 5). Thus no measurable contact resistance was found. The deviation can be explained by the insufficient accuracy of the electrical conductivity values given in the Liquid Metal Handbook [12].

Finally, the carry over of steam into the downflow of liquid metal passing the generator and its influence on generator performance was investigated. It is obvious that presence of voids in the generator channel is changing the internal resistance of the genrator and therefore the load factor is also changed. Fig. 5 shows the measured internal resistance values vs. the generator liquid metal velocity. It seems that the electrical resistance does not change with velocity up to 0.5 m/s, and from that velocity and up the internal resistance starts to increase due to the presence of bubble voids between the electrodes. The experimental values were compared with existing models for calculation of two phase conductivity. For low void fraction values up to 8%, the Maxwell [13] model was found to be the most suitable, and was adapted for calculating the generator theoretical load factor. In the Etgar-3 facility the theoretical value of the open circuit load factor is $K_{oc} = 0.971$. In order to obtain the experimental open circuit load factor the voltage between the electrodes was measured at different superficial velocities and magentic fields, as can be seen in Fig. 6. The experimental load factor for all runs was found to be 0.955 (1.6% lower than theoretical). The deviation can be explained by the fact that the magnetic field decay is not exactly exponential. It is supposed that the proportion between the theoretical load factor and the real value remains the same also in the closed circuit case, i.e., the correction factor for the load factor is 0.986.

The next comparison is related to the pressure drop in the genrator using the corrected value for load factor. Since calculations give only the pressure drop due to electromagnetic forces, friction pressure drop has to be subtracted from the measured values. Friction pressure drop was calculated using a semi-empirical formula suggested by Branover [14]. Results for open circuit conditions are compared in Fig. 7.

Comparison of output voltage with theoretical values for a loaded generator ($K_{cc} = 0.83$) is given in Fig. 8. Finally Fig. 9 presents the experimental electrical efficiency of the generator calculated as electrical power output divided by mechanical hydraulic power (calculated from full measured pressure drop minus frictional pressure drop).

The comparisons presented above demonstrate a very fair coincidence of experiment and theory. Figures 10-12 present the results obtained on the ER-4 facility. The process of data derivation was completely similar to that described above. Again, the experimental results fit the theory very fairly.

Since the MHD generator is only one of the components of the Etgar type system, it is interesting to compare also the performance of the entire system with corresponding theoretical predictions. To make calculations for such predictions it is necessary first of all to have full information on the characteristics of the two-phase flow in the riser pipe where the actual conversion of thermal into mechanical energy occurs. These characteristics were established in special detailed experiments at the Center for MHD Studies of Ben-Gurion University, as summarized by El-Boher et al. [11]. Using these results, together with approaches described above in this paper, in relation to single phase flow LMMHD generators, it is possible to predict Etgar systems performance quite satisfactorily, as shown in Fig. 13, and in Fig. 14 for Etgar-3 and ER-4, respectively.

5.   CONCLUSIONS

The comparison of experimental and theoretical values of output voltages, pressure drops, efficiencies, etc., for two single flow liquid metal MHD generators, working within systems with different liquid metals and operated at both open electrical circuit and closed circuit regimes, given in section 4, manifests very fair agreement of experimental and theoretical results. Since the experimental program as presented above addressed also a number of additional problems related to the MHD generator performance - contact electrical resistance, geometry of the decaying magnetic field in the end zones, carry over of voids into the generator channels - it can be concluded that the presently available theory is sufficient for calculating and designing of this type of generators with confidence. This also gives the reason to assume that for large-sized generators, where viscous friction losses are relatively low, a 85-86% generator efficiency is practically attainable as predicted in the numerical studies quoted above [7]. One comment is in place here. The present experimental program did not include testing of generators with insulating vanes in the end zones. Although theoretical calculations have been verified experimentally in a number of different cases and therefore it is very unlikely that the same theory would prove insufficient for the case with vanes, it is still desirable to test generators with vanes directly.

REFERENCES

[1]   ELLIOTT, D., 1962, "Two Fluid Magnetohydrodynamic Cycle for Nuclear-Electric Power Conversion," ARS Journal,
[2]   PETRICK, M. and BRANOVER, H., 1985, "Liquid Metal MHD Power Generation - Its Evolution and Status," Single- and Multi-Phase Flow in an Electromagnetic Field, Progress in Astronautics & Aeronautics, vol. 100, AIAA, New York, pp. 371-400,
[3]   BRANOVER, H., 1986, "Liquid Metal MHD," Proceedings of the 9th Inter. Conference on MHD Elec. Power Generation, Tsukuba, Japan, Vol. V, pp. 1735-1743
[4]   SUTTON, G.W., HURWITZ, H. and PORITSKY, H.Jr., 1962, "Electrical and Pressure Losses in a MHD Channel Due to End Current Loops," Trans. AIEE, Comm. & Electronics, Vol. 80, pp. 687-695.
[5]   BARNES, A.H., 1953, "Direct Current Electromagnetic Pumps" Nucleonics, Vol. 11, p.16.
[6]   BLAKE, L.R., 1956, "Conduction and Induction Pumps for Liquid Metals," Proc. I. Mech. A., p. 49.
[7]   GHERSON, P. and LYKOUDIS, P., 1979, "Analytical Study of End Effects in Liquid Metal MHD Generators," Final Report prepared for ANL, Purdue University, Indiana,.
[8]   BIRZVALK, YU.A., 1968, "The Theory and Calculations of D.C. MHD Pumps," Zinatne Press, Riga (In Russian).
[9]   PETRICK, M., FABRIS, G., PIERSON, E.S., FISCHER, A.K., JOHNSON, C.E., GHERSON, P., LYKOUDIS, P.S. and LYNCH, R.E. 1979., "Experimental Two-Phase Liquid Metal Magnetohydrodynamic Generator Program," Office of Naval Research, ANL, Final Report, ANL/MHD-79-1, U.S.A.
[10]   EL-BOHER, A., LESIN, S., UNGER, Y. and BRANOVER, H,. 1986, "Testing of OMACON Type Liquid Metal MHD Power Facilities," Proc. of the 9th Int. Conference on MHD Elec. Power Generation, Tsukuba, Japan, Vol. 2, pp. 892-902
[11]   EL-BOHER, A., LESIN, S., UNGER, Y. and BRANOVER, H. ,1988, "Experimental Studies of Liquid Metal Two-Phase Flows in Vertical Pipes," Proc. of the First World Conference on Experimental Heat Transfer, Fluid Mechanics and Thermodynamics, Dubrovnik, Yugoslavia.
[12]   LYON, R.N. 1954., Liquid Metals Handbook, Atomic Energy Commission, Office of Naval Research, 2nd Ed., US Government Printing Office, Washington DC, pp. 42-43,
[13]   MAXWELL, J.C., 1904, A Treatise on Electricity and Magnetism, Oxford University Press, Vol. 1, 3rd Ed..
[14]   BRANOVER, H., 1978, Magnetohydrodynamic Flow in Ducts, John Wiley & Sons, New York, N.Y., pp. 133-164,.

MODELLING OF MAGNETOHYDRODYNAMIC TWO-PHASE FLOW IN PIPE

J.P. THIBAULT & B. SECK
Institut de Mécanique de Grenoble
B.P. 53 X
38041 - GRENOBLE Cedex
France

ABSTRACT. One of the possibilities offered by Liquid Metal MagnetoHydroDynamic (LMMHD) conversion to convert heat into electricity is based on the interaction between a conducting Two-Phase flow and electromagnetic fields. Due to the very strong electromagnetic forces created within the flow, the physical behaviour of the Two-Phase MHD flow is completely changed comparatively to an ordinary Two-Phase flow. Consequently an accurate modelling of this original problem (based on the coupling between thermohydraulics and Maxwell equations) is needed. The first part of the paper, devoted to theoretical considerations concludes on the proposition of an averaged form of these equations. The second part presents some of the interesting results of the previous model and their comparison with experimental data.

1. INTRODUCTION

In a liquid-metal magnetohydrodynamic (LMMHD) conversion loop, the generator is the component where the mechanical energy of the flow is converted into electricity. There are two basic options depending on either the gas is injected upflow the generator (two-phase generator) or the gas is injected downflow the generator (one-phase generator). This paper concerns the first solution which offers, in comparison with the second one, the advantage of a reduced difference between the extrema of pressure and velocity in the complete conversion loop. In addition most of the transfer of heat into mechanical energy, which is realized by the expansion of the two-phase flow, takes place directly in the generator. But there are huge difficulties in predicting the physical behaviour of two-phase magnetohydrodynamic (2PMHD) flows, which is needed for performance prediction and design computations. As a matter of fact the very strong J x B forces created within the flow, by the interaction with electromagnetic fields, change completely the dynamic of the flow comparatively to an ordinary two-phase flow. Considering that theoretical and experimental studies of liquid-metal two-phase flow have not been yet very much explored, we have developed a theoretical analysis in order to express the basic equations in

119

J. Lielpeteris and R. Moreau (eds.), Liquid Metal Magnetohydrodynamics, 119–125.
© *1989 by Kluwer Academic Publishers.*

proper form for their numerical solution. Thus the models are able to
simulate a 2PMHD flow to establish what are the measurements needed
for their calibration and extrapolation.

2. FORMULATION OF TWO-PHASE MHD FLOW EQUATIONS

A 2PMHD flow results from the interaction between electromagnetic
fields, obeying Maxwell equations, and a conducting two-phase flow.
The latter consists in two continuous media, obeying conservation
principles, separated by interfaces which can be considered as
discontinuity surfaces. Considering that the present work is mainly
devoted to the modelling of a two-phase flow in a Faraday generator
(see figure 1), some basic hypotheses are justified: 1) The magnetic
Reynolds number is low enough, to consider that the magnetic field B
is not convected by the flow, consequently it remains equal to the
applied magnetic field. 2) The electrodes (see figure 1) are parallel
and their electrical conductivity is large enough in order that they
are considered as equipotential surfaces, consequently the external
electric field E applied to the flow remains constant along the
generator. 3) The mean flow can be assumed as one-dimensional. Finally

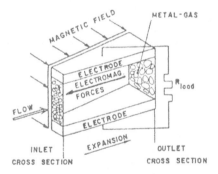

Fig. 1.   Two-phase flow Faraday generator

AVERAGING

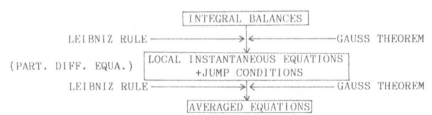

Fig. 2. - Summary of the establishment of averaged formulation of
         a two-phase flow model

based on these hypotheses the Maxwell equations become trivial and the electromagnetic problem is then reduced to what is following ; E the external electric field (constant between the electrodes), B the magnetic field and V the mean velocity of the flow form an orthogonal triedron. Thus the current density J, expressed by the Ohm's law is aligned with E and the J x B forces are aligned on the flow direction.

The classical procedure in establishing two-phase flow equations |1| |2|, consists in two main stages both based on Leibnitz rule and Gauss theorem (see figure 2). Firstly the integral balances (conservation equations for a continuous medium) are transformed to local instantaneous equations associated to local instantaneous jump conditions representing the interfaces. Secondly these equations are averaged. Thus the model is expressed in terms of two set of conservation equations governing the balance of mass, momentum and energy in each phase. However as the latter are not independent of each other, interfacial jump conditions appear as interaction terms in these balances. For 2PMHD flows two specific interaction terms appear, which represent the coupling between the flow and the electromagnetic fields, they correspond respectively to the J x B forces and to the Joule effect.

An example of the final mathematical formulation of an averaged model is following. This is a one-dimensional two-fluid model for two immiscible components (no mass transfer between the phases) :

Mass balances :
$$\frac{\partial}{\partial t}\left(S\alpha_k \rho_k\right) + \frac{\partial}{\partial Z}\left(S\alpha_k V_k \rho_k\right) = 0$$

Momentum balances :
$$S\alpha_k \rho_k\left(\frac{\partial V_k}{\partial t} + V_k \frac{\partial V_k}{\partial Z}\right) + S\alpha_k \frac{\partial P}{\partial Z} + S\alpha_k \rho_k g_z = Fmt_k + Fem_k + \mathcal{P}\tau_{w_k}$$

Thermal energy balances :
$$S\alpha_k \rho_k\left(\frac{\partial h_k}{\partial t} + V_k \frac{\partial h_k}{\partial Z}\right) - S\alpha_k\left(\frac{\partial P}{\partial t} + V_k \frac{\partial P}{\partial Z}\right) = Qi_k + Q_{J_k}$$

Constitutive Ohm's law : $J_k = \sigma_k C_c\left(E + V_k B\right)$

Constitutive state laws : $\rho_k = \rho_k\left(P, h_k\right)$

The subscript k refers to the medium : k = 1 for the liquid and k = g for the gas. Z is the axial coordinate, t the time, S the cross section of the pipe, $\mathcal{P}$ its wetted perimeter and P the mean pressure. The phasic variables are respectively the mean values of ρ the density, V the velocity, h the enthalpy and α the surfacic void fraction. The phasic interaction terms on the right hand side, are respectively Fmt the interfacial momentum transfer, Qi the interfacial heat transfer, τ_w the wall friction stress, Fem the unitary J x B force and Q_J the Joule effect (heat source). The two last terms need the Ohm's law which expresses J the phasic current density as a function of σ the phasic electrical conductivity, C_c the coefficient of conductivity (depending on the flow topology and on the external fields).

122

In the previous set of equations only the interaction terms contain some uncertainties because their formulation is necessarily semi-empirical. Most of these terms are well established for ordinary two-phase flow |3| and/or for liquid-metal MHD flow |4|, but not for 2PMHD flow. Thus considering the very rare works on this subject, we decided to introduce ordinary two-phase flow transfer laws with MHD modified coefficients (i.e. for wall friction or interfacial draging...)

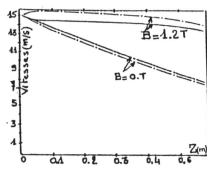

Fig.3 - Axial evolution of the velocities (liquid——; gas—.—.) with and without electromagnetic field

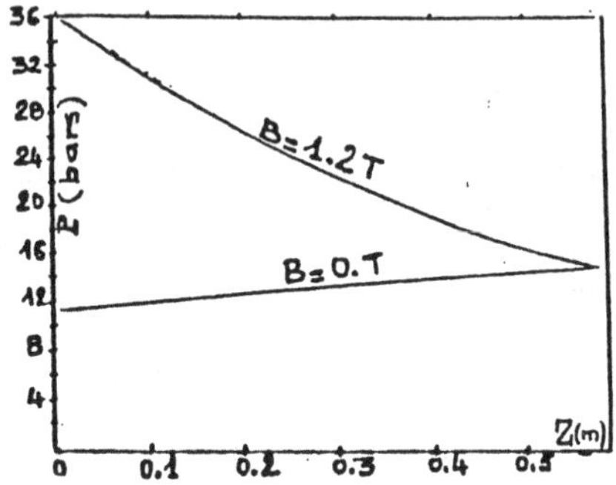

Fig.4 - Axial evolution of the pressure with and without electromagnetic fields

3. NUMERICAL TREATMENT

The numerical treatment of the previous set of partial differential equations is done by a code called FARADEX which includes in fact two models. The first one, called pre-solver, can be run optionally, it solves a simplified steady state homogeneous model |5|. It computes the idealized channel geometry based on criteria of both constant mixture velocity and electrical load factor. The second one, the main solver, computes the previous two-fluid model for an evolution of the channel geometry which is either the one computed with the pre-solver or the one which is fixed by the data file (i.e. the test section of an experimental facility). It is based on a fully implicit finite difference discretisation scheme developed for ordinary two-phase flow |6| using the well known ICE (Implicit Continuous-fluid Eulerian) method.

Presently FARADEX is capable of simulating the following pairs of components : tin + steam (Sn + H_2O), mercury + steam (Hg + H_2O), lead-bismuth + steam (Pb - Bi + H_2O) and mercury + nitrogen (Hg + N_2). This selection corresponds to the experimental facilities of the Ben Gurion University (Israel) |7| and of the Institut de Mécanique de Grenoble (France). Note that the addition of any new pair of components, only necessitates the introduction of new state laws. We plan to introduce lithium + cesium (Li + Cs) and sodium-potassium + nitrogen (NaK + N_2) in a near future.

4. RESULTS AND DISCUSSION

It is not possible in this paper, to present a complete selection of curves (i.e. pressure, void fraction, velocities, temperatures, current density,...) but, in most of the cases, pressure and/or velocities curves are enough significant. Thus corresponding to the following computation conditions : tin + steam flow, electric power 50 kW, inlet-outlet generator cross section 15 cm^2 – 26 cm^2 , electric gap 12 cm, inlet void fraction 0.4, temperature 570 K ; figures 3-4-5 show that the main controlling effect on the dynamic of the flow is due to the J x B forces. The comparison of predicted Velocities (see figure 3) and pressure (see figure 4) corresponding respectively to a two-phase flow diffuser (B = O.T) and to a two-phase flow MHD generator (B = 1.2 T) are significant from this point of view. On figure 3 two main differences appear : the sign of the relative velocity changes and the evolution of the mixture velocity is completely changed. On figure 4, note that the downflow pressure is needed as initial condition due to the numerical model, the pressure gradient is strongly changed since the inlet pressure jump from 11.5 10^5 Pa for B = OT to 36.2 10^5 Pa for B = 1.2 T. Note that for B = 1.2T the artificial annihilation of wall friction, nothing else being changed, reduces of about 4 10^5 Pa the inlet pressure |8|, that corresponds to a very moderate reduction of the pressure gradient in comparison with the effect of J x B forces. On the other hand figure 5 shows that changing from an idealized electrical conductivity coefficient (Maxwell law) to an empirical one (Petrick and Lee law) |9| has a very strong effect since the pressure gradients are about in the ratio of one to two.

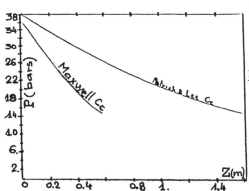

Fig. 5 - Axial evolution of the pressure with electromagnetic fields for two different conductivity coefficient

NB. the total length of the optimized Faraday generator, computed in both cases by the presolver, is 0.58 m for the Maxwell conductivity coefficient and 1.58 m for the Petrick & Lee conductivity coefficient

Based on the numerous closure laws and parametric sensitivity analyses that we done, and considering the hard difficulties of two-phase flow measurements we conclude that the first objective of the experimental analysis of 2PMHD flow is the calibration on the pressure evolution, with the aim of accurately expressing the electrical conductivity coefficient which controles mainly the pressure gradient. For that the first stage consists in verifying the accuracy of the model for ordinary liquid-metal two-phase flow. So figures 6 and 7 show the very good agreement between the measured pressures (data of the Ben Gurion University) and the computed one. Figures 6 and 7 have respectively been obtained on a mercury + steam (Hg + H_2O) facility and on a lead-bismuth + steam (PbBi + H_2O) facility |10|. These very good results allowed us to go to the next step which induced us to built an experimental "room temperature" facility using mercury as liquid-metal in a closed cycle and nitrogen as gas in open cycle. This facility is intented to produce a high velocity (about 10 m/s), high void fraction (up to 80 %) two-phase flow and to submit it to a magnetic field adjustable up to 1.2 T. It is now in test, we perform pressure and pressure gradient measurements on the test section (1 m length, 7 x 1 cm^2 cross section) through 2 x 9 measurements posts. The experimental program comprises four items which correspond to the measurements of: 1) pressure and pressure gradient (actual status), 2) temperature, 3) apparent electrical conductivity, 4) Void fraction, in any case with and without magnetic field. Two generations of test sections will be tested, the actual one is an ordinary constant rectangular cross section pipe and the next one will be a Faraday generator including segmented electrodes for electrical conductivity measurements.

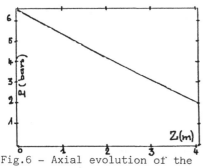

Fig.6 - Axial evolution of the pressure (computed ———— & experimental ___.___.) for a Hg + H_2O flow at about 1.2 m/s without electromagnetic fields

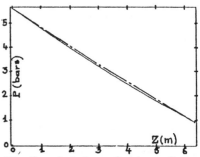

Fig.7 - Axial evolution of the pressure (computed ———— & experimental ___.___.) for a PbBi + H_2O flow at about 2 m/s without electromagnetic fields

CONCLUSION

The work done on modelling of 2PMHD flows abiding by the rules governing the construction of two-phase flow models, allowed us to propose a two-fluid monodimensional model and its numerical solution. The numerous analysises done with the latter show that the dynamic behaviour of the 2PMHD flow is mainly controlled by the J x B forces and that the pressure gradient prediction is strongly connected to the expression of the apparent electrical conductivity of the two-phase flow. On the other hand this advanced model constitutes a very good mean of experimental interpretation. In example the adjustment of computed pressure with experimental data gives a very good prediction of the evolution of the other flow parameters as void fraction or phasic velocities. Thus the achievement of a joined theoretical and experimental program can be rendered very powerful. The first step consists in a calibration on experimental data of extraneous laboratories and the second step consists in the running of our experimental 2PMHD facility. Finally all this knowledge on two-phase Faraday generator contributes to the accuracy of our performance evaluation for the potential applications of LMMHD conversion.

REFERENCES

|1| ISHII M. Thermo-fluid dynamic theory of two-phase flow , Eyrolles, Paris

|2| DOBRAN F., 1981, 'On the consistency conditions of averaging operators in two-phase flow and on the formulation of magnetohydrodynamic two-phase flow', Int. J. Eng. Sci., 19, 1353-1358

|3| HESTRONI G., 1982, Handbook of multiphase systems, Mac Graw Hill Book Comp.

|4| BROUILLETTE E.C. & LYKOUDIS P.S., 1967, Magneto-fluid-mechanic channel flow, (I & II), The Phys. of Fluids, 10-5, 995-1007

|5| THIBAULT J.P., 1983, 'Générateur de Faraday à métal liquide', Thèse de l'Institut National Polytechnique de Grenoble

|6| ROUSSEAU J.C., 1984, 'Module de base du code Cathare développements physiques et performances'. La Houille Blanche, 3/4, 199-208

|7| BRANOVER H., 1986, 'MHD national program of Israel'. 9th Int. Conf. on MHD Electrical Power Generation, Tsukuba, Ibaraki, Japan, 1771-1774

|8| SECK B., 1987, 'Contribution à la modélisation des écoulements diphasiques MHD : évaluation des pertes de pression par frottements pariétaux'. DEA d'Energétique physique, I.N.P. Grenoble

|9| TANATAGU N., FUJII-E. Y. & SUITA T., 1972, 'Electrical conductivity of liquid metal two-phase mixture in bubbly and slug flow-regime'. J. of Nucl. Sci. and Techn., 9-12, 753-755

|10| BRANOVER H., 1987, 'Table of experimental data', personal meeting at the Institut de Mécanique de Grenoble.

MELT MAGNETOHYDRODYNAMICS OF SINGLE CRYSTAL GROWTH

A.Bojarevics, Yu.M.Gel'fgat, L.A.Gorbunov
Institute of Physics, Latvian SSR Academy of Sciences
229021, Riga, Salaspils, USSR

ABSTRACT. The methods of mathematical and physical modeling are applied to study steady and time-dependent flows, and heat-mass transfer characteristics in the single crystal growth systems with MHD effects in the melt, and on the solidification front. A variety of MHD phenomena, occuring at closely similar conditions, and their possible effect on the growth conditions, and characteristics of the produced single crystals is considered and analysed consequentially. The theoretical and experimental results concerning effects due to the electromagnetic fields:suppression and (or) intensification of different kinds of convective flows in the melt, including specific flows caused by thermoelectric phenomena on the phase interface;thickness of the boundary layers, etc. are reported.

In recent years, a new branch of MHD applications - the semiconductor single crystal growth from a melt under the influence of an electromagnetic field - has been developed [1-3]. The similarity of thermal as well as electric properties of metal and semiconductor material melts makes feasible the use of practically all the known MHD-methods for the melt flow and heat & mass transfer control in these processes [4].

The majority of available publications dealing with the magnetic field effect on the single crystal melt growth pays attention to the quality of resulting materials [2,3]. The effect of the induced electromagnetic force on the melt motion and heat and mass transfer under peculiar conditions of a single crystal growth is not yet sufficiently studied. A review of some such studies are given below.

In the Czochralski process a crystal is pulled from a melt contained in a crucible-Fig.1. The crystal and the crucible are differentially rotated, crucible sidewalls are heated. The presence of temperature gradient, differential rotation, etc. causes the melt convection which, in its turn, governs the heat and mass transfer in the liquid and on the phase interface.

The basic consequences of the imposition of the steady axial magnetic field are: the reduction of the melt convection velocity as a result of an interaction between the magnetic field and the currents induced by the fluid motion; a peculiar boundary layer formation on the crystal interface and on the crucible walls; the suppression of various kinds of

J. Lielpeteris and R. Moreau (eds.), Liquid Metal Magnetohydrodynamics, 127–133.
© 1989 by Kluwer Academic Publishers.

velocity and, consequently, temperature and concentration pulsations cau-

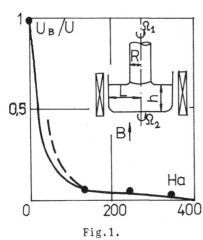

Fig.1.

sed by the vibration, time-dependent la-minar or turbulent flows, etc.; the ge-neration, in some cases, of additional kinds of melt convection due to field interaction with the thermoelectric cur-rents in the melt or with a medium mag-netization dependence on temperature.

Thorough analysis of the magnetohy-drodynamics in the magnetic Czochralski single crystal growth for heat and mass exchange requires to take into account all the forces in the melt responcible for: natural, thermocapillary, forced, thermoelectrical, etc. convections. Gene-ral solution of such a problem is extre-mely difficult. Therefore, we discuss in succession the characteristic effects of the magnetic field on every kind of the melt convection and its stability.

1. Natural convection. The natural convection of the semicon-ductor material melts (including the case in the magnetic field) has re-ceived till now the most of the attention. The main effect of the axial magnetic field is a substantial reduction of the melt velocity with a very slight change in a convective pattern of the melt.

As an example Fig.1 shows the results of the numerical simulation [5] of the model Czochralski natural convection in the axial magnetic field: $G_r = \beta g \Delta T L^3 / \nu^2 = 8,08 \times 10^5$; $Pr = \nu/a = 0,023$.

The dependence of the characteristic nondimensional velocity U_B/U (where U – the velocity when B=0 and $U_B - B \neq 0$) on the Hartmann number $Ha = BL(\sigma/\rho\nu)^{1/2}$ illustrates the obvious reduction of the melt convection.

As in [5] the criterion for the magnetic field effect on the cha-racteristic velocity is

$$U/U_B = Ha^2/Re = Ha^2/\sqrt{Gr} = N \qquad (1)$$

The values of the velocity, estimated from (1) are shown in Fig.1 by a dashed line. For $Ha > 50$ and $N > 2.8$ the discrepancy between the nume-rical and estimated results is negligible and, therefore, the latter can be used for an approximate evaluation of the magnetic field effect on heat and mass transfer in the melt. For example, the velocity evaluation from (1) shows that for the real Si crystal growth conditions in the magnetic field B=0.2 T the convective heat and mass transfer redu-ces$\approx$25 times [1].

2. Forced convection. There are detailed studies of such flows in the absence of the magnetic field [6]. The convection depends on ge-ometrical parameters of the crucible and the crystal, the angular velo-cities of the crucible and the crystal rotations, the melth depth, etc. Thorough analysis of the magnetohydrodynamics for the laminar axisymmet-ric flow in the axial magnetic field is reported in [7].

The experimental investigation of the forced convection in the axial magnetic field B<0.4 T [8] have been conducted on the model Czochralski set-up. As a working fluid the eutectic melt of In-Ga-Sn is used. Several crucibles and crystal models made from the materials with different electrical conductivity are used. The azimuthal component of the flow velocity is measured by electrical potential probes with two electrodes.

The characteristic azimuthal velocity distribution is shown in Fig.2. For the sufficiently high Hartmann number $Ha=BR(\sigma/\rho\nu)^{1/2}$ and the MHD-interaction parameter $N=Ha^2/Re$; $(Re=|\Omega_1-\Omega_2|R^2/\nu)$ in the case of a nonconducting crystal $(\sigma^*=\sigma_c/\sigma=0)$ and unrotated crucible $(\Omega_2=0)$ the convective cell with the uniform angular velocity in its core is formed under the crystal. Fig.2a demonstrates that the angular velocity in the core is a half of that of the crystal. The boundary layer with large velocity gradients $\delta v_\theta/\delta z$ is formed on the wetted crystal surface and on the bottom of the crucible, their length scale being $\sim Ha^{-1}$. Between the convective cell and the outer region a shear layer is formed with the length scale $\sim Ha^{-1/2}$. In the shear layer the angular velocity changes with the radius as well as with the melth depth, the width of the layer grows parabolically with the depth of the melt [8]. There is no distinguishable melt flow in the outer region.

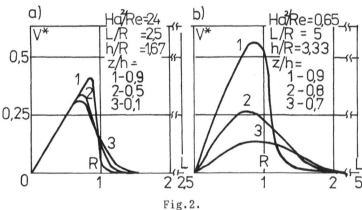

Fig.2.

With the decrease of the MHD-interaction parameter ($N\leqslant1$) the convective cell under the crystal is transformed - there are no more definite core and the shear layer region, Ω depending on r and z everywhere, and the motion is transferred to the region with much larger r (Fig.2b).

When the crystal has a good conductivity $\sigma^*=17$ (copper), the angular velocity in the core of the convective cell is practically that of the crystal - $\Omega=\Omega_1$. The boundary layer with the depth $\sim Ha^{-1}$ is formed only at the bottom of the crucible.

The forced convection in the case of $\Omega_2\neq0$ and $N\gg1$ is in general the same. There are the same boundary layers and the shear layer, the angular velocity in the outer region is uniform - $\Omega=\Omega_2$ and in the inner core is $\Omega=0,5(\Omega_1+\Omega_2)$ for $\sigma^*=0$ and $\Omega=\Omega_1$ for $\sigma^*=17$. It is proved that this type of the flow looses its stability with the increase of Re to the value of $Re_{cr}=f(Ha;R/H)$. This phenomenon will be discussed in p.4.

3. Thermocapillary convection. Till now, those flows have
been neglected and assumed unsignificant for the Czochralski crystal
growth. However, as shown in [1] the melt motion due to temperature gradient on the free surface in some cases can be quite comparable to natural convection, and affect the flow pattern as well as its stability.

Let us, following [1], choose the velocity scale from the equality
condition of the viscous and capillary shear stresses at the free surface. Assuming for simplicity the boundary layer type flow,

$$U=[(\delta a/\delta T)^2\Delta T/\rho^2\nu L]^{1/3} \qquad (2)$$

where a - the surface tension of the melt, ΔT - temperature difference
on the characteristic length scale L. Evaluating the electromagnetic
field effect for the Couette type flow in the transverse magnetic field
the equality condition of the electromagnetic and capillary forces gives:

$$U_B=(|\delta a/\delta T|\Delta T/\rho\nu L)/B\sqrt{(\sigma/\rho\nu}\quad)=(|\delta a/\delta T|\Delta T/\rho\nu)/Ha \qquad (3)$$

Then

$$U/U_B=Ha/Re^{1/3}, \text{ where } Re=|\delta a/\delta T|\Delta TL/\rho\nu^2 \qquad (4)$$

Thus, the reduction of the thermocapillary convection by the magnetic field can be evaluated by the parameter U/U_B which in this case is
proportional to the Hartmann number.

4. Stability of the melt motion. The melt convection in the
Czochralski crystal puller can be laminar as well as turbulent [1]. Yet,
even the laminar flow regime can be time-dependent and then the flow has
local velocity, temperature and concentration pulsations[1,8].

Ihe effect of a magnetic field on the stability of the natural and
forced convections. Have been investigated on the models using eutectic
In-Ga-Sn as a melt [8].

The natural convection is studied at the Grashof numbers $2 \times 10^6<$
$<Gr<5 \times 10^7$. The time-averaged temperature fields and the local
temperature pulsations are measured. For B=0 temperature pulsations
exist in all the region of the Gr numbers. Temperature pulsations are
observed in all points of the melt, the amplitude and frequency spectra
are strongly nonuniform in space.

The axial magnetic field as well as the transverse one effectively
suppresses temperature pulsations and at B=0.1 T (Ha=130) temperature
field becomes steady for all the region of Gr. The suppression of temperature pulsations by the magnetic field are illustrated in Fig.3, where
a nondimensionalized mean amplitude of temperature pulsations - $\delta T^*=$
$=\delta T/\Delta T$ (ΔT - temperature difference between the crystal and crucible
wall) are shown as a function of the MHD-interaction parameter $N=Ha^2/$
$/\sqrt{Gr}$ (for Gr characteristic temperature difference ΔT are taken at B=0).

The stability of the forced convection of an isothermal melt in the
axial magnetic field when the crystal and the crucible are differentially rotated, are studied by the measurements of local melt azimuthal

velocity. The unstability threshold of the axisymmetric flow is determined by the emergance of the velocity timedependence or by the flow visualisation on the free surface.

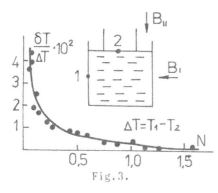

Fig.3.

It is established that the axisymmetric flow structure described in p.2 for Re>Re_{cr} and Ha>>1 looses its stability to the three-dimensional azimuthally periodic flow regime. The loss of the stability can be described as a Kelvin-Helmholtz instabbility of the free shear layer with the inflexion point in the azimuthal velocity distribution - $\delta^2 V_\theta/\delta r^2=0$ [9,10]. The evolving flow has an azimuthally periodic system of vortices with the vertical axes at r>R (Fig.4). This flow pattern has the angular velocity which is a function of HaR/h; Ω_1; Ω_2 and σ^*

The results of the experiment for 0.05 T<B<0.35 T and different values of R/h show that a stability of the axisymmetric flow is governed by the local Reynolds number $Re^*=|\Omega_1-\Omega_2|R\delta/\nu$ where the characteristic length scale is a maximal width of the shear layer - $\delta=C(hR/Ha)^{1/2}$ [7]. Then

$$Re^*=\nu^{-1}|\Omega_1-\Omega_2|R(Rh/Ha)^{1/2}=Re(HaR/h)^{-1/2} \qquad (5)$$

The experimental results - Fig.5. - show that Re_{cr} is approximately linear function of a nondimensional parameter $(HaR/h)^{1/2}$; number Re^*_{cr} has a constant value $\sim 90\pm15$.

Fig.4.

5. Thermoelectromagnetic convection (TEMC). This type of convection is due to the interaction of the external magnetic field with the thermoelectric currents caused by the differences in the absolute thermoelectric power α at the phase interface [11]. TEMC can be an additional type of the melt convection in the magnetic Czochralski growth [11].

For the majority of semiconductor material melts the values of α are large and can reach, for example, $\alpha=30\mu V/K$ for the melt of InSb. On the other hand, temperature difference in the real crystal growth is 20+40 K, and the difference of α between the phases is also large. Then the thermoelectric potential differences can initiate thermoelectric currents at the phase interface and, consequently, TEMC, if B≠0.

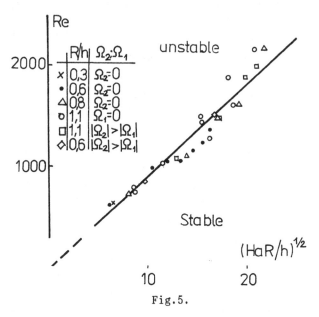

Fig.5.

Consider now the characteristics of TEMC. The thermoelectric current is $j_T=\sigma(-\alpha\, gradT - grad\, \Phi)$, where Φ - an electrical potential. It is shown [12] that the thermoelectric currents can be generated if the distribution of α and σ in the melt are nonuniform, or if the current paths are completed through external regions. In the latter case such outer regions can be the crystal or the conducting crucible walls. Otherwise, if the thermoelectric potential difference is induced in the solid phase the current paths may be completed through the melt.

The evaluation of the electromagnetic force induced in the melt (for shortcircuited thermoelectric currents) are:

$$f_T=j_T \times B=\alpha\sigma\, gradT \times B \qquad (6)$$

By analogy with the Grashof number for the free convection $Gr=\beta g \Delta T L^3/\nu^2$ the characteristic number for TEMC can be found - $Te=\alpha\sigma B\Delta T L^2/\rho\nu^2$. Then $Te/Gr=\alpha\sigma B/\rho g\beta L$ characterizes the intensity of TEMC in terms of the intensity of the natural convection. E.g., for a number of semiconductor melts (B=0.2 T and L=2 x 10^{-2} m):

* * * *	Ge	InSb	GaSb
$\alpha(\mu V/K)$	-3.5	-20	-61
Te/Gr	19	16	66

The results of the experimental study of the axial magnetic field effect on the melt dynamics in the Czochralski model are reported in [11], and confirm the criterial evalutions mentioned above.

This review proves that even for the simplest case for the single crystal Czochralski growth in the magnetic field-i.e.,in the steady axial field-the MHD phenomena can be manifold, and a careful analysis is needed to rationalize the crystal growth conditions.

REFERENCES

[1] GEL'FGAT Y.M., GORBUNOV L.A., SORKIN M.Z., PETROV G.N., 1985, 'MHD Effect on Semiconductor Material Melt in Process of Czochralski Monòcristal Growth', Magnitnaya Gidrodinamika, no. 1, 81.
[2] HOSHIKAWA K., KOHDA H., HIRATA H., 1984, 'Homogeneous Dopant Distribution of Silicon Crystal Grown by Vertical, Magnetic Field Applied Czochralski Method', Jpn .J. Appl. Phys., Vol. 23, L37.
[3] TERASHIMA K., FUKUDA T., 1983, 'A New Magnetic Field Applied Pulling Apparatus for LEC GaAs Single Crystal Growth', J. Cryst. Growth'Vol. 63, 423.
[4] GEL'FGAT Y.M., LIELAUSIS O.A., SCHERBININ E.V., 1976,'Liquid Metal Under Action of Electromagnetic Forces, Riga, "Zinatne", 248.
[5] GEL'FGAT Y.M., GORBUNOV L.A., STARSHINOVA I.V., SMIRNOV V.A.,FRYAZINOV I.V., 1985, 'Axial Magnetic Field Effect on Czochralski Natural Convection', Applied Problems of Mathematical Physics, Riga, Latvian State University, 127.
[6] SHASHKOV Y.N., 1982, 'Single Crystal Growth by a Pulling Method', Moscow, "Nauka", 312.
[7] HJELMING L.N., WALKER J.S., 1986, 'Melt Motion in a Czochralski Puller with an Axial Magnetic Field: Isothermal Motion', J.Fluid Mech.,Vol. 164, 237.
[8] BOJAREVICS A., GORBUNOV L.A., 1987, 'An Axial Magnetic Field Effect on Melt Motion in a Model Czochralski', 12-th Riga Conference on MHD, Salaspils, Vol.II, 151.
[9] LEHNERT B., 1955, An Instability of Laminar Flow of Mercury Caused by an External Field', Proc. Roy. Soc., Vol.A233, 299.
[10] KLYUKIN A.A., KOLESNIKOV Yu.B.,LEVIN V.B., 1980, 'Experimental Study of a Free Rotating Layer in an Axial Magnetic Field', Magnitnaya Gidrodinamika, no.4, 65.
[11] GORBUNOV L.A., 1987,'The Effect of Thermoelectromagnetic Convection upon Monocrystal Growth Process from Semiconductor Alloys in a Constant Magnetic Field', Magnitnaya Gidrodinamika, no.4, 65.

THE EFFECT OF ELECTROMAGNETIC FIELD ON SEMICONDUCTOR SINGLE CRYSTAL GROWTH AND CHARACTERISTICS

El.P.Bochkarev*, V.M.Foliforov**, Ju.M.Gelfgat**,
L.A.Gorbunov**, V.S.Gorovic**, O.V.Pelevin*, G.N.Petrov*
* State Research Institute for Rare-Metal Industry (GIREDMET),
Bol'shoj Tolmachevskij per., 5, 109017 Moscow, USSR
** Institute of Physics, Latvian SSR Academy of Sciences,
229021 Riga, Salaspils, USSR

ABSTRACT. The effect of constant (axial and transverse) magnetic field on the Czochralski semiconductor single crystal growth is studied. Relationship between the changed melt dynamics and the distribution of dopants and contaminants is considered. It is found that a combination of the MHD- and conventional growth control means enables a purposeful change of the semiconductor single crystal characteristics.

A wide variety of advanced microelectronic devices with exceptional functional capabilities in many ways is the result of the achievements in semiconductor materials technology and, primarily, of a dramatic improvement in the quality of starting monocrystalline materials.

As it is known [1] the quality of single crystals obtained is largely determined by hydrodynamics and heat and mass transfer processes occurring in melt and at the crystallization front. In this context, the nature of convective flows (thermogravitational, thermocapillary and forced convection), the fluctuation of rate, temperature and doping concentration, as well as the laminar or turbulent flow condition of the medium determine heavily the uniformity of dustribution of doping and background impurities, the presence and type of macro- and microdefects in crystals, etc [1].

One of the main problems encountered in monocrastalline materials growth processes is the improvement of the uniformity of distribution of doping and background impurities. As regards the tasks of obtaining silicon single crystals to be used for producing LSI and VLSI, this problem involves, in particular, the production of single crystals with a specified and uniform oxygen distribution throughout the entire bulk of the ingot [2.3].

One of the ways for achieving the objective in question are the recently proposed MHD-methods of acting on semiconductor melts [4,5] by constant longitudinal and transverse fields [6,7], rotating electromagnetic field [8] and others, that enable to actively control hydrodynamics and heat and mass transfer in the entire bulk and at the crystallization front.

J. Lielpeteris and R. Moreau (eds.), Liquid Metal Magnetohydrodynamics, 135–143.
© 1989 by Kluwer Academic Publishers.

The given paper discusses the results of studying the impact of constant magnetic field, vertical relative to the growing axis, on the growth processes of semiconductor silicon, germanium and indium antimonide single crystals using the Czochralsky method. As an example, the analysis covers mainly the data obtained in growing silicon single crystals under conditions where the magnetic field modifies the basic factors that determine the uniformity of distribution of doping additives, the concentration and distribution of oxygen, the presence of imperfections, streakiness caused by impurities, etc. Such factors are known to include thermal gradients in melt and at the crastallization front, temperature rate and fluctuation level, temperature field asymmetry, melt stirring intensity both throughout the entire bulk and at the crystallization front, etc.

Investigations were carried out on single-crystal growing installations equipped with special magnetic systems to produce vertical magnetic fields B with an induction up to B < 0.35 T in the zone of the crucible with melt.

Let us start discussing investigation results with the analysis of the averaged field and melt temperature fluctuation measurement data obtained in the experiments on a crucible with a diameter of 152 mm and a loading of 2 kg. As the experiments have shown, the application of longitudinal magnetic field produces a more pronounced increase in the vertical thermal gradient, as against the radial one.

The effect in question manifests itself most strongly under the conditions where no rotation of the crystal and crucible takes place ($n_1 = 0$, $n_2 = 0$) and there is no forced convection in melt. Thus, for example, for $n_1 = 0$, $n_2 = 0$, crystal diameter $D_1 = 60$ mm, and for melt's depth H = 40 mm, temperature difference between the crystallization front and the crucible bottom for B = 0.2 T amounts to $T_z = 23°C$, which is about twice as much as temperature difference for B = 0, when $T_z = 13°C$. At the same time, radial temperature difference between the crystallization front and the crucible wall at a depth of 5 mm beneath melt level amounted to $T_r = 15°C$ for B = = 0.2 T and $T_r = 11°C$ for B = 0, respectively.

The presence of forced convection in melt ($n_1 \neq 0$, $n_2 \neq 0$) somewhat decreases thermal gradient both for B = 0 and for B $\neq$ 0, however, on the whole, the application of magnetic field under all investigated conditions resulted in the increase of thermal gradients in melt. This result is in good agreement with [9,5] where it is shown that the application of magnetic field decreases the intensity of convective motion of melt.

Convective flows in melt under the actual single crystal growth conditions, even in the case of laminar condition, are characterized by various kinds of instabilities [1,5,9] that manifest themselves in the experiments as temperature, rate and concentration fluctuations. As the investigations have demonstrated, in the absence of forced convection in melt ($n_1 = 0$, $n_2 = 0$), the application of a relatively weak field B = (0.05-0.06) T causes a complete suppression of temperature fluctuations. These data are confirmed by the results in [9,10] where practically complete suppression of temperature fluctuations with the nonrotating crystal and crucible took place for the value of MHD-action parameter $N = Ha^2 / Gr = 1 \div 2$, where $Ha = BR_2$ — is the Hartmann number and $Gr = g\ T_r R^3_2 / \ ^2$ is the Grashof number. Indeed, for a 152 mm diameter crucible with characteristic temperature difference $T_r = 15°C$ the Grashof number $Gr = 2 \cdot 10^8$; the Hartmann number $Ha = 1.7 \cdot 10^2$ and $N = 2$.

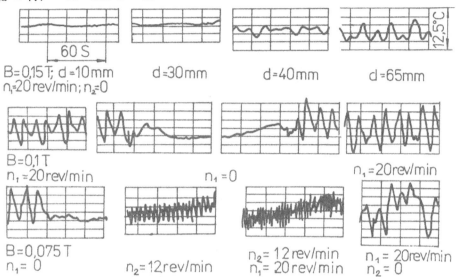

Figure 1. Measurement results of temperature fluctuations

The above results are illustrated in Fig.1. which shows temperature fluctuation data obtained directly in the process of diameter increase to a specified value and a subsequent growth of the single crystal in magnetic field with an induction of B = 0.15 T.

As it is evident from the thermograms presented, a noticeable temperature fluctuation amplitude appears when the crystal is rotating, starting with the diameter of the crystal being grown D_1 = 10 mm. As the diameter of the growing portion of crystal increases, so does temperature fluctuation amplitude, and at D_1 = 65 mm (which is close to the nominal diameter) at attains the value T = 5-6°C.

Note the periodical, close to sinusoidal, nature of variation of T over time. When crystal rotation ceases, temperature fluctuations disappear.

A similar effect, i.e. the appearance of temperature fluctuations in magnetic field, results from rotation of the crucible at n_1 = 0, as well as from simultaneous rotation of the crystal and crucible. It should be stressed that, as folloes from [9], the presence of such temperature fluctuations at $n_1 \neq 0$ is due to the instability of forced convective flows in vertical magnetic field.

Single crystal growth process is markedly influenced by the degree of asymmetry of temperature field in melt [11,12]. Under the

138

conventional growth conditions in the absence of magnetic field the rotation of the crucible and crystal is commonly used to improve temperature field symmetry. The application of vertical magnetic field with B = 0.05-0.2 T even at n_1 = 0 and n_2 = 0 fairly effectively symmetrizes temperature field in melt.

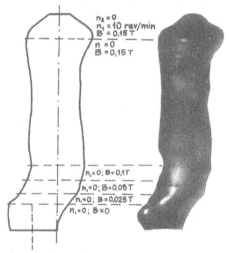

Figure 2. A form of monocrystal at asymmetric thermal field in crucible.

To illustrate this effect, Fig.2a shows the photograph, and Fig.2b - the pattern of longitudinal section of the single crystal grown under the specially provided conditions of clearly pronounced temperature field asymmetry. The first portion of single crystal has been grown at n_1 = 0, n_2 = 10 rev/min and B = 0.15 T. Subsequently, when crystal rotation was switched on in the course of growth process, the single crystal somewhat shifted toward the cold crucible wall, and portion 2 characterizes the position of thermal centre in the crucible for B = 0.15 T. Gradual decrease in magnetic field induction resulted in the shift of thermal center in melt toward the side wall and in the curvature of single crystal shape. Portion 3 of the single crystal grown at n_1 = 0 characterizes the position of thermal center in melt at B = 0.

As we see, even with the artificially introduced, clearly pronounced asymmetry of temperature distribution in melt, the application of magnetic field improves temperature field asymmetry to a great extent. Clearly, this effect is largely determined by the suppression of convective motion of melt which is due to thermogravitational convection. Here the situation is encountered in which a decisive contribution to heat transfer in melt is due to molecular heat transfer.

As it is known, the convective to molecular heat transfer ratio is determined by the Peclet number Pe = Pr Re = Pr Gr. In the presence of magnetic field the Peclet number can be written [5] as Pe = Pr Re/N = PrGr/Ha². Characteristic estimate for Pe at T_r = 15°C and B = 0.2 T in our case gives Pe = 4.3. Thus, as follows from the experimental data, the symmetrizing action of magnetic field manifests itself at Pe = 4÷5.

Let us now discuss the results of the impact of vertical magnetic field on the impurity structure of single-crystal ingots. Here phosphorus doping up to =1-50 Ohm·cm was used (the coefficient of phosphorus distribution in melt k_{ef} = 0.3).; Hydrodynamic growth conditions were varied as a function of magnetic induction value.

Characteristic oxygen concentration data for one of the crystal growth conditions (n_1 = const, n_2 = const.) with the variation of magnetic field induction B in the growth process are presented in Fig.3

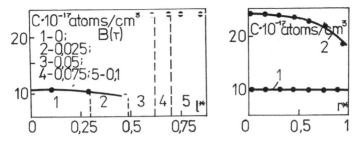

Figure 3. Change of oxygen level at various values of magnetic field.

[13]. A sharp increase in oxygen concentration to $C = 26 \cdot 10^{17}$ atoms/cm³ at $B > 0.05$ T (Fig.3) is due, presumably, to the appearance of instability of melt motion in vertical magnetic field [9] and to the increase in oxygen transfer to the subcrystal area.

Fig.3b demonstrates the data on oxygen distribution along washer radius $C(r)$ for the first (curve 1) and fifth (curve 2) zones of the single crystal - Fig.3. As we see, an increase in C in this case has resulted in the increase in radial distribution nonuniformity to $C = 16\%$ [14].

Typical relationships for oxygen concentration along the ingot length (g/G - crystallized portion of melt, G - loading weight) grown under completely stable melt flow conditions [9] are given in Fig.4. Radial scatter C in this case amounted to $C = 5 \div 9\%$.

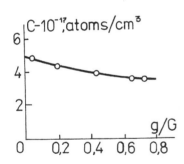

Figure 4. Oxygen distribution at stable melt flow conditions.

On the whole, the experiments performed with 2 kg loading have shown that oxygen and dopant concentration in silicon single crystals is dependent on magnetic field and growth conditions (n_1, n_2, B, V) and can be varied within the limits of $C = (1-2) \cdot 10^{17}$ atoms/cm³ to $(20-30 \cdot 10^{17}$ atoms/cm³.

Let us also dwell upon dopant distribution microheterogeneity data. Here, in the absence of MHD-action, minimum microheterogeneity in check ingots has amounted to 4.4%. In the experiments involving electromagnetic action, microheterogeneity depends also on growth conditions and lies within the limits of $= 2.8-10.7\%$, its minimum value having been attained in the crystals with the minimum macroheterogeneity value $-$ Fig.5 [14].

The results of the experiments concerned with the growth of silicon single crystals 100 mm in diameter from the crucible 280 mm in diameter and a loading of 16-30 kg are similar in principle to the

140

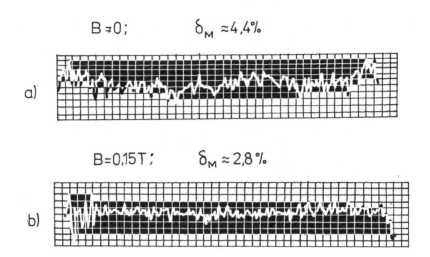

Figure 5. Measurement results of distribution of microheterogeneous alloyed admixture

above mentioned ones [15]. Fig.6a shows the relationships for oxygen distribution along the length of the crystals 100 mm in diameter that have been grown at B = 0 (curve 4) and B ≠ 0 (curves 1-3, 5-6). In the latter case different combinations of magnetic field induction, frequency and rotation sense of the crystal and crucible were varied. The shape of curves for B ≠ 0 testifies to the fact that, depending on the specific growth conditions, single crystals have different oxygen concentrations C (ranging from $2 \cdot 10^{17}$ atoms/cm³ to $1.6 \cdot 10^{18}$ atoms/cm³) and uniformity of its distribution along the length of the ingot. It should be stressed that the single crystals grown under relatively low magnetic field strengths (Fig.6a, curves 5,6) have oxygen concentrations along the entire length within the limits of $(2-5) \cdot 10^{17}$ atoms/cm³.

Different oxygen content and distribution of single crystals naturally affect the distribution of resistivity . As an example, Fig.6b displays the relationships for distribution of   along the length of a boron-doped single crystal (p-type conductivity) for B = 0 (Fig.6b, curve 4) and B ≠ 0 with C = $(6-16) \cdot 10^{17}$ atoms/cm³ and C = = $(2-5) \cdot 10^{17}$ atoms/cm³ (curves 1, 6), respectively. In the crystals with low oxygen level, obtained in magnetic field, the values of even in the absence of stabilizing heat treatment (under thermodonor elimination condition) do not fall outside the limits of admissible values. The scatter of   throughout the bulk of the crystal amounts to 15%, whereas throughout the section it does not exceed 5%. It means that in the process of growth of such crystals at B ≠ 0 no high thermodonor concentration therein occurs, capable of markedly changing the value of   , which enables to abandon the use of heat treatment, compulsory for conventional single crystals.

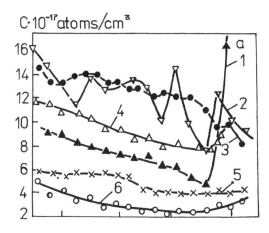

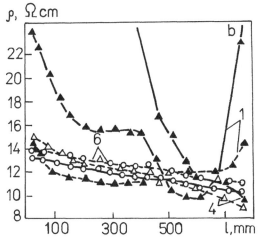

Figure 6. The results of distribution of C and in silicon single crystals of 100 mm diameter.

In the single crystals grown in the absence of magnetic field (Fig.6b, curve 4) and at B ≠ 0 with high oxygen content (Fig.6b, curve 1), a high level of thermodonors generated in the growth process results in a sharp resistivity variation, right up to the conductivity type inversion.

Thus, the above data testify to the fact that by employing an appropriate combination of magnetic field parameters, crystal and crucible rotation, pulling rate, etc., it is possible to control the quality and uniformity of distribution of doping and background impurities in single crystals so that under the optimum conditions fairly satisfactory results in terms of C and can be obtained.

The above discussed uniformity of distribution of doping and background impurities in single crystals, as it is known, is determined by the effective coefficient of distribution k_{ef} at the crystal-melt interface. At B = 0 it is found from the Burton-Prime-Slichter formula in which one of the variables represents the thickness of boundary diffusion layer d at the crystallization front. MHD-action on melt modifies the structure of velocity flows therein and, accordingly, affects the values of d and k_{ef}.

Criterion-based estimation of the variation of k_{ef} in constant magnetic field [16] and experimental check of the obtained relationship for $k_{ef}(B)$ in the growth of tellurium-, zinc- and cadmium-doped indium antimonide single crystals in transverse magnetic field [16] yields a good agreement between the experimental and theoretical data-Fig.7.

142

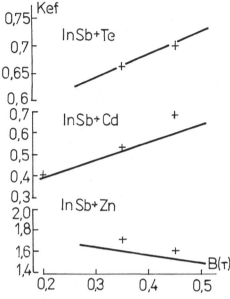

Figure 7. Change of k_{ef} at various
values of magnetic field induction.

In conclusion it should be stressed that investigations of the application of MHD-action means in single crystal growth processes are at present only at their initial stage. Further development of activities in this direction will undoubtedly lead to the emergence of practically efficient methods for actively controlling hydro-dynamics and heat- and mass trans-fer in obtaining bulk single crystals.

REFERENCES

[1] SHASHKOV Yu.N., 1982, *Single Crystal Pulling*, M., 312 (in Russian).

[2] RAIVY C., 1984, *Defects and Impurities in Semiconductor Silicon*, M., Mir Publishers, 475 (in Russian).

[3] BUZYNIN A.N., Koroleva E.A., 1985, 'Recent advances in the field of silicon single crystal growth', *Zarubezhnaya Elektronnaya Tekhnika*, No.2, 68 (in Russian).

[4] FIEGL C., 1983, 'Recent advances and future directions in CZ-silicon crystal growth technology', *Solid State Technol.*, Vol.26, No.8, 121.

[5] GELFGAT Yu.M., GORBUNOV L.A., SORKIN M.Z., PETROV G.N., 'On the MHD-actions on semiconductor materials melt in the Czochralski single crystal growth processes, *Magnitnaya Gidrodinamika*, 1985, No.1, 81 (in Russian).

[6] HOSHIKAWA K., 1982, 'Czochralski silicon growth in the vertical magnetig field', *Jap.J.Appl.Phys.*, Vol.21, No.9, L545.

[7] TORASHIMA K., FUKUDA T., 1983, 'A new magnetic field applied pulling apparatus for LEC GaAs single crystal growth', *J.Crys. Growth*, Vol.63, No.3, 423.

[8] SHASHKOV Yu.N., SHUSHLEBINA N.P., 1972, 'On the impact of electromagnetic mixing of melt on silicon single crystal growth', *Fizika i khimiya obrabotki materialov*, No.1, 34 (in Russian).

[9] BOYAREVICH A.V., GORBUNOV L.A., 1987, 'Investigating the impact of longitudinal magnetic field on hydrodynamics of melt in a Czochralski process model', *12-th Riga Conf. on Magnetic Hydrodynamics*, Salaspils, II, MHD-Devices, 151 (in Russian).

[10] BOYAREVICH A.V., GORBUNOV L.A., 1985, 'Investigating temperature fluctuations in semiconductor melts in the magnetic fields with various orientations', *Thes. 3rd All-Union Conf. Single Crystal Groeth Methods: Current Status and Development Prospects*, Kharkov, 23 (in Russian).

[11] HIRATA H., INOUE N., 1986, 'Study of thermal symmetry in Czochralski silicon melt under vertical magnetic field', *Jap.J.Appl.Phys.*, Vol.23, No.8, L527.

[12] HIRATA H., INOUE N., 1986, 'Improvement of thermal symmetry in CZ silicon melts by the application of a vertical magnetic field', *J. Crept. Growth, 330*.

[13] GORBUNOV L.A., REMIZOV O.A., 1985, 'Magnetic field impact on the Czochralski silicon single crystal growth process', *6th All-Union Conf. on Crystal Growth*, Tsakhadzor, Vol.1, 200 (in Russian).

[14] GELFGAT Yu.M., GORBUNOV L.A., PETROV G.N., POPOV A.I., REMIZOV O.A., 'Peculiarities of the Czochralski silicon single crystal growth in the vertical magnetic field', *12-th Riga Conf. on Magnetic Hydrodynamics*, Salaspils, II, MHD-Devices, 143 (in Russian).

[15] GELFGAT Yu.M., LEVSHIN E.S., POGODIN A.I., SALNIK Z.A., SMIRNOV B.V., EIDENZON A.M., 1987, 'Properties of the silicon crystals grown under the impact of vertical magnetic field', *Tsvetnye metally*, No.7, 72 (in Russian).

[16] ZEMSKOV V.S., RAUHMAN M.R., MGALOBLISHVILI D.P., GELFGAT Yu.M., SORKIN M.Z., 1986, 'Coefficients of impurity distribution in the growth of indium antimonide single crystals under magnetic field impact on melt', *Fizika i khimiya obrabotki materialov*, No.2, 64 (in Russian).

GAS BUBBLES MOTION DURING VACUUM TREATMENT OF LIQUID ALUMINIUM IN
MDV-TYPE DEVICES

Yu.Gelfgat*, V.Polischuk**, L.Puzhaylo**, M.Sorkin*
* Institute of Physics of Latvian SSR Academy of Sciences,
229021 Salaspils, Latvian SSR, USSR
** Institute of Casting Problems of the Ukranian SSR Academy
of Sciences, Vernadsky St. 34/1, 252142 Kiev, USSR

ABSTRACT. Vacuum degassing of liquid metal is possible both by evapo-
ration and bubbling, but in case of fast metal flow in vacuum devices
the bubble degassing does not occur. In working zones of MDV devices
where the liquid metal is affected by electromagnetic forces the
bubble degassing may take place in certain conditions. These conditions
were found by numerical modelling of bubble motion in working zone of
MDV device where hydrodynamic structure and the distribution of
electromagnetic forces were known, so the effective dimensions of the
working zone and the parameters of metal treatment in MDV were determined.

Modern metallurgy requires high quality production at low
labour consumption and short production cycle, all of which can be
received by introducing continuous technologies.
As far as pouring and simultaneous refinement of aluminium
and its alloys is concerned the problem of continuous technology
development can be solved by using MDV-type magnetodynamic devices
(MDV) designed at the Institute of Casting Problems of the Ukranian
SSR Academy of Sciences [1]. Some of these devices are able to pump
liquid metal through the vacuum chamber (the residual gas pressure is
2-5 mm Hg), where metal layer of about 60 mm is treated by vacuum and
pour the metal out through a pouring tube with a discharge up to 1 l/s
without stopping the vacuum treatment.
Available data on vacuum degassing kinetics of aluminium [2]
show that degassing is possible both by diffusion and bubbling under
the conditions of vacuum treatment in the vacuum chambers of MDV.
Considerable rate of metal flow in vacuum chambers (1-3 m/s) would,
however, make it impossible for the small gas bubbles inside the liquid
metal body to rise to the surface before they are carried away by the
flow, and the contribution of bubbling to the whole effect of degassing
would be negligible. At the same time, in MDV there exist working zones
(w.z.) where liquid metal is affected by three-dimensional electromagnetic forces (EMF) of considerable density (up to 5 MN/m^3). These
forces result in essential apparent increase in the weight of molten

145

metal. Thus intensification of degassing by means of bubbling becomes possible. To this end, the rate of the metal flow and the EMF density in the w.z. should be combined to ensure the rising of bubbles to the surface of the metal when it is in the vacuum chamber. The aim of the present study is qualitative determination of both the effective dimensions of the w.z. and of the parameters for metal treatment in MDV on the basis of a theoretical analysis of the single gas bubble motion in the upper T-piece zone of the MDV.

The hydrodynamic flow structure in the w.z. in question is rather complicated, but for approximate analysis it may be represented

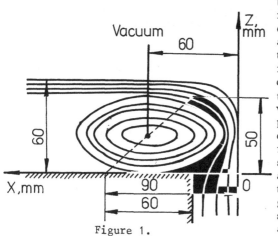

Vacuum

Figure 1.

by the model structure shown in Fig.1. The latter is a closed ellipse-shaped vortex above which there flows a thin layer of liquid bordering on vacuum. This flow is orginally rectilinear and then falls down into the vertical channel of the T-piece zone (Fig.1). The dotted line shows the distribution of EMF maximum density of which was 5 MN/m³ and the EMF density fall was taken to be linear with the gradient of 100 MN/m⁴. The symmetrical half of the zone and its main dimensions are shown in Fig.1.

The bubble motion in the rectilinear metal flow region was determined by numerical computation of the following system: the bubble motion equation (1), inertia forces, resistance forces, the attached mass, the force of gravity and pressure gradients in the liquid being taken into consideration; and the equation of the dimensional change of a spherical bubble (the Rayleigh's equation) (2), surface tension and viscosity being taken into consideration. At the initial moment the bubble was taken to be in equilibrium with the liquid, the process of the bubble growth to be isothermal, and the penetration of gas from the liquid into the bubble to be absent:

$$\frac{d^2x}{dt^2} = - \frac{F_0}{1+ \rho/2\rho_1} \frac{1}{R} (\frac{dx}{dt} - u);$$

$$\frac{d^2z}{dt^2} = - \frac{F_0}{1+ \rho/2\rho_1} \frac{1}{R} \frac{dz}{dt} + \frac{g(\rho/\rho_1 -1)-f_e}{1+ \rho/2\rho_1};$$

$$\frac{d^2R}{dt^2} = - \frac{3}{2} \frac{1}{R} (\frac{dR}{dt})^2 - \frac{4\nu}{R^2} \frac{dR}{dt} - \frac{2\sigma}{\rho R^2} +$$

$$+ \frac{0.25}{R} |v-v_1|^2 + \frac{1}{\rho R} [P_0 + \rho g(h_0 - z_0)] \frac{R_0^3}{R^3} -$$

$$- [P_0 - (\rho g - f_e) (h_0 - z)]$$

Here $F_0 = 3/8 \, C_d(\rho/\rho_4) |v-v_1|$, v and v_1 are the velocities of motion
of the liquid and the bubble, x and z are the coordinates (current ones)
of the bubble, z_0 is the initial ordinate of the bubble, R_0 and R are
the initial and the current values of the bubble radius, ν is the kinematic viscosity, σ is the coefficient of the surface tension, ρ and
ρ_4 are the densities of the liquid and the bubble, P_0 is the pressure
above the free surface, h_0 is the height of the liquid metal layer, f_e
is the density of EMF the vector of which and the one of the gravity
acceleration have the same sense of direction, C_d is the coefficient of
the hydraulic resistance of the bubble which was taken to be equal C_d
for solid spherical particle of radius R. The values of C_d were calculated by expressions taken from [3]. The initial conditions for Ed.(1),
(2) were given in the following form:

$$dx/dt \underset{t=0}{= 0}; \quad dz/dt \underset{t=0}{= 0}; \quad x \underset{t=0}{= x_0}; \quad z \underset{t=0}{= z_0};$$

$$dR/dt \underset{t=0}{= 0}; \quad R \underset{t=0}{= R_0}$$

Eq.(1), (2) were solved by means of the BESM-4M computer by
Runge-Kutta method with automatically chosen step at the accuracy of
10^{-3}%. As described, the problem does not take into consideration
either spacing apart of the bubbles by turbulent pulsations or EMF
pulsating character in real devices: $f_{e,max} = f_0 \sin|314t|$. As was
shown by control computations, taking into account the pulsating
character of f_e gives 5-10% difference in the results and is negligible
as far as qualitative results are concerned. As to spacing the bubbles
apart by turbulent pulsations the results obtained should be considered
as describing the probable tendency of bubble motion. A bubble which
happened to get into the downward flow (crossing ordinate $x = - 60$ mm
before rising to the surface) was assumed to be trapped by the flow
and stay in the bulk of the liquid.

The typical results of the computation are presented in
Figs.2-4. Fig.2 demonstrates the trajectories of the motion of the
bubbles having initial radius of 100 μm at different rates of metal
flow and different initial abscissae of the bubble nucleation point.
Bubbles come to the surface from the depth of 2 mm at residual pressure
of 262 N/m², $f_{e,max} = 5$ MN/m³. It can be seen that independently of the
initial location of the bubble its rising is practically determined by
the region of bubble motion in the zone of EMF action, the place
where the bubble comes to the surface of the liquid being determined
by the rate of the flow almost not depending on the initial point of
the bubble motion. It should be noted, however, that the computation
of the bubble motion close to the surface may be considered to be

148

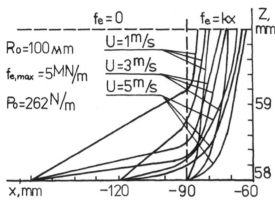

Figure 2.

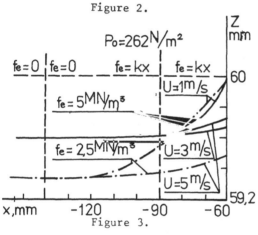

Figure 3.

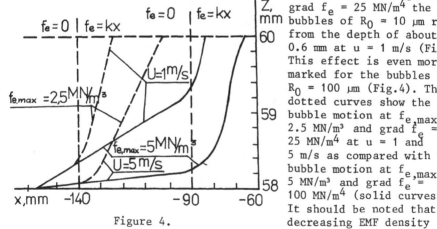

Figure 4.

qualitative only, as the Rayleigh equation for varying bubble dimensions is in this case in error. The results presented in Fig.2 show that to increase multiplicity of metal vacuum treatment the metal flow rate can be increased to 5 m/s, in which case the bubbles of R_0 = 100 μm will rise from the depth of 2 mm.

The data for the bubbles of R_0 = 10 μm are presented in Fig.3. Here the rising of the bubbles can be observed only in the region of EMF action, the bubbles of R_0 = 10 μm rising from the depth of about 0.3 mm only at the rate of the liquid of u < 1 m/s. At u = 5 m/s they rise from the depth of no more than 0.05 mm. Decreasing density EMF gradient in the region of its fall results in a considerable intensification of bubble of removal even if the value of $f_{e,max}$ is decreased. For example, at $f_{e,max}$ = 2.5 MN/m³ and the gradient grad f_e = 25 MN/m⁴ the bubbles of R_0 = 10 μm rise from the depth of about 0.6 mm at u = 1 m/s (Fig.3). This effect is even more marked for the bubbles of R_0 = 100 μm (Fig.4). The dotted curves show the bubble motion at $f_{e,max}$ = 2.5 MN/m³ and grad f_e = 25 MN/m⁴ at u = 1 and 5 m/s as compared with the bubble motion at $f_{e,max}$ = 5 MN/m³ and grad f_e = 100 MN/m⁴ (solid curves). It should be noted that decreasing EMF density gradient results in expanding the zone of EMF action, i.e. the zone of increased hydrostatic pressure where the bubble nucleation can be impeded.

Thus, the effective conditions of vacuum treatment are determined by 1/ the dimensions of the bubbles in which the bulk of gas is carried and 2/ the depth of the layer in which the majority of bubbles nucleate. If the bulk of gas is carried by the bubbles of the average dimension of about 100 μm or if the depth of the layer where the majority of smaller bubbles nucleate does not exceed 0.1 mm, it is worth-while increasing the rate of metal flow through w.z. up to 5 m/s in order to increase the multiplicity of the vacuum treatment and to combine degassing by diffusion and bubbling more effectively. Otherwise, it is worth-while decreasing the EMF density gradient in the region where the density falls and, correspondingly, expanding the zone of EMF action provided the metal surface which is in the vacuum chamber and outside the zone mentioned is preserved.

Since in the zone of vortical motion of the liquid there are regions of upward flow as well as the ones where the pressure gradient in liquid results, due to the effect of the centrifugal force, in additional buoyant force, degassing by bubbling is possible in the zone mentioned.

To analyze this possibility the bubble motion inside the elliptical vortex of liquid was computated. The following equations were solved numerically:

$$\frac{d^2x}{dt^2} = \frac{B_1}{R}\frac{dx}{dt} + B_2x + \frac{B_3}{R}z + B_4\frac{dz}{dt};$$

$$\frac{d^2z}{dt^2} = \frac{B_1}{R}\frac{dz}{dt} + B_2z + \frac{B_5}{R}x + B_6\frac{dx}{dt} + B_7;$$

$$\frac{d^2R}{dt^2} = -\frac{3}{2}\frac{1}{R}\left(\frac{dR}{dt}\right)^2 - \frac{4\nu}{R^2}\frac{dR}{dt} - \frac{2\sigma}{\rho R^2} + \frac{0.25|v - v_1|^2}{R} +$$

$$+ \frac{1}{\rho R}\ [P_0 - \rho g(\frac{h_0}{2} - z)]\ \frac{R_0^3}{R^3} - [P_0 + (\rho g - f_e B)(\frac{h_0}{2} - z)] +$$

$$+ 2\rho a_1 a_2(x^2 + z^2)\ ;$$

$$B_1 = -\frac{F_0}{1 + \rho/2\rho_1}\ ;\qquad B_2 = \frac{4a_1 a_2 \rho}{(1 + \rho/2\rho_1)};$$

$$B_3 = \frac{2a_2 F_0}{1 + \rho/2\rho_1}\ ;\qquad B_4 = \frac{a_2 \rho}{(1 + \rho/2\rho_1)}\ ;\qquad B_5 = \frac{2a_1 F_0}{1 + \rho/2\rho_1}\ ;$$

$$B_6 = -\frac{a_1 \rho}{(1 + \rho/2\rho_1)}\ ;\qquad B_7 = \frac{g(\rho/\rho_1 - 1) - f_e B/\rho}{1 + \rho/2\rho_1}\ ;$$

$$a_1 = ub/a;\qquad\qquad a_2 = u/2b;$$

where a and b are the half-axles of the typical elliptical current line described by the equation $\psi = a_1 x^2 + a_2 z^2$, and u is the rate of vortical motion at the point x = 0, z = b on that line. The pressure distribution was established as for the ideal liquid. The coordinates of the bubble center at t = 0 and the initial radius of the bubble were taken as the initial conditions. The initial rate of radius change was taken to be zero.

The computation showed (Fig.5) that the bubbles of 100 μm dimension move upwards and may be carried out of the vortical flow provided the EMF is present. At the same time the bubbles of 10 μm dimension are trapped by the metal stream and move practically following the current lines, the bubble trajectory being slichtly divergent (or convergent) despite the effect of EMF. The computation also showed that the bubbles of initial dimensions of about 1 mm which nucleate outside the EMF action zone and get into it during their further motion may collapse, i.e. behave like the cavitational ones. Analogous results have been obtained for the motion of the bubbles of 1 mm dimensions in recti-linear flow provided they nucleate at the depth of more than 2 mm. The change of the bubble radius and ordinate depending on time at R_0 = 1 mm and Z_0=0.03 m is shown in Fig.6. It can be seen that the rate of fluctuations of R is incre-ased sharply when entering the EMF zone action and the moment comes eventually at which R → 0.

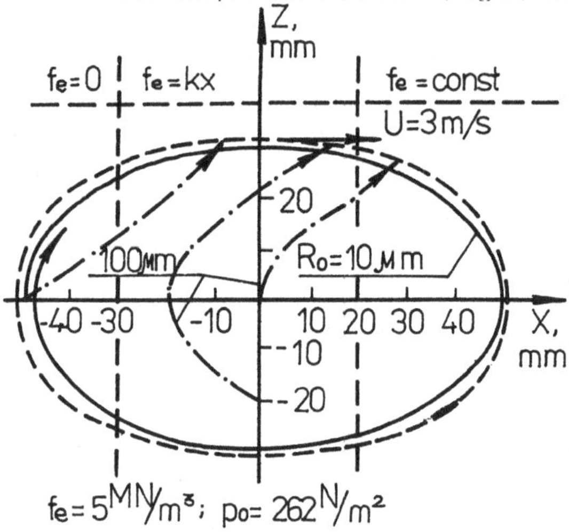

Thus, degassing is only slightly affected by the presence of vortical flow. Yet the possibility of cavi-tational phenomena in the active zones of MDV should be taken into consideration.

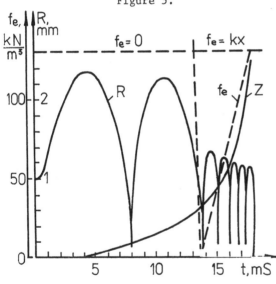

Figure 5.

Figure 6.

REFERENCES

[1] POLISCHUK V., PUSHAYLO L. et al., 1978, 'Investigation on Magneto-dynamic Device for Vacuum Treatment and Continuous Pouring of Wrought aluminium Alloys in Industry', *IX Riga Conference on Magnetohydrodynamics*, Vol.3, Zinatne, 123.

[2] AL'TMAN M., GLOTOV E. et al., 1977, *Vacuum Treatment of Aluminium Alloys*, Metallurgy, 240.

[3] BRUYATSKY E., 1965, 'Determination of the Particle Fall Rate Inside viscous Conducting Liquid, *Technical Electromagnetic Hydrodynamics*, Metallurgy, 366.

MHD TURBULENCE DECAY BEHIND SPATIAL GRIDS

A.A.Kljukin, Ju.B.Kolesnikov
Institute of Physics, Latvian SSR Academy
of Sciences, 229021, Riga, USSR

ABSTRACT. Turbulence decay after spatial grid in a transverse magnetic
field has been investigated. The grid consists of two rows of cylinders
orientated parallelly and perpendiculary to the magnetic field. The correlation factor distribution and velocity pulsations energy were measured for two positions of grid in the flow: i) cylindrical elements perpendicular to the field were located in front of elements parallel to
the field; ii) behind them. In the weak magnetic field for both grid positions the laws of turbulence decay are the same, at the begining with
a power of -2, but then with a power of -I. In a relatively strong magnetic field turbulence energy along the flow in the first case of grid
position changes initially with law -I, but then reaches a constant value. In the second grid position decay law is characterised by a power
of -0.66. The physical interpretation of the results obtained is presented.

I. INTRODUCTION

In the previous works [1-4] using liquid metals in a magnetic field it
was found experimentally that in a relatively strong magnetic field,or-
thogonal to the flow of metal, turbulence becomes anisotropic, and its
features tend to the features of two-dimensional turbulence. The flows
behind a cylindrical body, behind grid and in the channel in the transverse magnetic field were studied. Specifically, it was shown [3] that
in the presence of relatively strong magnetic field in the flow after
grid transfer of a passive addition, being introduced at the axis of
channel, takes place mainly in the orthogonal to the field plane. In the
absence of the magnetic field distribution of the passive addition is
homogeneous in the flow cross-sections. In the case of plane jet flow
in the presence of a strong magnetic field, simultaneously with the formation of two-dimensional structure of turbulence, the effect of turbulent "negative viscosity", i.e. the energy transfer from pulsations to
the mean flow, was found [5]. The formation of k^{-3} spectrum, which ca-
racterizes the quasi two-dimensional MHD homogeneous turbulence, was also studied in [6].

In these investigations flows with two-dimensional MHD-turbulence
features were realized and some main turbulence features were studied.

153

J. Lielpeteris and R. Moreau (eds.), Liquid Metal Magnetohydrodynamics, 153–159.
© 1989 by Kluwer Academic Publishers.

But question about possibility of transition at the presence of strong
uniform transverse magnetic field from initially three-dimensional turbulence to two-dimensional turbulence, in the orthogonal to the magnetic
field plane, was still open. The work [7] was the first attempt to ans-
wer this question on the example of a simple flow behind grid, formed as
a set of mutually orthogonal cylinders. This work is a further investigation of the problems discussed in the work [7].

2. EXPERIMENTAL TECHNIQUE

Experiments were carried out on the horizontal mercury loop, described
in [8]. In an insulating channel with a cross-section of 70 x 70 mm,
placed between poles of an electromagnet, two-plane spatial grid formed

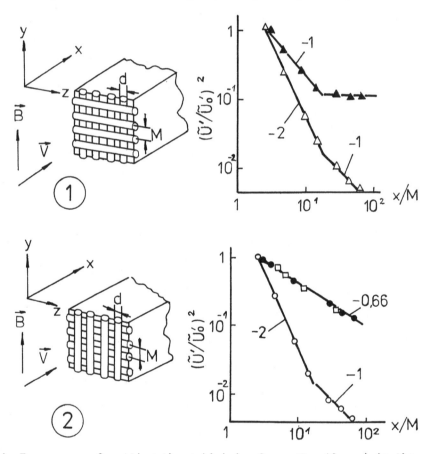

Figure 1. Two manner of setting the grid (d = 5 mm, M = 10 mm) in the
channel and the laws of turbulence decaying. B, Tesla : $\triangle$,O - 0.1;▲,
●- 0.62; □ - [7].

from mutually orthogonal cylindrical rods was arranged. A set of parallel rods with a distance between them equal to the rod diameter was put on the same orthogonal set of rods. Grid was arranged in the channel in such a way, orthogonal to the field elements were, in one case, in front of parallel to the field elements, and in the other, behind them (Fig.1). In the latter case, grid must generate more isotropic turbulence in comparison with the first one. Grid was put in the region with a uniform magnetic field.

In order to eliminate forming of M-shape velocity profiles, a honeycomb from insulating material was arranged at the initial section with nonuniform magnetic field. All measurements were carried out at the constant mean velocity V = 12.2 cm/s and intensity of magnetic field up to 0.62 T. In experiments, conduction anemometer with four electrodes sensor was used. Longitudinal velocity pulsations at different cross-sections and electrical current component pulsations were measured and distribution of spatial correlation factor of these pulsations in the field direction was measured using two probes.

3. EXPERIMENTAL RESULTS

In the weak and strong magnetic fields turbulent flow behind grid at a small distances from it is sufficiently homogeneous.

As seen from Fig.1, turbulence behind grids with different orientation in the weak magnetic field, B = 0.1 T, decays in the same way. By the way, one can trace two specific decay regions with laws: $(U'/U_0')^2 \approx (x/M)^{-2}$ and $(U'/U_0')^2 \approx (x/M)^{-1}$. When magnetic field B = 0.6 T, initial turbulence appears different for different grid orientation. In the first case, Fig.1-①, at the initial region decay law is $(U'/U_0')^2 \approx (x/M)^{-1}$ but then perturbation energy comes to quasistationary level. In the second case, Fig.1-②, decay law is near $(U'/U_0')^2 \approx (x/M)^{-0.66}$.

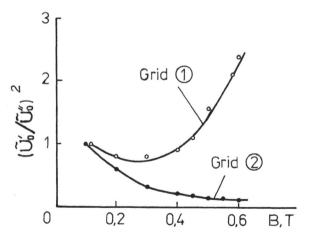

Figure 2. The magnetic field intensity influence on the initial turbulence energy. x/M = 4.5. U_0''-perturbations intensity at B = 0.1 T.

156

Given in Fig.2 relationships of velocity pulsation energy and magnetic field intensity show that with increasing magnetic field in the
second case of grid orientation turbulence is strongly depressed, while
in the first case of grid orientation turbulence intensity sharply increases.

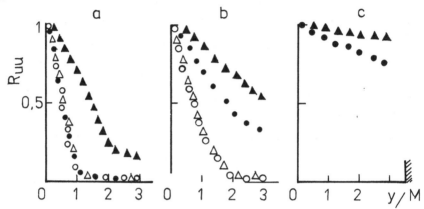

Figure 3. The correlation factor distribution spaced in the field direction. x/M : a - 2.5; b - 13.5; c - 39.5. B, Tesla: △ , ○ - 0.1; ▲,
● - 0.62.

In Fig.3, distribution of spatial correlation factor in the different cross-sections of the flow obtained by direct measurements is shown.
The results illustrate the dependence of initial turbulence on grid orientation in a relatively strong magnetic field B = 0.6 T and, particu-
lary, more rapid transition to two-dimensional parturbations behind grid
with elements parallel to magnetic field placed behind orthogonal elements.
    It should be noted that initial turbulence intensity at B = 0.6 T
is four time larger in the first case than in the second one. At B =
0.1 T, intensities in both cases are approximately the same.

4. DISCUSSION

To explaind the obtained results, we put forward the following considerations.
    In the weak magnetic field, B = 0.1 T, grid orientation affects
weakly initial turbulence structure. The both cases breakaway from grid
elements short, ≈d, oriented in the field direction perturbations become more elongated in the field direction due to electromagnetic diffusion and soon reach scales ≈2d. Vortex diffusion time at a distance
about its transverse dimensions in accordance with evaluations from work
[9] is τ_{diff} ≈1 s at B = 0.1 T, in our case this corresponds to x/M =
= 13.5. Electromagnetic diffusion process occurs with big Joulean dissipative losses and accounts for rapid intensity drop with law -2. If vortices breaking away from grid elements are sufficiently coherent, then

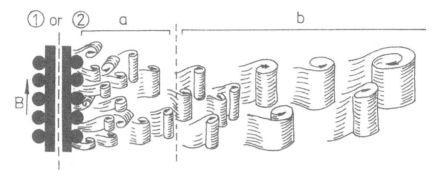

Figure 4. Transformation of turbulence under different initial conditions ① or ② . a - ancoherent brakaway, dissipation and diffusion ; b - interaction and diffusion. B = 0.1 T.

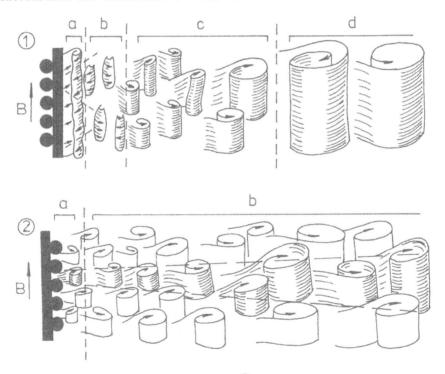

Figure 5. Transformation of turbulence. ① : a - coherent brakaway (intensity is large); b - three-dimensional instability; c - interaction and diffusion; d - two-dimensional state. ② : a - ancoherent brakaway (intensity is small); b - interaction is large in the direction ortogonal to field, diffusion is suppressed in the direction of field. B = 0.62 T.

158

during time τ_{diff} vortices breaking away from neighbour elements placed one after another in the field direction are unified into one vortex with a scale in field direction ≈ 3 - 4d, (see Fig.3 and 4). Larger vortices were not formed due to three-dimensional instability. The law of decay for such a strongly anisotropic turbulence is -1.

In the strong magnetic field, in the first case of grid orientation strongly correlated in the field direction perturbations with high intensity breakaway from grid and they are unstable and decay instantly onto scales ≈ 3 - 4d, (see Fig.3 and Fig.5-①) and they are damping as in the case of weak field at x/M > 13.5 with law -1. As intensity along flow decreases, perturbations with larger longitudinal scale become more stable and after perturbation scale in the field direction become equal to the distance between channel walls, turbulence becomes fully two-dimensional, and reaches quasistationary level of perturbations energy.

In the strong magnetic field, in the second case of grid orientation its elements, orthogonal to the field, disturb coherence of vortices breakaway from elements lying in the field direction (see Fig.3). As decay rate appears minimal, we may suppose that interaction between vortices in the orthogonal to the field plane is the main factor in the decay process of such turbulence. Strong interaction of vortices in orthogonal to the field plane prevents vortices diffusion in the field direction. In this case, decay law - 0.66 appears intermediate between the cases of fully two-dimensional turbulence and anisotropic turbulence. Turbulent flow corresponding to decay law - 0.66 is a "sandwich" in each layer of which turbulence is as quasi two-dimensional. Interaction and mutual penetration of these layers down the flow leads to slow transition of all turbulent flow to two-dimensional (Fig.5-②).

It should be noted, that the grid generated turbulences correlation factor in the second case of grid orientation, practically does not dif-

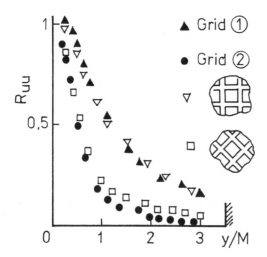

Figure 6. The correlation factor comparison in the flow behind different grids (x/M = 4.5) in the strong magnetic field. B = 0.62 T.

fer from that of the slanting plane grid generated turbulences. Comparison shows, that at small distances from both grids (Grid ② and slanting grid on Fig.6) correlation factor of longitudinal velocity pulsations in the field direction quickly drops to zero. Velocity pulsations intensities in both cases are approximately equal.

So, peculiarity of investigations of turbulence decay behind grids in the presence of magnetic field consists of uncertainty of turbulence generation initial conditions due to influence of magnetic field on the flow breakaway from the grid elements. This influence may act in two ways: first, due to Joulean dissipation magnetic field affects intensity of perturbations breakaway process; second, magnetic field produces anisotropy in initial turbulent structure, and at sufficiently strong magnetic fields may lead to the coherency of perturbation breakaway from grid elements lying in the field direction. Both these factors, in combination with electromagnetic diffusion of velocity perturbations in the field direction and unstability of such anisotropic structures to three--dimensional disturbances, may basically affect all the turbulence decay processes - the turbulence structure and the laws of decay.

REFERENCES

[1] VOTSISH A.D. & KOLESNIKOV Ju.B., 1976, 'The space correlation and vorticity in two-dimensional homogeneous turbulence', Magn. Gidrodin., Vol. 3, 25
[2] KOLESNIKOV Ju.B. & TSINOBER A.B., 1972, 'A two-dimensional turbulent flow bihind a round cylinder', Magn.Gidrodin., Vol. 3, 23
[3] KOLESNIKOV Ju.B. & TSINOBER A.B., 1974, 'Experimental investigation of two-dimensional turbulence behind a grid', Izv. A.N.S.S.S.R. Mekh. Zh. i Gaza, Vol. 4, 146
[4] PLATNIEKS I., 1972, 'Correlation study of turbulent velocity perturbations in MHD channel', VII Riga Conference on MHD, Vol. IA, 31
[5] VOTSISH A.D. & KOLESNIKOV Ju.B., 1976, 'The anomalous impulse transfer in MHD shear flow with two-dimensional turbulence', Magn. Gidrodin., Vol. 4, 47
[6] CAPERAN Ph. & ALEMANY A., 1985, 'Homogeneous MHD turbulence at weak magnetic Reynolds number', J.Mec.Theor.Appl., Vol. 4, n. 2, 175.
[7] VOTSISH A.D. & KOLESNIKOV Ju.B., 1976, 'Three - to two-dimensional turbulence transformation in the external magnetic field', Magn. Gidrodin., Vol. 3, 141
[8] SELYUTO S.F., 1984, 'Influence of a magnetic field on turbulence structure formation in flow past grids of various configurations', Magn. Gidrodin., Vol.3, 55
[9] SOMMERIA J. & MOREAU R., 1982, 'Why, how and when, MHD turbulence becomes two-dimensional', J.Fluid Mech., Vol. 118, 507.

THE EFFECT OF A UNIFORM MAGNETIC FIELD ON STABILITY, TRANSITION AND TURBULENCE AS A CONTROL MEANS FOR LIQUID METAL FLOW MIXING

V.M. IEVLEV, A.S. KOROTEEV, V.B. LEVIN
Phys. Techn. Inst.
G. Dolgoprudny
Moscow Region I4I700
USSR

ABSTRACT. The effect of a magnetic field on the flow regime provides a means of control of mixing in liquid metal flows. The paper presents results of the following experimental investigations : laminarization of pipe and jet flows by a coaxial magnetic field : decay of a homogeneous MHD turbulence with the near-isotropic initial state ; transverse magnetic field effect on Rayleigh-Taylor instability in thin films.

1. Laminarization of Pipe Flows by a Coaxial Magnetic Field

In the experiments [I, 2], in a range of Reynolds number values up to $Re = Ud/\nu \approx 10^4$ (where U is the average velocity, d is the pipe I.D.) the full suppression of turbulence, entering the magnetic field, and reduction of the fully developed turbulent flow hydraulic resistance coefficient, λ_0, up to the fully developed laminar value λ_l, are obtained. When $Re = 10^4$, $\lambda_0/\lambda \approx 2$. For the critical Reynolds number of the turbulence suppression, Re_1, an experimental relationship $Re_1 = Re_{01}(1 + 0.4 \, N)$ is obtained, where $Re_{01} = 2250$ is the value in the absence of the magnetic field, $N = \sigma B^2 d/\rho U$ is the interaction parameter. For $N \gg 1$ this relationship reduces to $M/Re = 0.033$ where the Hartmann number $M = (\sigma/\rho\nu)^{1/2}Bd$.

As Re and M values increase, the laminar flow entry length, l_1/d, required for achieving a parabolic velocity profile, also increases. With the rise of Re value, for obtaining a fully developed laminar flow it is necessary to increase values of M and l_1/d, which is practically difficult to realize. That is why the values of λ_l were not achieved in experiments with $Re > 10^4$. Thus, at $Re = 10^5$ the ratio $\lambda_0/\lambda_l \approx 30$; in the experiment [3], when $Re \approx 10^5$ and $M/Re \approx 0.035$ the reduction of λ_0 approximately by an order of magnitude was obtained. During further increase of M, the coefficient λ remained practically at the constant level what was in agreement with the assessment of the value of λ for the partially developed laminar flow corresponding to the parameters of the experiment. Taking into account the experimental data [I, 2], the result obtained in [3] can be explained by the suppression of turbulence at $M/Re \approx 0.035$ with the realization of a partially developed laminar flow. Thus the expansion of Reynolds number value range by an order of magnitude did not lead to a considerable change of parameter M/Re critical value.

J. Lielpeteris and R. Moreau (eds.), Liquid Metal Magnetohydrodynamics, 161–166.
© *1989 by Kluwer Academic Publishers.*

2. Jet Laminarization by Coaxial Magnetic Field

2.1. Introduction

In spite of sufficiently high values of the parameter N, the suppression of turbulence in the experiments [4-6] involving jets was considerably weaker compared with the experiments involving pipe flows ; the flow regime in jets remained very far from being laminar. In this experiments the flow regime in the central pipe and coflowing stream was turbulent. Boundary layers on the pipe walls and the edge wake formed the initial velocity profile with two inflexion points and large local gradients of velocity, $U \gg U/d$. The local gradients reduce considerably the value of local interaction parameter, $N_l = \sigma B^2 d/\rho U' \ll N$. Under these conditions, the magnetic field effect is weakened and on the initial region of the jet there develops a MHD analog of jet turbulence with large-scale eddy structures. At some distance, as a result of turbulent mixing, the longitudinal shear layers become blurred and the local interaction parameter increases, but large disturbances, developed upstream, prevent the flow laminarization.

2.2. Experimental facility and procedure

In the present study, the jet laminarization was accomplished by transition delay with the joint effect of a coaxial magnetic field and an inlet device forming in the initial cross-section a velocity profile with a smooth change at the jet boundary and a low disturbance level (corresponding experiments in the absence of a magnetic field is presented in [7-10]).

The experiments were carried out on a liquid metal loop filled with an In-Ga-Sn alloy. The first results are reported in [II]. The test section consisted o f a 15 cm I.D. cylindrical tube mounted in the solenoid ; a magnetic field was uniform to within 3 % over the length of 1.3 m. Flow straightening and initial disturbance reduction were carried out with a uniform honeycomb, fine-pored grids and a confuser. To eliminate the magnetic field influence on the initial velocity profile, the confuser wall was arranged close to the corresponding line of force of the solenoid leakage field. To obtain a smooth velocity change on the jet boundary, the confuser was butted to the 2 cm I.D. cylindrical pipe with the length of 35 cm. The experiment was performed with a submerged jet. According to the linear theory, unstable disturbances develop in a submerged jet faster than in a coflowing jet, thus reducing the length of transition and making laminarization more difficult. Measurements of Pitot and static pressure drop, performed with a U-shaped spirit piezometer, were accurate to 0.2 mm. Using a four-electrode sensor of the electrical potential difference probe (EPDP) with 2 mm distance between electrodes and a three-channel amplifier the potential difference corresponding to radial and azimuthal components of fluctuating velocity and fluctuating current in the axial direction was recorded.

2.3. Results and discussion

The experiments were carried out with the velocity on the jet axis in the initial cross-section $U_{mo} = 10.5$ cm/s, the boundary layer width at the pipe exit $\delta \approx 0.5$ cm and the Reynolds number Re = $U_{mo}d/\nu = 3300$. The velocity profile measured at a distance of x = 3 mm from the pipe exit is shown in Fig. 1. Solid symbols correspond to B = 0, open symbols correspond to the maximum in the present experiment value of symbols correspond to the maximum : in the present experiment B = 0.2 T($M \approx 120$). In a number of values of B and x, mean velocity distributions were measured, and during the sensor displacement across the jet at a speed of 0.5 mm/s EPDP signals were recorded.

The transition length, x^*, was determined by the jet boundary break. The dimensionless length of transition, $x_1 = x^*/x^*_o$ (where x^*_o is a value of x^* for B = 0) is shown in Fig. 2 as a function of the Hartmann number. Near $M \approx 116$ (M/Re $\approx 3.5.10^{-2}$) the enhancement of x_1 increase is so great that it indicates the approach to stable regime, evidenced by results of EPDP signal recording. Fig. 3 is a montage of EPDP signal distributions across the jet for $B = 2{,}8 \cdot 10^{-2}$ T ($M \approx 17$). Figures indicate values of x(cm) and the dotted line indicate the jet boundary. Fig. 4 illustrates typical records for $M \approx 83$ (denoted by figures I-8) and for $M \approx 106$ (denoted by figures 9-15). The pattern of velocity pulsations in Fig. 4, enlarged ~ 10 as compared with Fig. 3. According to Fig. 2, for $M \approx 83$ (M/Re $\approx 2.5 \ 10^{-2}$) we obtain $x_1 \approx 6.3$; for the inlet device used $x^*_0 \approx 4$ cm, so $x^* \approx 25$ cm. At $x > 25$ cm a rather weak development of disturbances is observed. At $M \approx 106$ (M/Re $\approx 2.5 \ 10^{-2}$) EPDP records obtained during repeated passage of the sensor across the jet at $x = 50$ cm, correspond to the amplifier background where sometimes weak bursts appear, connected with random variations of initial conditions. At $M \approx 106$, the above bursts were not observed up to $x = 100$ cm (at $x > 100$ the solenoid magnetic field non uniformity increased abruptly and so the measurements were not made).

At the transition length the jet expansion is close to laminar level. For M/Re $\ll 3.5 \ 10^{-2}$, the jet expansion after transition grows abruptly. With the M/Re increase, this growth slackens and at M/Re $\approx 3.5 \ 10^{-2}$ the jet becomes stable.

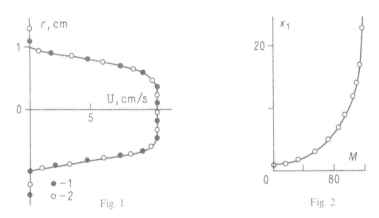

Figure 1. Velocity profile in the jet initial cross-section.

Figure 2. Dimensionless length of transition in Hartmann number.

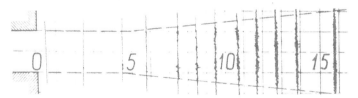

Figure 3. Qualitative pattern of the distribution of pulsating velocity at the weak action of a magnetic field (figures indicate x [cm]).

3. Experimental Investigation of the Turbulence Behind the Grid Suppression by a Magnetic Field

As in Ref [12], in the experimental investigation [13-15] turbulence was generated by a grid movement in a cylindrical vessel filled with an In-Ga-Sn alloy and mounted in the solenoid. In contrast to [12] : a) a transverse component of the pulsating velocity v was measured with fixed sensors ; (b) measurements were made both during the movement of the grid and after grid stopping ; c) grid movement took place both in the presence of the magnetic field (solid dots in Figs. 3.1, and 3.2) and in the absence of the magnetic field, with solenoid switched off after grid stopping (open dots in Figs 3.1 and 3.2). In the latter case the generated turbulence is nearly isotropic and the process of field growth practically does not disturb its isotropy. For each regime, the grid movement and signal recording were repeated many times. Statistical processing of nonstationnary random signals was performed by the combined time interval and realization ensemble averaging.

Comparison of the parameters of experiments [12] and [13-15] is given in Table I (M, U - grid mesh and velocity, respectively, $N = \sigma B^2 M/\rho U$). The time t, counted from the moment of grid passage past the sensor, corresponds to $x = Ut$.

TABLE I

Parameter	Ref. [12]	Ref. [13-15]
B_{max}, T	0.25	0.4
M, m	0.02	0.02
U, m/s	0.05 - 0.8	0.1 - 1.0
$(x/M)_{max}$	1.8	100
N_{max}	1.8	7
Generation	In a magnetic field	In a field and without field

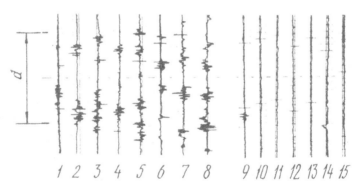

Figure 4. Qualitative pattern of pulsating velocity development under strong action of a magnetic field.
$M \approx 83$; x [cm] : 1-20 ; 2-25 ; 3-27 ; 4-32 ; 5-35 ; 6-40 ; 7-45 ; 8-50
$M \approx 106$; x [cm] : (9-15)-50.

Th results of the three-dimensionality coefficient measurements, $\kappa = 2 < e_x^2 > / < e_y^2 >$, where e_x and e_y are electric field pulsations along and across magnetic field direction, respectively, are given in Fig. 5. In the process of practically full decay of the component v, the decrease of κ from its initial value, $\kappa_0 \approx 1$, did not exceed 50 %. Hence, it follows that : (a) for the present experiment conditions, the accuracy of measurements by an EPDP is satisfactory ; (b) if MHD turbulence initial state is close to isotropic, then during the turbulence decay the pulsating current and, therefore, Joule dissipation remains great in spite of the longitudinal scale of the turbulence strong growth. This essential property of the three-dimensional MHD turbulence is maintained right up to practically full decay of the component v, which is confirmed by the power law $< v^2 > / < v^2 >_0 \sim [(t - t_0) U/M]^{-n}$ (where v_0 is the initial value of v, n = n(N)), which is valid over the whole range of x/M and N values investigated, Fig. 6.

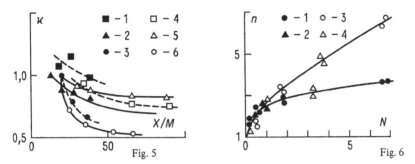

Figure 5. Change in the three-dimensionality coefficient during the decay.
Generation in a fields ; N : 1-0.43 ; 2-1.7 ; 3-6.6.
Generation in the absence of a field ; N : 4-0.43 ; 5-1.7 ; 6-6.6.

Figure 6. Power dependence exponent for a transfer component of pulsating velocity.
Generation in a field ; Re : 1-1.47 : 10^4 ; 2-2.94 : 10^4
Generation in the absence of a field ; Re : 3-1.47 : 10^4 ; 4-2.94 : 10^4.

4. Rayleigh-Taylor Instability in Thin Liquid-Metal Films in the Presence of Transverse Magnetic Field

In the experiment, an annular specimen was placed in a chamber, filled with inert gas, and was fixed to a heated flat plate mounted on the lower end face of a solenoid [16]. The general picture of instability development was similar to that observed in the absence of a magnetic field. However with induction growth (in the range of 0-0.8 T) a considerable decrease of the disturbance development rate was found both at linear and non linear stages.

References

/1/ FRAIM F.W., HEISER W.H., 1968, "The effet of a strong longitudinal magnetic field on the flow of mercury in a circular tube", J. Fluid Mech., vol. 33, n° 2, 397.
/2/ KRASIL'NIKOV Ye. Yu., LUSHCHIK V.G., NIKOLAENKO V.S., PANEVIN I.G., 1971, "Experimental investigation of the conducting fluid flow in a circular tube in a coaxial magnetic field", Izvestiya AN SSSR, Mekhanika zhidkosti i gaza, n° 2, 151.
/3/ LEVIN V.B., CHINENKOV I.A., 1970, "Experimental investigation of coaxial magnetic field effect on Hydraulic resistance with conducting fluid turbulent flow in tube", Magnitnaya Gidrodinamika, n° 3, 145.

/4/ SAJBEN M., FAY J.A., 1967, "Measurement of the growth of a turbulent mercury jet in a coaxial magnetic field", J. Fluid. Mech., vol. 27, n° 1, 81.

/5/ PREOBRAZHENSKY S.S., CHINENKOV I.A., 1970, "Experimental investigation of coaxial magnetic field effect on turbulent jets on conducting fluid", Magnitnaya Gidrodinamika, n° 2, 65.

/6/ BAUSHEV B.N., KRASIL'NIKOV Ye. Yu., LUSHCHIK V.G., PANEVIN I.G., 1972, "Coflowing jet mixing in a coaxial magnetic field", Isvestiya AN SSSR, Mekhanika shidkosti i gaza, n° 5, 33.

/7/ VULIS L.A., ZHIVOV V.G., YARIN L.P., 1969, "Flow transition region in a free jet", Inzhenerno-frizichesky zhurnal, n° 2, 239.

/8/ NAVOZNOV O.I., PAVEL'EV A.A., 1969, "On turbulence transition in coflowing jets", Izvestiya AN SSSR, Mekhanika zhidkosti i gaza, n° 6, 131.

/9/ NAVOZNOV O.I., PAVEL'EV A.A., YATSENKO A.V., 1972, "On turbulence transition in submerged and coflowing jets", Izvestiya AN SSSR, Mekhanika zhidkosti i gaza, n° 4, 148.

/10/ NAVOZNOV O.I., PAVEL'EV A.A., 1980, "Influence of initial conditions on the flow of axisymetric coflowing jets", Izvestiya AN SSSR, Mekhanika zhidkosti i gaza, n° 4, 18.

/11/ KOLOMEETS V.N., LEVIN V.B., FOMENKO V.P., 1978, "Transition delay in a jet by a coaxial magnetic field", in : Riga Conference on MHD, vol. I, 24.

/12/ MOREAU R., ALEMANY A., 1987, "Experimental results on MHD homogeneous turbulence", Lect. Notes Phys., n° 76, 369.

/13/ VORONCHIKHIN V.A., GENIN L.G., LEVIN V.B., SVIRIDOV V.G., 1984, "Experimental study of turbulence decay behind the grid in a magnetic field", in XI Riga conference on MHD, vol. I, 39.

/14/ VORONCHIKHIN V.A., GENIN L.G., LEVIN V.B., SVIRIDOV V.G., 1985, "Experimental investigation of grid turbulence decay in a uniform magnetic field", Magnitraya Gidrodinamika, n° 4, 131.

/15/ VORONCHIKHIN V.A., GENIN L.G., LEVIN V.B., 1987, "Experimental investigation of the decay of MHD turbulence with the initial state close to isotropic", in : XII Riga Conference on MHD, vol. I, 55.

/16/ KOROTEEV A.S., REY I.N., 1984, "Rayleigh-Taylor instability in thin liquid metal films in the presence of magnetic field", Izvestiya Sibirskogo Otdeleniya AN SSSR, seriya technicheskikh Nauk, n° 10, Pt 2, 113.

Session C:
Current Carrying Melts

ELECTRICALLY INDUCED VORTICAL FLOWS

Ed. V. SHCHERBININ
Institute of Physics, Latvian SSR Academy of Sciences
229021 Riga, Salaspils
USSR

ABSTRACT. This paper considers the origin of electrically induced vortical flow theory as one aspect of magnetohydrodynamics. We consider the physical principles of theory associated with the presence of the critical value of electrically induced vortical flow parameter, the possible origin of this critically, the properties of both converging and diverging electrically induced vortical flows, the flow structure in internal magnetic fields, the characteristics of heat transfer and the behaviour of bodies and gaseous inclusions in current-carrying liquid. The results of experimental study of electrically induced vortical flows including those at multi-electrode current supply have been presented. Promising developments have been outlined.

Introduction

The first reference concerning a current-carrying fluid behaviour is dated 1907, when Edwin Northrup published the results of his two experiments on an electric current passage through a liquid conductor of variable cross-section*.

Northrup also presented the first theory of phenomena in a current-carrying flow based on a concept of pressure increase at the axis of cylindrical liquid conductor as the result of compression by the electromagnetic pinchforce due to the interaction of the electric current and its self-magnetic field.

On the basis of this concept intense jets in electrical arcs were explained, which were widely investigated in the 30 s ~ 50 s, along with other phenomena. Only in 1960 did Zhigulev notice that the electromagnetic force generated within the electrical discharge could be compensated by the pressure force, and, two years later, Uberoi correctly explained the origin of motion in current-carrying fluid : at certain conditions the electromagnetic force becomes rotational and should drive a fluid motion.

However, substantial development of current-carrying fluid hydrodynamics began in early 70 after the publication, in a short period, of three independent investigations of the Lundquist problem of a point current source on a plane on the basis of full equations of motion. This can be explained by relating the problem to electrometallurgical processes and natural phenomena. Since 1977 the theory of effects in a current-carrying fluid has been called "Electrically Induced Vortical Flows" (EVF). The origin of the term is related to the generation of these flows i) due to electric current passage within the fluid and ii) owing to a vortical nature of the electromagnetic force arising when the electric current interacts with its self-magnetic field.

At present several aspects of the theory are being developed. Historically one of the first aspects is related to periodic EVF in tubes, which has been initiated by Uberoi and extended by Freiberg and others. The second aspect, based on studies by Chow, concerns flows bodies in current-carrying fluid, and has been applied by Oreper to mass transfer. In the Institute of Physics, we are concerned with behaviour of bodies in current-carrying fluid.

The bibliography can be found in [1]. In the following we will make references only for the works published after 1985 or not included in [1].

J. Lielpeteris and R. Moreau (eds.), Liquid Metal Magnetohydrodynamics, 169–178.
© 1989 by Kluwer Academic Publishers.

The third aspect is related to the problem of an electrode , small in respect to the flow domain, and at the limit, a point electrode. Many papers on this subject have been published by Sozou and co-workers, and others, a principal solution was proposed by Bojarevics.

Most useful for technological purposes are studies of EVF in closed volumes. As with the directions listed above, we cannot mention here all the researchers involved in these studies. We would like to mention the works by Szekely & Dilawari on turbulent EVF numerical simulations, and by Vlasyuk and others on laminar (and turbulent) EVF and the effect of these on a heat mass transfer.

A separate direction is a control of EVF by external magnetic field, which is of special interest for applications and for presenting a number of new physical effects in moving fluids.

The experimental investigations are in a special position, most of them are based on visual observation and measurements of pressure distribution (Sharamkin et al.). Recently an experiment on velocity field measurements [2] has been made permitting the comparison of different numerical methods for the turbulent flow predictions.

1. Physical Principles of Electrically Induced Vortical Flows

For homogeneous boundary conditions an electromagnetic force $f_e = j \times B$ can drive a fluid motion, from an initial rest state, if the force is rotational, i.e. curl $f_e \neq 0$. It is instructive to determine the conditions in an axisymetric situation in cylindrical coordinates (z, r, Φ).

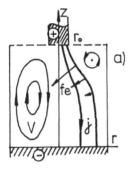 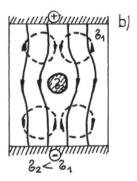

Fig. 1

For an axisymetric electric current, accounting for $j = \text{curl } B/\mu_0$, it is easy to show that the self-magnetic field contains only the azimuthal B_Φ - component related to the electric current stream function $(j_z = r^{-1}\partial\psi/\partial r, j_r = -r^{-1}\partial\psi/\partial z)$ by the equation $\psi = rB_\Phi/\mu_0$. The rotationality condition for the electromagnetic force then is of the form :

$$\text{curl } j \times B = -i_\Phi 2\mu_0 r^{-3} \psi \, \partial\psi/\partial z = 0 .$$

Considering the expression for j_r current component, one can conclude that the fluid would be set in motion by any radial current component. The second conclusion derived from the expression for curl f_e : the flow driven always lies in meridional planes containing the symmetry axis, the same planes where the electric current lines are.

Several typical situations are shown in Fig. 1. The most common case, when electrodes are immersed in a container with liquid conductor of a constant cross section at least of one of these is not coinciding with the section of container. The most common case is shown in Fig. 1 a by following the line of electric current from the smaller electrode, it is easy to see that the current density falls inversely to the square of distance from the axis, and the magnetic field-inversely to the distance. The electromagnetic force decreases like r^{-3} with a maximum value at the small electrode. This distribution of the force causes, first, the current line to move as a whole under the action of a resultant force, and, second, to be rotated to the nonuniform force distribution. Since the liquid is bounded by rigid walls, then the fluid should move along the symmetry axis from the small electrode and return along the bottom and side walls of container.

This and numerous other examples allow us to formulate a fairly general rule ; in a region of nonuniform current density the fluid moves towards the decrease of current density. Applying this rule, one may predict a fluid flow in the container of varied cross section and equal electrode surface areas. A flow is set up if within the fluid, while passing an electric current of uniform density, a body is introduced, the electric conductivity σ_2 of which is different from the conductivity of the fluid (σ_1), and the flow direction at the body with $\sigma_2 > \sigma_1$ (Fig. 1b) is opposite to the flow with $\sigma_2 > \sigma_1$.

All the examples are not merely theoretical, since these are directly related to specific problems in electrotechnology. Thus, the situation shown in Fig. 1a is typical for electroslag and electricalarc processes, that shown in Fig. 1b for a problem of drop, gas and rigid inclusion behaviour in electrical smelting devices.

Often the processes of welding, smelting, or remelting are done with alternating current. In such cases, a mechanism related to skin layers is added to the above mechanisms. So, even for the plane interface between two fluids of different electrical conductivities, the flow is driven at the boundary due to the difference of skin-layers.

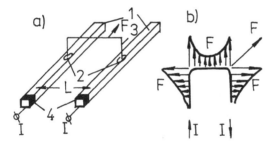

Fig. 2

A significantly more complex electrically induced flow is set up for a multiple electrode electric current supply to the fluid (this situation is also typical for electrometallurgy). The flow configuration in this case loses the axial symmetry, becomes essentially three-dimensional, and depends on a variety of parameters : the number of electrodes, their immersion depth, the sort of current, etc.

It is instructive to consider a simple schema of two electrodes held at different potentials by direct current supply. We shall restrict ourselves to two limiting cases : a small and a great immersion depth of the electrodes.

The basic mechanism for the small immersion of electrodes, when the electric current is assumed to flow along a π-shaped shortest path between the ends of electrodes, i.e. in the plane including the electrode axes, may be explained by the following experiment [3]. Two rectangular channels are filled with mercury and placed parallel to each other at a distance L between the axes (Fig. 2a). Two floats 2 on the mercury surface hold a copper wire bridge 3 electrically connecting the mercury channels. After the electric current source is connected to the electrodes 4, the bridge moves in the direction of force F shown in Fig. 2a.

As seen in Fig. 2b, in the sites of electric current curves the integral force acts on each current-element in the direction of external surface of the angle $3\pi/4$ to the element. This explains the flow configuration with the two-electrode current supply : two diverging fluid jets are ejected from the electrodes, these reach the side walls of container, converge along the wall, and then return along the z-axis penetrating the volume between the electrodes. The flow is opposite when the electrodes are of equal polarity and the second pole is the end wall of the container. Here the electrode jets attract each other like two electric currents of the same direction, merge at the z-axis, and then return along the bottom and side of the container.

A more general way to predict a flow configuration is based on the curl of electromagnetic force, which can be found by numerical methods or by simplified analytical models.

To analyse the consequences of electric flow in the volume of bath, let us assume the electrodes be of hemispherical shape, and the electrical field of the individual electrode in the spherical coordinates centered at the centre of the hemisphere is expressed by $j_R = + I/2\pi R^2$ (the + sign means that the current enters the fluid, (-) exists it). The magnetic field of the current then is $B_\Phi = +\mu_0 I (2\pi R)^{-1} (1 - \cos \theta) / \sin \theta$.

Applying a superposition principle and Cartesian coordinates shown in Fig. 3a, one can express the electromagnetic force and its curl. The result shows that z-component of the force curl is generated, where as it is absent for a single electrode current supply. This means that the circulation of melt in the planes $z = Cst$ normal to the axes of electrodes should occur. Corresponding to the signs of (curl $f_e)_z$ in different quadrants of the plane x-y one should expect 4-contour circulation with two jets penetrating along the line connecting the projections of electrodes (Fig. 3b), which agrees with experimental observations.

A different flow configuration takes place for a greater immersion of electrodes within the fluid. To analyse the situation, assume that the two electrodes with distance 2a (Fig. 4a) between their axes, are immersed to the whole depth 1 of the liquid layer which is assumed unbounded in the horizontal direction. For each electrode, at which we impose a condition of uniform current density on the surface,

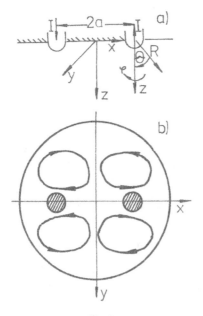

Fig. 3

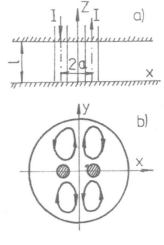

Fig. 4

the solution for the current density and magnetic field in the cylindrical coordinates fixed to the electrode are of the form :

$$j_r = + I / 2\pi l r, \quad B_\Phi = + Iz / 2\pi l r.$$

By using the Cartesian coordinates shown in Fig. 4a and applying superposition, we can construct the electric current and magnetic field lines. With this information, it is straightforward to construct the electromagnetic force containing only z-component, and its curl having only x and y components in comparison to the case of unimmersed electrodes.

Let us predict the flow in two planes : $y = 0$ and $x = 0$. The result shows, that in the plane $y = 0$, the electromagnetic force swirls the fluid in such a way that it should move downwards along the electrode surface and return upwards along the z axis.

Within the plane $x = 0$ the maximum of force $(f_e)_z$ is situated at the z-axis, and (curl $f_e)_x$ determines that the fluid should move downwards along the z-axis and return in the region remote from the plane $y = 0$.

Thus, at the z-axis two effects oppose each other. Direct experimental velocity measurements V_z show that the second tendency is present at the immediate vicinity of the container's bottom. This is also evident in the motion at the free surface of mercury if the electrodes are fixed in the bottom of the container and protrude within the fluid almost to the free surface : now the diverging jets can be observed at the plane $x = 0$ (Fig. 4b), i.e. the motion is opposite to that of unimmersed electrodes.

Our experience observing electrically induced flows in more complex situations shows that the observed flow configuration principally agrees with the above flow generation mechanism.

2. Behavior of Bodies in a Current-Carrying Fluid

The flow configuration of a body in a current-carrying fluid depends on the relative electrical conductivities of the body and fluid. Of course, these flows intensify mass transfer between the body and the fluid. These flows also change the hydrodynamic drag of body when it is in an external flow. In particular, an interaction of EVF with an external flow at a nonconducting sphere leads to the reversed flow region at the rear critical point even for Re ≈ 1. This results in an increase , although small, of the hydrodynamic drag. At a good conducting body the reversed flow region takes place at the front critical point, which decreases the hydrodynamic drag. Therefore, in a flowing current-carrying fluid a separation of particles with different conductivities occurs.

If the body is a current-carrying volume bounded by walls, then a direct effect of electromagnetic force on the body can be observed. It is known that, when an electric current is passed along a cylindrical conductor, the electromagnetic force is radial and leads to a parabolic pressure distribution

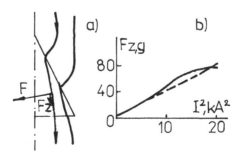

Fig. 5

174

with the maximum pressures on the axis of conductor. The non-conducting particle is affected by the radial pressure gradient and should move to the wall of the tube. The resulting radial force depends on the radius of the spherical particle "a", its distance from the tube's axis, and the density of the unperturbed electric current j_1 : $F_r = \pi\mu_0 j_1{}^2 a^3 r$.

If a particle is electrically conducting, an electric current of density j_r penetrates the particle. In this case, in addition to the pressure gradient pushing the particle to the wall of the tube, an electromagnetic force is applied to the volume of the particle and directed to the axis of the tube. The resulting force can be quite easily estimated for the particle of the cylindrical shape of volume V :

$$F_r = r\,j_1(j_1 - j_2)\,V\mu_0/2.$$

The formula may be used to estimate the force on particles of different shape. However, even for a sphere of extremely high conductivity, the inner current density j_2 exceeds by only three times density within the fluid, i.e. $j_2 = 3j_1$. Analysis of the motion shows that, independently of total current magnitude, two regimes of motion are possible : an asymptotic, its motion to the tube's axis (for small currents) and dying oscillations relative the axis (for high currents). This means for the electroslag remelting process that drops of liquid metal passing the slag bath will concentrate close to the axis of bath, which is extremely unfavourable for the process.

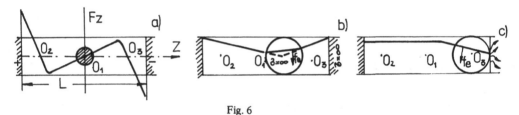

Fig. 6

Apart from the radial, an axial electromagnetic force acts on a highly conducting particle. The effect is most easily observed for highly conducting bodies of nonsymmetric shape (drop, cone, etc.). If a conical body is oriented by its axis parallel to the direction of unperturbed electric current, then the cone is set in motion with its blunt end forward. This phenomenon could be explained in the following way. If the electrical conductivity of the cone is higher than the fluid's, the electric current is concentrated at the apex of the cone and its density increases within the cone with the increase of its cross-sectional area (Fig. 5). In other words, the lines of electric current within the cone are not parallel, and the field is still azimuthal. The electromagnetic force is as usual, normal to the lines of electric current and, consequently, acquires an axial component. The axial component of the integrated force applied to the body moves the cone.

If applied to the electroslag process, this means that a drop at the melting electrode is drawn by the force exceeding the gravity force within the slag. The acceleration acts on the metallic drops while it travels through the slag bath.

It was found, unexpectedly, that bodies of symmetric shape (sphere, spheroid) were also subjected to an action of axial force, yet for the condition of a limited length of the surrounding cylindrical liquid conductor. Consider the experimental data.

In a semicylindrical channel, the end walls of which are the electrodes of diameter 7 cm, and filled with mercury, a copper sphere of diameter 3 cm is inserted. The interelectrode distance L is variable (Fig. 6). When the current is swithched on, the sphere first remains on the central axis of the channel, which is obvious because the electrical conductivity of copper is 50 times the conductivity of mercury. Further, for L < 3.5 cm, the sphere stays exactly at the midpoint of the electrodes and remains stationary. With an increase of L the sphere moves from the mid-position, reaches one of the electrodes, then it is repelled from it, approaches the other one, etc. Finally, L > 10 cm, the sphere does not reach the second electrode, and starts oscillating in a vicinity of the electrode relative to a point at a distance from the electrode approximately equal to the diameter of the sphere.

The direct measurement of axial force gives evidence (Fig. 6 a), that for L > 3.5 cm there are three positions of equilibrium ($F_z = 0$), two of which (O_2 and O_3) are stable (the force is directed opposite to a deviation) and one (O_1) unstable (the force increases a deviation). The origin of unstable position could be explained in the following way. Assume that the electrodes and sphere are made of infinitely conducting material (then an electric current enters the material normal to its surface) and the sphere is of equal diameter with the channel (Fig. 6, b). Obviously, when the sphere deviates from the middle position, the concentration of current in the left of the figure will be higher than in the right part of the sphere. Consequently, the electric current lines within the sphere are not parallel, and similarly to the cone, an axial force increases the deviation.

When the sphere is positioned at the electrode, and for a finite conductivity of the electrode's material, the electric current is redistributed within the electrode itself, and the concentration of current increases in the part of sphere close to the electrode. The resulting inclination of the current lines within the sphere is opposite to the previous one and the axial electromagnetic force repels the sphere from the electrode (Fig. 6, c).

3. Numerical Studies of Electrically Induced Flows

Intensity of electrically induced flows is determined by the parameter $S = \mu_0 I^2/4\pi^2 \rho \nu^2$ and by a degree of electric current nonuniformity. For a comparatively small magnitude of total current $I = 10^3$ A, the value of parameter is $S = 10^8$ for mercury. From this estimate it is obvious that a problem of electrically induced flow computation is similar to general hydrodynamics, i.e. the problem of a large parameter in the Navier-Stokes equation.

The first solutions of EVF, based on a similarity approach, have led to unexpected results : the problem of flow at a point electric current source gives a regular solution only for a modest value $S_{cr} = 150$ (the source on a rigid plane) and $S_{cr} = 47$ (the source on a free surface). The following efforts have been directed to improve the model and to increase the available magnitude of the parameter.

Sozou & Pickering introduced a bounded flow region within a hemisphere. However the result was surprising : the problem could be solved up to $S \approx 9.5$ (instead of 47 for the problem with the free surface). An analysis of the results led the authors to the conclusion that the reason of flow breakdown is the steep velocity gradient growth in a region where the hemisphere is crossed by the symmetry axis. Bojarevics introduced a finite size electrode retaining the infinite flow region, yet it was found that for large radial distances the solution tends to the self-similar one, with all its difficulties. Attempts were made to account for the effect of electric currents induced by the flow. Even assuming for Batchelor number $\beta = \mu_0 \sigma \nu = 1$ (yet for real media $\beta = 10^{-6}, 10^{-7}$), the value of S_{cr} is merely increased.

At the present time two ways are proposed to resolve the paradox. The first is based on the property of extreme sensitivity of the converging flow at a small electrode to azimuthal perturbations. The solution shows that a small stationary perturbation of azimuthal motion is increased in the converging meridional flow up to a magnitude leading to a decrease of the axial velocity incremental rate where the parameter S and the problem of S_{cr} is absent. In natural conditions, the stationary perturbation is the axial component of terrestrial magnetic field, which interacting with the radial current component gives a stationary small azimuthal force, which is realy demonstrated in the physical experiment [1].

The second way is related to a limited flow region and a finite size of the electrode. Thus, the solution for a hemispherical container with a hemispherical electrode yields a stable result up to $S = 6.3 \times 10^3$ for the ratio of hemispheres 0.1, and up to $S = 1.5 \times 10^5$ for 0.5. For a cylindrical container with the depth equal to the radius, reliable results can be obtained up to $S = 10^8$ for the ratio of small electrode to the large (usually the bottom of container) equal to 0.2, and up to $S = 10^{12}$ for 0.8 [4]. This speaks in favour of the solutions for high values of S parameter when the relative current density at the small electrode is not too high, or the nonuniformity of the current is low (in the spherical container it is higher than in the cylindrical) ; from this finally depends the density of electromagnetic force and its curl.

As a whole, the flow configuration within a closed container is sufficiently standard : the flow takes a form of toroidal vortex with the motion directed from the small electrode along the central line (Fig. 7 a). This seems to be the only information obtained from the hydrodynamic streamlines. The nuances of the flow require a different approach.

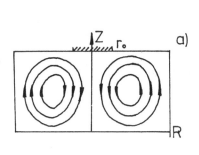
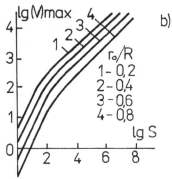

Fig. 7

Let us consider for example, a maximum magnitude of the nondimensional axial velocity v_{zmax} ($V_z = v_z \nu / L$) while varying the parameter S (the results by Vlasyuk are applied). As evident in Fig. 7 b, this dependence clearly contains three intervals. The first, approximately up to $S \approx 10^3$, $V_{zmax} \approx S$; the third, also linear for $S > 10^6$, $V_{zmax} \approx \sqrt{S}$; the second, intermadiate, is more complex. It is easy to see from the equations of motion, that the first regime corresponds to Stokes flow. In this region the dimensional velocity is proportional to the square of electric current and inversely to the dynamic viscosity $V \approx \mu_0 I^2 / \rho \nu L$.

The most significant for applications is the third region, since the values of S typical in electrometallurgy exceed 10^6. In this region the dependence $V \approx \sqrt{S} \nu / L$ follows from the equality of typical inertial and electromagnetic terms in the equation of motion, and the regime is non linear.

In the nonlinear regime $V \approx \sqrt{(\mu_0/\rho L^2)}$, i.e. the velocity grows proportional to the first power of current, and, what is noticeable, does not depend on the viscosity of medium.

In addition to this, the numerical results and experiments give evidence that within the main volume of the bath (excluding boundary layers) the velocity field becomes self-similar, i.e. the velocity normalized by $\sqrt{S}$ is a function of merely the nondimensional coordinates. This means the data of a physical or numerical model can be related to the geometrically similar object simply by multiplying the shape function by the coefficient $aI \sqrt{(\mu_0/\rho L^2)}$. The coefficient a depends on the shape of bath, geometric conditions for the current flow, and it should be determined in every special case.

Useful information can be obtained by employing the concept of the boundary layer for flows in closed volumes, introduced in [5], as the line where $w = \text{curl } V = 0$.

As an exemple we refer to the computational results of Vlasyuk (Fig. 8, a) for the flow in a cylindrical container of two small electrode sizes $r_0 = 0.2$ and 0.8. In both cases the thickness of the boundary layer, determined by the zero vorticity line, in the region diverging from the axis flow (i.e. at the bottom of the container) decreases with S, which corresponds to the behaviour of ordinary boundary layer with R_e number growth. However in the region of converging flow (near the free surface or at small electrode) the thickness of the boundary layer grows with S.

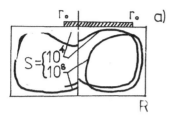
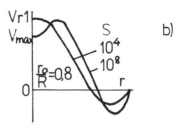

Fig. 8

Evidently, velocity perturbations, generated at the wall, are convected by the converging flow to the near electrode region, and since the perturbations grow with velocity (or what is the same, S) then the volume occupied by these also grows. Moreover, with the increase of S, the domain of perturbation begins to expand into the axial zone even reaching the bottom of the bath, and the zero vorticity is deflected from the symmetry axis. Since the perturbations result in a retardation of fluid, then the zero vorticity line separation from the axis means the maximum velocity is located off the axis. There should be a deflection in the velocity profile of V_z - component (Fig. 8, b). The theoretical result is supported by the experiment.

Most of the EVF computations are for laminar flows. Nevertheless, the computational results satisfactorily agree with the experimental not only qualitatively, but also quantitatively. Here a single fact calls for special attention : the coefficient a in the velocity expression, determined in the experiment, is approximately 25 % higher than predicted. This is quite unusual, since predicted velocity according to the laminar theory is, as a rule, higher than the experimental, which is commonly explained by the turbulence of the real flow.

At present attempts are being made to compare the laminar and turbulent prediction. There are several works predicting turbulent flows according to k-ε model electroslag remelting flows. In [6] the predicted flow is compared to the experimental within a cylindrical container with a conical electrode. The numerically computed velocity is approximately 20 per cent higher than the experimental. The value of laminar predicted velocity higher than the experimental of the turbulent predicted could be explained by the special turbulence properties in converging EVF.

4. Control of Electrically Induced Flows. Heat and Mass Transfer

Since the flows considered are driven by electromagnetic forces, the control aslo can be made by an external magnetic field.

We should consider the flows in an axial magnetic field, because the analysis is simpler due to the conserved axial symmetry.

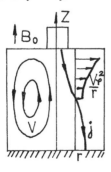
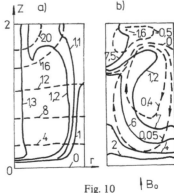

Fig. 9

Fig. 10

The axial magnetic field interacting with the radial current component drives an azimuthal motion of fluid. A special property is that the liquid layers closer to the small electrode rotate faster than the more distant, i.e. the rotation is differential. This leads to the following effect. Along the cylindrical surface r = Cst (Fig. 9) the centrifugal force grows approaching the small electrode and its curl > 10. Consequently, the differential rotation drives a secondary meridional flow in reverse to the electrically induced flow. By increasing the magnitude of axial magnetic field, the electrically induced flow can be suppressed and even reversed.

This property can be used to control the electrostag welding process. As it is well known, in the absence of the field the EVF jet from the small electrode causes a significant deformation of the interface slag-liquid metal. The resulting surface between the liquid metal and the weld takes the shape of a crater with the maximum depth on the weld axis. The experiments show that the magnetic field magnitude can be chosen so that the pressure distribution along the interface is uniform, and the crystallization front is practically a plane.

The heat and mass transfer is also affected in the external magnetic field. So, in the absence of field, the axial zone is overheated (Fig. 10 a). By choosing a certain value of the magnetic field, one can obtain a practically uniform temperature on the interface 1.6 (Fig. 10 b). As the numerical solution shows, this occurs for N/S = 0.6, where N is the parameter responsible for the rotation intensity, B_0-external magnetic field induction.

In the same manner an additive can be distributed uniformly over the bath, introducing it through the slag free surface [7].

There is one other intriguing effect observed in the rotating fluid. It has been found that by keeping the value for N constant and increasing S, the fluid total angular momentum and the kinetic energy grow, i.e. the energy of meridional flow goes to the rotational motion. The mechanism of energy transfer is nonlinear, and the phenomena are typical of flow converging to the axis of symmetry.

The effect has been explained by the amplification of an azimuthal perturbation in the converging flow, and it could be related to a number of flows observed in nature and technology. Thus, the effect has been included within the tornadic vortex model as one of the mechanism maintaining the intense rotation [1].

References

[1] BOJAREVICS V., FREIBERGS Ya.Zh., SHILOVA Ye.I., SHCHERBININ E.V., 1985, Electrically Induced Vortical Flows, Zinatne (In Russian).

[2] ZHILIN V.G. et al., "An experimental investigation of the axisymetric electrovortex flow velocity field in a cylindrical container", Magnitnaya Gidrodinamika, N3, 110.

[3] BOJAREVICS V., CHAIKOVSKY A.I., CHUDNOVSKY A.Yu., SHCHERBININ E.V., 1986, "On a force acting on moving bridge in electromagnetic railgun launchers", Magnitnaya Gidrodinamika, N2,105.

[4] VLASYUK B.Kh., SHARAMKIN V.I., 1986, "Numerical study of heat and mass transfer in an electrovortex flow in a longitudinal magnetic field. I. Problem formulation and calculation of heat transfer in a cylindrical volume", Magnitnaya Gidrodinamika, N3, 78.

[5] SHCHERBININ E.V., YAKOVLEVA Ye. Ye., 1986, "An electrovortex flow in a spheroidal container", Magnitnaya Gidrodinamika, N4, 64.

[6] MEDOVAR B.I., EMELJANENKO Yu. G., SHCHERBININ E.V., SZEKELY D., DILAWARY A., 1982, "A comparison of results of physical and mathematical modelling of flow velocity field in a slag bath at electroslag remelting", Problemi specialnoi electromerallurgii, vol. 17, 9.

[7] VLASYUK B. Kh., SHARAMKIN V.I., 1987, "Numerical study of heat and mass transfer in an electrorotational flow in a longitudinal magnetic field. 11. Mass transfer in a cylindrical container", Magnitnaya Gidrodinamika, N1, 86.

LIQUID METAL FLOW NEAR MAGNETIC NEUTRAL POINTS.

Richard G. Kenny,
D.A.M.T.P.,
Silver Street,
Cambridge,
United Kingdom.

ABSTRACT. In this study the possibility of employing a magnetic neutral point as an electro-magnetic brake is explored. We consider flows along the tapering part of a magnetic neutral point rather than perpendicular to it as the problem then becomes very intractable for gener-alised flows (Alemany [a]). A full scale nonlinear approach to the problem has yet to be realised but the principal characteristics can be illustrated by considering a simpler model which yields analytic solutions. Pai [b] and Regirer [c] found expressions for the velocity profiles in both ax-isymmetric and two dimensional geometry. The fact that the flow rates in both dimensions are $O(1/M^2)$ (M=Hartmann number) in the domain of large M suggests a means of controlling the mass flow rate. Intuition would anticipate jet-like structures for both geometries at large M but this is only found to be the case in 2D. The contrasting behaviour of the two flows results from the fact that curl(Lorentz force) equals zero in the core of the flow for large M and this has two differing effects in axisymmetry and 2D.

1. Introduction

Liquid metal flows in non-uniform fields have been the attention of much study due to their im-portance in both fusion technology and casting processes. The analytical approach to such flows has been quite extensive [d]. However, there remains plenty of scope for a combination of both analytical and numerical techniques to study the fully nonlinear flow problems which are more readily observed. In this article the initial analytical component to such a study is presented and it should be borne in mind that this has yet to be incorporated into a complete description of the fully developed flow.

The resulting steady state flow structure is examined for a conducting liquid near a magnetic neutral point in both a cylindrical pipe and a two dimensional duct with a possible view to em-ploying the arrangement as an electromagnetic brake.

The ultimate problem which requires investigation concerns a fully developed flow which im-pinges upon this region of non-uniform field and re-emerges fully developed, assuming that the pipe is sufficiently long. There is also a fixed pressure drop across the ends of the pipe which pro-vides the driving force for the flow and is balanced by both viscous and electromagnetic forces in the steady state. However the full problem is nonlinear in nature and requires numerical tech-niques to make any significant headway. In order to gain some understanding of the more salient

179

J. Lielpeteris and R. Moreau (eds.), Liquid Metal Magnetohydrodynamics, 179–185.

features therefore we consider a simplified problem amenable to an analytic approach. For example in the present case we consider liquid metal encountering a tapering magnetic field of the form $\mathbf{B}_0 = (\alpha r, 0, -2\alpha z)$ in axisymmetry or $\mathbf{B}_0 = (-\alpha x, \alpha y, 0)$ in two dimensions in either a pipe or duct respectively. The ends of the respective channels are held at a specified pressure difference which provides a scaling for this linear problem. The idealisation of the model arises in the simplified description of the neutral point which appears to extend throughout space.

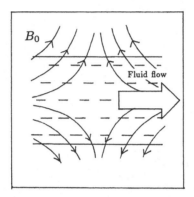

Figure 1. Fluid flow near a neutral point.

Up to the time of the Riga conference the author was unaware of the extent of prior analytic work in this area especially by S.I. Pai (1954) [b] and S.A. Regirer (1974) [c] who found solutions for the simplified flows in both axisymmetry and two dimensions respectively. However, there are certain characteristics of the two flows which require emphasis and comparison making it valuable to sketch an outline of the solutions.

2. Axisymmetric flow through a pipe.

A conducting fluid of conductivity σ and viscosity μ flows through a cylindrical pipe of radius a under the influence of an applied pressure difference and through an imposed field of the form, $\mathbf{B}_0 = \nabla \times (0, \frac{\psi(r,z)}{r}, 0)$, in terms of cylindrical coordinates (r, θ, z). Unidirectional solutions to the flow exist if $\psi(r, z) = zf(r) + g(r)$, which arises from inspection of both the steady state equation of motion and the induction equation shown in a paper by Regirer [e]. Consequently the velocity profile has the form $\mathbf{u} = (0, 0, u(r))$. The induced field $\mathbf{b}$ arises from the azimuthal currents induced in the flow and since these are independent of z it follows that $\mathbf{b} = (0, 0, b(r))$. The magnetic neutral point is described by an applied magnetic field of the form $\mathbf{B}_0 = (\alpha r, 0, -2\alpha z)$ and since there are no free currents it follows that $\nabla \times \mathbf{B}_0 = 0$.

The steady state equation of motion of the fluid is,

$$0 = -\nabla p + \mathbf{j} \times \mathbf{B} + \mu \nabla^2 \mathbf{u}, \tag{2.1}$$

which must be solved in tandem with the steady state induction equation,

$$(\mathbf{u} \cdot \nabla)\mathbf{B} = (\mathbf{B} \cdot \nabla)\mathbf{u} + \eta \nabla^2 \mathbf{B}. \tag{2.2}$$

By integrating the components of the pressure equation we find that the pressure distribution becomes,

$$p(r, z) = -kz - \frac{b^2}{2\mu_0} + \frac{2\alpha}{\mu_0} zb(r). \tag{2.3}$$

Two interesting features emerge from this expression, firstly the constant $-k$ which is the longitudinal pressure gradient typical of Poiseuille flow scaling both the velocity and mass flux and secondly, the fact that the Lorentz force terms give a radial variation which will be seen to lead to unexpected results.

The curl of (2.1) results in the vorticity equation which is a third order d.e. relating u and b namely,

$$u''' + \frac{u''}{r} - \frac{u'}{r^2} = \frac{\alpha}{\mu\mu_0} b' - \frac{\alpha}{\mu\mu_0} rb''. \tag{2.4}$$

The induction equation links the velocity u and induced field b by $b' = -\frac{\alpha}{\eta} ru(r)$ and substitution in (2.4) yields,

$$u''' + \frac{u'}{r} - u'\left(\frac{1}{r^2} + M^2 r^2\right) = 0, \tag{2.5}$$

where r is scaled so that $0 \leq r \leq 1$ and $M^2 = \frac{\sigma a^2 a^4}{\mu}$ (Hartmann number). The boundary conditions are that the velocity vanishes at the wall i.e. $u(1) = 0$ and that the flow is symmetrical, so that $u'(0) = 0$. The flow is completely specified by the pressure gradient $-k$ of (2.3) which scales both the velocity and induced magnetic field. Analogously, the induced field must satisfy $b(1) = 0$ at the wall using the usual electromagnetic boundary conditions at a non-conducting wall and also that $b'(0) = 0$ from the symmetry of the flow.

Equation (2.5) can be transformed into a second order d.e. whose solution for u exhibits the form,

$$u(r) = a_0(M) \int_{M^{\frac{1}{2}} r}^{M^{\frac{1}{2}}} \frac{\sinh(\frac{\xi^2}{2})}{\xi} d\xi, \tag{2.6}$$

where M is the Hartmann number and $a_0(M)$ is a constant determined by the z component of the pressure equation. The induced field is related to the velocity distribution by the induction equation and is given by,

$$b(r) = \frac{\alpha a^2}{\eta} \frac{a_0}{2M} \left(\cosh(\frac{M}{2}) - \cosh(\frac{Mr^2}{2}) - Mr^2 \int_{Mr^{\frac{1}{2}}}^{M^{\frac{1}{2}}} \frac{\sinh(\frac{\xi^2}{2})}{\xi} d\xi \right). \tag{2.7}$$

If we examine the result obtained for the velocity profile we can quickly see that no anticipated jet structure emerges for large Hartmann number. This is graphically illustrated in figure 2 where the profiles are observed to flatten (see Pai [b]) for increasing M and where the velocity assumes unit value along the axis of the pipe. In terms of the absolute velocity, $\lim_{M \to \infty} u(r) = O(\frac{1}{M^2})$, and the fluid is braked. The calculations are consistent and will for example yield the familiar Poiseuille flow as $M \to 0$.

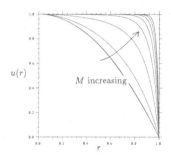

Figure 2. Axisymmetric velocity profiles at various values of the Hartmann number.

3. Flow through a two plane slot

The physical setup is similar to that described for the axisymmetric flow. The imposed magnetic field assumes the form, $\mathbf{B_0} = (-\alpha x, \alpha y, 0)$, which permits unidirectional solutions to the equations of motion. This point can be justified using a similar analysis to that employed in the cylindrical case. Once again we seek unidirectional solutions of the form $\mathbf{u} = (u(y), 0, 0)$ and $\mathbf{b} = (b(y), 0, 0)$. The resulting pressure distribution includes Lorentz terms which vary across the width of the duct in contrast to hydrodynamic flow with no magnetic field i.e.

$$p(x,y) = -kx - \frac{b^2(y)}{2\mu_0} + \frac{\alpha}{\mu_0}xb(y), \tag{3.1}$$

and where once again $-k$ is the imposed pressure gradient. The velocity and magnetic field are further related by the steady state magnetic induction equation (2.2) to give, $b'(y) = -\frac{\alpha}{\eta}yu$. Again there is no electric field present in (3.7) due to the symmetry of the flow. Finally eliminating b' from (3.6) and (3.7) results in,

$$u''' - M^2 y(yu)' = 0, \tag{3.2}$$

where $-1 \le y \le 1$ and M is the Hartmann number defined by $M^2 = \frac{\sigma\alpha^2 a^4}{\mu}$. The boundary conditions on the flow are exactly analogous to the axisymmetric case namely that, $u'(0) = 0, u(\pm1) = 0$ and, $b'(0) = 0, b(\pm1) = 0$.

Equation (3.2) can be transformed to a second order d.e. whose solutions for the velocity profiles can be expressed in terms of modified Bessel functions so that,

$$u(y) = A(\beta)y\{I_{\frac{1}{4}}^2(\beta)I_{-\frac{1}{4}}^2(\beta y^2) - I_{-\frac{1}{4}}^2(\beta)I_{\frac{1}{4}}^2(\beta y^2)\}, \tag{3.3}$$

which are shown in figure 3 and where $\beta = M/4$ with $A(\beta)$ determined from the z component of the equation of motion (2.2) i.e.

$$A(\beta) = \frac{k}{2\beta^2}/\{I_{\frac{1}{4}}^2(\beta)I_{\frac{3}{4}}^2(\beta) - I_{-\frac{1}{4}}^2(\beta)I_{-\frac{3}{4}}^2(\beta)\}, \tag{3.4}$$

(Antimirov 1978 equation (31) [f]).

The consistency of the solutions for all values of M is demonstrated by the limit $M \to 0$ in which domain the Poiseuille profile is recovered. Interesting behaviour arises when the Hartmann

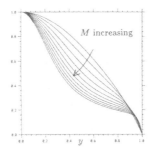

Figure 3. Two dimensional velocity profiles for various values of the Hartmann number.

number M tends to infinity. In accordance with the geometry we anticipate that a jet structure will emerge in this limit and this is indeed verified since, $\lim_{\beta \to \infty} u(y) \propto \frac{1}{\beta^2 y}$. Consequently for $y = O(1)$ the fluid flows along the field lines and is effectively squeezed up the middle of the duct. However the total flow is braked according to $O(\frac{1}{\beta^2})(= O(\frac{1}{M^2}))$ in this limit analogously to the axisymmetric flow.

4. Comparison of Results

We are now in a position to compare the chief characteristics of the two dimensional and axisymmetric flows (especially in the limit of large M). At large Hartmann numbers the flows of both geometries are braked according to $O(\frac{1}{M^2})$. Intuitively one might anticipate that the fluid would tend to flow along the magnetic field lines in order to reduce the Lorentz forces, as observed in the 2D problem. The comparable axisymmetric case exhibits no such tendency and in fact reduces to a slug or uniform flow. In order to explain this discrepant behaviour consider the curl of the steady state equation of motion (2.1),

$$M^2 \nabla \times ((\mathbf{u} \times \mathbf{B}_0) \times \mathbf{B}) + \nabla^2 \omega = 0, \tag{4.1}$$

which is the dimensionless vorticity equation for ω and where $\mathbf{j} = \sigma(\mathbf{u} \times \mathbf{B}_0)$. Consequently in the limit $M \to \infty$ and away from any boundary layers we have to a good degree of approximation,

$$\nabla \times ((\mathbf{u} \times \mathbf{B}_0) \times \mathbf{B}) = 0. \tag{4.2}$$

The curl of the Lorentz force in the core of the flow vanishes for large M and so permits flows which are not necessarily parallel to the magnetic field lines. This explains the nature of the pressure distributions present in the two flows which possess variation transverse to the flow and allow it to develop according to (4.2).

Despite the two independent studies of the flows by Regirer and Pai and their apparent contrasting behaviours they are essentially similar. The two cases studied are extreme examples of flows through a channel with an elliptical cross section. In order to further emphasize the similarity of the two neutral point flows studied we can alter the configuration. For example if we solve the problem for $a \leq r \leq c$ in axisymmetry and $d \leq y \leq e$ in 2D then let $a, d \to \infty$ respectively it is found that Hartmann type solutions arise for both flows (c.f. Shercliff, J., 1965, 'Magnetohydrodynamics', Pergamon Press).

5. Discussion of Stability

It is shown by Antimirov [f], [g] that the unsteady solutions to the two flows relax monotonically to the given solutions of Pai [b] and Regirer [c]. However, an important question arises from the preceding work concerning the stability of the jet structure observed in the two dimensional flow at large M. One might suspect with reference to 'Rayleigh's Inflection point theorem' that the flow would be unstable from inspection of figure 3. However, this criterion derives from inviscid theory in the absence of a magnetic field. In order to address this problem therefore a stability approach is formulated by introducing perturbations of the form, $\mathbf{u}'(x, y, t) = (\phi_y, -\phi_x, 0)$, and, $\mathbf{b}'(x, y, t) = (\psi_y, -\psi_x, 0)$. The ensuing linearised equations of motion and induction become,

$$\rho \frac{D}{Dt}(\nabla^2 \phi) - \rho u''(y)\phi_x = \frac{1}{\mu_0}\left\{(\mathbf{B}_m \cdot \nabla)\nabla^2 \psi - b''(y)\psi_x\right\} + \mu \nabla^2 \nabla^2 \phi, \tag{5.1}$$

and,

$$\frac{D\psi}{Dt} = (\mathbf{B}_0 \cdot \nabla)\phi + \eta \nabla^2 \psi, \tag{5.2}$$

respectively where $\mathbf{B}_m = \mathbf{B}_0 + \mathbf{b}$. In the spirit of the study presented so far an analytic limiting case is examined for which the perturbations have a long wavelength. The full stability problem is quite intractable and requires a numerical approach due to the fact that the coefficients depend on both x and y. The system (5.1)-(5.2) can however be considerably simplified if we examine the limit of long wavelength perturbations. The linearised equations reduce to,

$$\phi_{yyt} + M^2(\mathbf{B}_0 \cdot \nabla)^2 \phi - \phi_{yyyy} = 0, \tag{5.3}$$

where the Prandtl number $\frac{\mu}{\rho\eta}$ is assumed to be negligible for the liquid metals used.

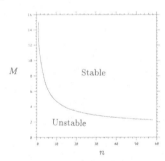

Figure 4. Neutral stability curve in $M - n$ space.

Seeking solutions to (5.7) which are products of functions of x, y and t so that, $\phi(x, y, t) = x^n f_n(y)e^{-\lambda_n t}$ yields a fourth order d.e. for $f_n(y)$. The boundary conditions are that $f(\pm 1) = f'(\pm 1) \doteq 0$ corresponding to no slip at the wall of the duct and that $f(0) = 0$ indicating there is no net pressure disturbance along the duct. The neutral stability curve $(\lambda = 0)$ is depicted in $M - n$ space and clearly shows that the flow is unstable for all M *. For most practical purposes

---

* Including $M = 0$ but requires an infinite length of pipe for the instability to show itself.

the length of channel is not long enough for the perturbations considered to manifest themselves. Indeed in a real situation the magnetic field is only approximated by $\mathbf{B}_0$ in the vicinity of the neutral point.

6. Concluding Remarks

The analytical studies of the neutral point flows have served to illustrate the consequences when the curl of the Lorentz force was required to vanish at large Hartmann numbers. As emphasised earlier the unidirectional solutions to the flow in this specific case provide a good initial understanding of the resulting behaviour when a fully developed nonlinear flow is established in this region. It would be reasonable to assume that a possible jet structure coupled with some form of eddy circulation would result. Such generalised flows have been reviewed by Hunt and Holroyd (1978) [h] but it remains to develop the full nonlinear problem and explore the potential use as an electromagnetic brake.

The author would like to thank Dr. Molokov and Dr. Antimirov for their guidance on the background of 'Neutral point flows'.

References

[a] Alemany, A., 1975, 'The flow of conducting fluids in a circular duct under rotating magnetic fields with several dipoles', Proc. of Bat-Sheva International Seminar on MHD Flows and Turbulence, Halstead Press.

[b] Pai, S., 1954, 'Laminar flow of an electrically conducting Incompressible fluid in a circular pipe', Journal of Applied Physics , Vol. 25, 1205.

[c] Regirer, S., 1975, 'Magnetohydrodynamic flow in a tapered magnetic field', Journal of Soviet Technical Physics , Vol. 19, 874.

[d] Vatazhin, A., Lyubimov, G. & Regirer, S., 1970, Magnetohydrodynamic Flows in Channels, Nauka, Moscow.

[e] Regirer, S., 1960, 'On exact solutions of the equations of Magnetohydrodynamics', Prikladnaia Matematika i Mekhanika (USSR), Vol 24, No.2, 383, [Journal of Applied Mathematics and Mechanics, Vol 24, 556, (1960)].

[f] Antimirov, M., 1978, 'Time-independent flow of a conducting fluid in a half-space and a two dimensional channel in a nonuniform external magnetic field', Magn. Gidrodin., Vol. 14, No.1, 59, [Magnetohydrodynam., Vol. 14, 49, (1978)].

[g] Antimirov, M., 1978, 'Nonstationary flow of conducting fluid in a round channel in a nonuniform magnetic field', Magn. Gidrodin., Vol. 14, No.2, 47, [Magnetohydrodynam., Vol. 14, 173, (1978)].

[h] Hunt, J.C.R. & Holroyd, R.J., 1978, 'Theoretical and experimental studies of liquid metal flow in a strong non-uniform magnetic field in ducts with complex geometry', Proc. of 2nd Bat-Sheva Seminar on MHD-flows and Turbulence, Israel University Press.

MODELLING OF ELECTRICALLY INDUCED FLOWS FOR STUDYING CURRENT CARRYING MELTS OF ELECTROMETALLURGICAL DEVICES

A.Ju.Chudinovskij*, S.B.Dement'ev**, E.V.Shcherbinin*,
V.Kh.Vlasjuk*, L.A.Volokhonskij**
* Institute of Physics of Latvian SSR Academy of Sciences,
Riga, USSR
** Institute of Electro-heating Equipment, Moscow, USSR

ABSTRACT. A new experimental technique and theory 'Electrically Induced Vortical Flows' enabled to develop a new method for studying
hydrodynamics of current carrying melts, i.e. a physical modelling of
velocity and pressure fields. The findings of the experiments correlat-
ing with theoretical models afford a possibility to safely determine
peculiarities of any three-dimensional flows, to verify calculated
data, to determine the existence of various flow régimes and the
respective critical values of the parameters. The paper cites some
examples for the modelling in question.

Vacuum-arc , electro-slag, arc steel-melting, thermal ore-
processing and other technological processes play an important role
among the modern techniques for high-quality steel production.
In terms of the magnetohydrodynamics all of them are characterized by high currents (from 50 A up to 150 kA) passing through a
system of contacting conductive media having different physical properties (slag, metal and flux melts; steel and graphite electrodes;
archs).
The interaction of spatially spreading current (of the $\vec{J}$ density) with the own magnetic field $\vec{B}$ induces the Lorentz electro-
magnetic force $\vec{F} = \vec{J} \times \vec{B}$, and if rot $\vec{F} \neq 0$ then vortical motions
are induced in liquid conductors – Electrically Induced Vortical Flows
(EIVF) [1].
Investigations on hydrodynamic processes in current carrying
melts started in the second half of the 70-ies. Among the first
studies there are numerical calculations of the electro-slag remelting
(Szekely et al. [2], Kreyenberg et al. [3]),theoretical estimation
for flows in the crucible induction furnace (Moreau [4]), the EIVF
experimental studies at the electro-slag welding by means of the physical modelling (Shcherbinin et al., Review [1]).
The last decade exhibits a growing need for the studies of
the kind. The power of electrical devices tending to an augmentation
leads to the hydrodynamic processes intensification in melts and,
consequently, to new technological problems, e.g. one of them – an

187

J. Lielpeteris and R. Moreau (eds.), Liquid Metal Magnetohydrodynamics, 187–193.
© 1989 by Kluwer Academic Publishers.

irregular lining wash-out in large baths with multielectrode current supply. So, nowadays efficient technologies for the EIVF hydro-dynamic structures are developed considering newly designed and re-constructed devices. The physical modelling on mercury models proves serviceable for the forced convection study.

In the majority of cases the modelling can be carried out only by a single similarity criterion $S = \mu_0 I^2 / \varrho \nu^2$, where μ_0 is the magnetic constant, ϱ and ν - the density and kinematic viscosity of the conducting liquid, I - the strength of current. Here the labo-ratory model scale should be arbitrary, since parameter S is indepen-dent of the specific dimension L. Modelling by S only is possible if subject to the restriction that: 1) the medium is homogeneous; 2) the deformation of the melt free surface is negligible when compared to the bath depth; 3) the Batchelor number is small: $Bt = \mu_0 \sigma V_0 L \ll 1$ where V_0 is the specific velocity, σ - the medium electroconductivity; 4) the thermal convection is insignificant; 5) parameter $\omega = 2\pi f \mu_0 \sigma L^2$ is such that $\delta = \sqrt{2/\omega} > 1$ at the current frequency $f = 40$-50 Hz [5].

As an example, consider (Fig.1a) an axisymmetrical EIVF in a cylindrical container 1 (H - the depth, D - the diameter) with non-conducting walls and the current supplied by two electrodes 2 and 3, their radii - $R_2 = D/2$ and $R_1 = kR_2$, $0 < k < 1$ [6]. Such a flow simulates a forced convection in slag bath 4 in the process of electro-slag remelting.

The impulse transfer equation integration by the ABCDA-con-tour (Fig.1a), limiting the meridional cross-section, leads to an integral half-empirical estimate enabling to neglect the effect of the thermal convection on the EIVF

$$\Delta T \ll \mu_0 I^2 M / \varrho g \beta H R_2^2 , \qquad (1)$$

as well as to the specific velocity expression

$$V_0 = \nu S M / R_2 \alpha . \qquad (2)$$

Here: ΔT is a mean (bath depth) difference between the symmetry-axis and the side-wall temperatures, ΔT being evaluated in reality; $M = (1 - k^2) / 2\pi^2 k^2$ - the flux of the electromagnetic force vector through the ABCDA cross-section calculated in case the current density in the electrodes is constant and $Bt \ll 1$; α is the empirical coef-ficient which depends on S, M and the bath configuration; β is the volumetric expansion coefficient; g is gravity acceleration.

The experimentally measured typical dependence of the axial velocity V_{ax} on the current strength I at the distance $\bar{z} = z/H \approx 0.48$ from the small electrode at $k = 0.2$ and $H = R_2$, is presented in Fig.1b (a scheme of the experimental device is given in [6], the velocity is measured by monocomponent optical-fibre sensor [7]). If $S < S_1$, then $V_{ax} \sim I^2$, i.e. the viscosous and electromagnetic members of the impulse transfer equation are of the same order ($\varrho \nu \nabla^2 \vec{V} \sim \vec{J} \times \vec{B}$, or $\varrho \nu V_0 / L^2 \sim \mu_0 I^2 / L^3$), that means that the EIVF exist in the Stokes regime. At $S_1 < S < S_2$ the regime is transitional, and at $S > S_2$ it is

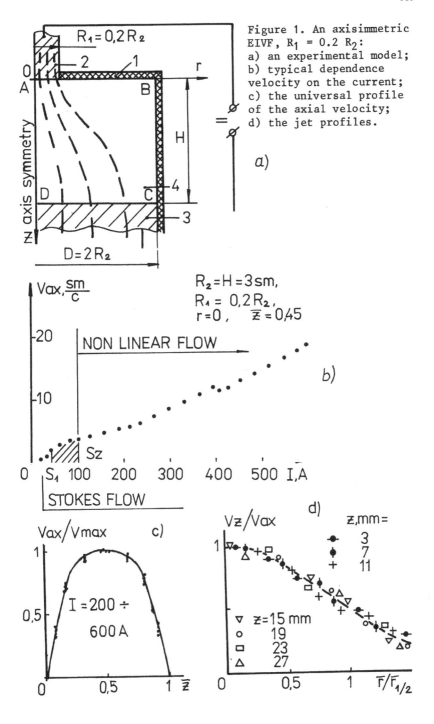

Figure 1. An axisimmetric
EIVF, $R_1 = 0.2 R_2$:
a) an experimental model;
b) typical dependence
velocity on the current;
c) the universal profile
of the axial velocity;
d) the jet profiles.

nonlinear, developed, since here $\varrho(\vec{V}\nabla)\vec{V} \sim \vec{J}\times\vec{B}$ and $V_0 \sim V*$, $V* = \sqrt{S}\nu/R_2 = I\sqrt{\mu_0/\varrho}/R_2$.

Similar measurements of other axial points of the bath prove that if $S > 2.8\cdot10^8$, then the axial V_{ax} velocities normalized by $V*$ are described (Fig.1c) by the universal function $V_{ax}/V* = 0.9\,\bar{f}$, $\bar{f} = f(\bar{z}) = 1 - (2\bar{z} - 1)^4$.

Fig.1d presents the results of the jet profile measurements, i.e. the V_z component of the $\vec{V} = V_r\,\vec{1}_r + V_z\,\vec{1}_z$ velocity at various distances $\bar{r} = r/H$ from the axis in various cross-sections. The data for $I = 600$ A are practically identical to those for $I = 400$ A and $I = 200$ A. These data are generalized by the formula $V_z/V* = 0.9\,\bar{f}(1 + 0.41(\bar{r}/\bar{r}_{1/2})^2)^{-2}$ where $\bar{r}_{1/2} = r_{1/2}/H = 0.34\,\bar{f}$ is a half-width of the jet. Considering the relations of $V_z = -r^{-1}\partial\psi/\partial r$ and $V_r = r^{-1}\partial\psi/\partial z$ the hydrodynamic stream-function ($\psi_0 = 0.21\ V*R_2^2$) could be restored

$$\psi/\psi_0 = \bar{f}^3 t\,(1+t)^{-1}, \quad t = 4.1\,(\bar{r}/\bar{f})^2 \qquad (3)$$

So, for a nonlinear developed EIVF, the velocities in the centre of the jet (viscous boundary layers excluded) are proportional to $V*$ and independent of the medium viscosity, and the velocity field normalized by $V*$ is a universal function of nondimensional coordinates of the point. It should be indicated that for the bath of the given configuration the developed regime sets in at $S > 2.8\cdot10^8$.

The results of the V_z velocity measurements at $k = 0.8$ and $I = 1,000$ A are shown in Fig.2 by arrows (solid lines - $\psi = $ const). With k increase, the integral velocity M of the electromagnetic force decreases in such a way that the critical value of the S_2 parameter, which initiates the developed regime, augments, and the Reynolds number critical value experimentally proved to remain the same for $k = 0.2$: Re$\approx$14,000.

In general, the EIVF preserves the structure of a toroidal vortex, but at $k = 0.8$ the axial jet is considerably wider, the jet profiles have a gap at the axis (the velocity defect is $\approx$8%), and the returning flow velocities are by 10-12% higher than those of the jet.

A comparison of $V*$ values at various k enables to conclude that in (2) the friction coefficient α with accuracy to constant $c*$ is equal to $c*\sqrt{SM}$ at Re $= \sqrt{SM}/c*> 14,000$; and the constant depends only on H/R_2. Numerical calculations and the measurements of the EIVF jet pressure upon the cylindrical bath bottom confirm the above [1]. The dependence of $c*$ on H/R_2 is experimentally stated in [6]. All this results give a formula for the maximum velocity V_{max} calculations (and the Reynolds number Re $= V_{max}R_2/\nu$) in a current carrying medium with the density of ϱ_c and the melting current I_n

$$V_{max} = \gamma I_n\sqrt{\mu_0\,(1-K^2)/\varrho_c}/KR_2, \qquad (4)$$

where for a shallow bath ($0.5 < H/R_2 < 1$) - $\gamma = 0.11 + 0.07\ H/R_2$, and for a deep one ($1 < H/R_2 < 2$) - $\gamma = 0.25-0.07\ H/R_2$.

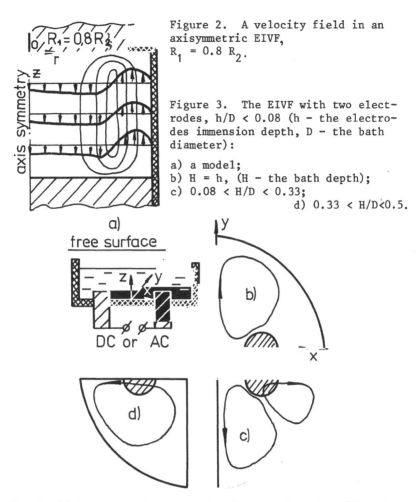

Figure 2. A velocity field in an axisymmetric EIVF, $R_1 = 0.8 \, R_2$.

Figure 3. The EIVF with two electrodes, h/D < 0.08 (h - the electrodes immension depth, D - the bath diameter):

a) a model;
b) H = h, (H - the bath depth);
c) 0.08 < H/D < 0.33;
d) 0.33 < H/D < 0.5.

It should be stressed once again that the formulae (3) and (4) are meaningful if (1) and Re > 14,000. (4) allows to estimate applicability limits of the electrodynamical approximation: $B_t \ll 1$ gives $I_n \ll k/\mu_0 \sigma \gamma \sqrt{\mu_0 (1 - k^2)/\rho_c}$.

As an example, let us then discuss hydrodynamic structures emerging in baths with a nonconducting case at the two- or three-electrode current supply to the melt free surface [8,9]. The problem is constrained by the structure dependence on the bath depth H when the bath diameter is 120 mm, diameter of the electrodes is 22 mm, the electrode opening is 52 mm, the depth of the electrode immersion h. The experimental model scheme is shown in Fig.3a. In all the cases the EIVF is symmetrical to the electrode plane y = 0 and the x = 0 plane, that is why Fig.3-d presents the flows observed on the surface in one quadrant only.

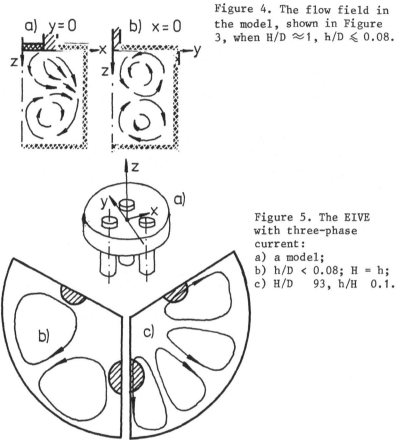

Figure 4. The flow field in the model, shown in Figure 3, when $H/D \approx 1$, $h/D \leqslant 0.08$.

Figure 5. The EIVE with three-phase current:
a) a model;
b) $h/D < 0.08$; $H = h$;
c) H/D 93, h/H 0.1.

Fig.3b demonstrates the EIVF in a flat layer $h = H = 3-8$ mm, which is discussed in [10]. H growing up to 40 mm, one more pair of vortices appears above the end-face of the electrodes (Fig.3c) (see [10]).In the range of 40 mm < H < 60 mm, the other pair of vortices prevails (Fig.3d).At H > 60 mm, it is impossible to make a conclusion concerning the whole picture of the flows, if observing the bath free surface.

The measurements of velocities in the x = 0 and y = 0 cross-sections with the sensor [7] lead to a conclusion that the EIVF in these planes have the form shown in Fig.4a,b. So, it follows that, in general, two large toroidal vortices are set on the z axis. In one of them (0 < z/H < 0.5) the velocities are periodic by the azimuthal co-ordinate , so that at the side wall of the bath $V_z > 0$ at y = 0, and $V_z < 0$ at x = 0; the second vortex is close to axisymmetrical, its axial jet is directed opposite to the z axis.

At a three-phase current supply (Fig.5a), between each pair of the electrodes the same EIVF structure is formed as is depicted in Fig.3. Thus, at h = H = 3-5 mm the EIVF has the form of three pairs of vortices (Fig.5b); at h = 3 < H < 25 mm – of six pairs of vortices (Fig.5c). At H > 40 mm the jets formed above the end-faces of the electrodes prevail, the EIVF turn to one large toroidal vortex having V_z > 0 near the wall and V_z < 0 in the near-axial region.

REFERENCES

[1] BOJAREVICH V.V., FREJBERG Ja.Z., SHELOVA E.I., SHCHERBININ E.V., 1985, *Electrically Induced Vortical Flows*, Riga, Zinatne, 315.

[2] DILAVARI A.H., SZEKELY J.A., 1977, 'A mathematical model of slag and metal flow in the ESR process', *Met.Trans.*, Vol.8B, 227.

[3] KREYENBERG J., SCHWERDTFEGER K., 1979, 'Stirring velocities and temperature field in the slag during electroslag remelting', *Arch. Eisenhüttenwessen*, Bd.50, No.1, 1.

[4] MOREAU R., 1978, 'MHD flows driven by alternating magnetic field', *Proc.2 Bat-Sheva Int. Sem.*, Beersheva, 165.

[5] CHUDNOVSKIJ A.Ju., 1988, 'About the modelling of EIVF', *Magnitnaja Gidrodinamika*, Vol.4.

[6] ZHILIN V.G.,IVOCHKIN Yu.P., OKSMAN A.A., LURINS G., CHAIKOWSKY A.I., CHUDNOVSKIJ A.Ju., SHCHERBININ E.V., 1986, 'An experimental investi-gation of the axisymmetrical EIVF velocity field in a cylindrical container', *Magnitnaja Gidrodinamika*, Vol.3, 110.

[7] ZHILIN V.G., ZVJAGIN K.V., IVOCHKIN Yu.P., OKSMAN A.A., 1988, 'Diagnostics of liquid metal flows using fibre-optical velocity sen-sors', *IUTAM Symp. on Liquid Metal MHD*, Riga.

[8] DEMENT'EV S.B., CHAIKOWSKY A.I., CHUDNOVSKIJ A.Ju, 'Forming of the EIVF in baths with many current-carrying electrodes', *Magnitnaja Gidrodinamika*, Vol.1, 95.

[9] DEMENT'EV S.B., ZHILIN V.G., IVOCHKIN Ju.P., OKSMAN A.A., CHUD-NOVSKIJ A.Ju., 1988, 'For a question about forming of the EIVF with many current-carrying electrodes', *Magnitnaja Gidrodinamika*, Vol.4.

[10] SHCHERBININ E.V., 1988, 'EIVF', *IUTAM Symp. on Liquid Metal MHD*, Riga.

Session D:
Aluminium Reduction Cells

AMPLITUDE EVOLUTION OF INTERFACIAL WAVES IN ALUMINIUM REDUCTION CELLS

René J. MOREAU, Sylvain PIGNY and Sherwin A. MASLOWE (*),
Laboratoire MADYLAM, ENSHMG, BP 95
38402 ST MARTIN D'HERES Cedex, FRANCE
(*) Permanent address : Mc Gill University, MONTREAL, CANADA

ABSTRACT. We consider instabilities at the interface of two liquids (representing cryolite and molten aluminium) which arise due to electromagnetic forces, possibly combined with the Kelvin-Helmholtz mechanism. Linear stability boundaries are computed and physical interpretations are given of the results. A multiple-scales analysis is then discussed which describes the amplitude evolution of a wave-packet.

I. Introduction

Instabilities at the interface of two liquids (cryolite above and molten aluminium below) representative of aluminium cells have recently been studied using two approaches. Sneyd [1] has developed a linear analysis with the simplifying assumption that the electric current is purely vertical through the liquids as well as through the electrodes. Among the effects neglected in [1], are the mean motion of the fluids, friction, and surface tension. Sneyd, none the less, demonstrates that the non-uniformity of the magnetic field created by external conductors can destabilize the interface.

Moreau and Ziegler [2], on the other hand, underline the importance of the horizontal component of the electric current which takes place in the aluminium and show that it can also destabilize the interface. Their analysis is local in the sense that the cell is assumed to be infinite and all the main parameters of the cell are taken to be uniform. Mean motion is taken into account as well as friction and surface tension.

The purpose of this paper is to proceed further toward a description of those phenomena responsible for instabilities and waves in aluminium cells. Special care is therefore taken to model consistent distributions of the electric potential and the magnetic field. In particular, these distributions allow basic ingredients such as the horizontal component of the current density and the gradient of the horizontal magnetic field. For the sake of brevity simplifying assumptions are still made :

a) The cell is supposed to be infinite in the direction y perpendicular to the plane of the electric current, so that the problem remains two-dimensional (in the following : ky <<1).

b) The two fluid layers are assumed to have the same thickness H as well as the same friction coefficient κ, (see Moreau and Evans [3]). This last hypothesis eliminates the differential pinch effect mentioned in [2], which is a minor stabilizing or destabilizing mechanism.

In section 2, a linear analysis leads to a characteristic equation and to neutral curves which illustrate the influence of the main parameters. Section 3 introduces an equation for the amplitude of a wave-packet based on a multiple scales technique. Electromagnetic forces are assumed to be of order

197

J. Lielpeteris and R. Moreau (eds.), Liquid Metal Magnetohydrodynamics, 197–204.
© 1989 by Kluwer Academic Publishers.

ε^2 (ε = (kL)$^{-1}$ where k is the wave number and L a typical length of the wave-packet) and the evolution equation is derived on this basis. For the reader who is familiar with interfacial stability problems, but not in the present context of aluminium reduction cells, some background material which may be helpful is presented in Moreau and Evans [3].

2. Stability analysis

2.1 Undisturbed state

In the domain -L $<$ x $<$ L, -H $<$ z $<$ H of the (x,z) plane, the electric potential is modelled by the following expressions, which satisfy the Laplace equation and the conditions of continuity of $\varnothing$ and $\sigma \frac{\partial \varnothing}{\partial z}$ at the interface (indexes 1 or 2 respectively refer to cryolite O$<$z$<$H or to aluminium -H$<$z$<$0) :

$$\varnothing_1 = \frac{J}{\sigma_1} \left[z + \frac{a}{H} \frac{\sigma_1}{\sigma_2} (x^2 - z^2) \right]$$

$$\varnothing_2 = \frac{J}{\sigma_2} \left[z + \frac{a}{H} (x^2 - z^2) \right] .$$

(1)

The factor a is a small non-dimensional parameter allowing variation in the shape of the electric current lines. Depending on its value the horizontal component of the current density may be positive, as when the cathodic bars collect the current partially near the border of the cell, or negative, as when the current lines are distorted by a peripheral ledge of frozen cryolite.

The magnetic field **B** is a sum of two contributions. One, the field induced by the external conductors, is a pure gradient. As recognized by Sneyd [1] it is important to take into account its variation ; it is therefore modelled by **B** = (-γy, B$_0$- γx, O). The other, the field of the internal current, is easily derived from Eqs. (1) through ampère's law curl **B** = μ**J**. Finally, we model the magnetic field as :

$$\mathbf{B_1} = (-\gamma y, \quad B_0 - \gamma x - \mu Jx, \quad 0)$$

$$\mathbf{B_2} = (-\gamma y, \quad B_0 - \gamma x - \mu Jx (1 - 2a z/H), \quad 0).$$

(2)

The undisturbed forces **JxB** are rotational and drive some motion in the (x, z) plane, as well as some deformation of the interface. This deformation, which could be calculated as in Moreau and Evans [3], is assumed to be small enough to be neglected. However, the stirring motion, as well as the preexisting turbulence, is represented only via the friction coefficient κ. In the framework of this two-dimensional analysis the main horizontal motion of the two liquids cannot be represented. However, in order to allow some interaction between Kelvin-Helmholtz instability and electromagnetically driven instabilities, uniform velocities U_1 and U_2, both in the x-direction, are assumed to be present. They are considered as independent parameters, whereas in actual cells the two horizontal velocity fields are controlled by the **JxB** forces (see [3]).

2.2 Disturbances

Because the relevant magnetic Reynolds number is much smaller than one, the disturbances of the electromagnetic quantities are independent of the velocity field. They depend only on the geometry of the interface and they instantaneously adjust themselves to any change in that geometry. Introducing a disturbance of the interface of the form

$$\eta = A \exp \left[i \, (kx - \omega t) \right] \tag{3}$$

expressions are easily obtained for perturbations of the electric potential satisfying Laplace's equation $\Delta \varphi = 0$ and the following boundary conditions

$$\varphi_1(H) = \varphi_2 (-H) = 0,$$
$$\emptyset_1(\eta) + \varphi_1(0) = \emptyset_2 (\eta) + \varphi_2(0), \tag{4}$$

$$\sigma_1 \left[\frac{\partial \emptyset_1}{\partial z}(\eta) - ik\eta \frac{\partial \emptyset_1}{\partial x}(0) + \frac{\partial \varphi_1}{\partial z}(0) \right] = \sigma_2 \left[\frac{\partial \emptyset_2}{\partial z}(\eta) - ik\eta \frac{\partial \emptyset_2}{\partial x}(0) + \frac{\partial \varphi_2}{\partial z}(0) \right]$$

It is straighforward to deduce expressions for the disturbances of the current density **j**, the magnetic field **b**, and for the electromagnetic forces **f** = **jxB** + **Jxb**.

It is remarkable that these forces are irrotational, except for the terms involving the gradient γ (see Eq. (2)) of the far magnetic field. This suggests expressing them, as well as the velocity field as (i=1, 2) :

$$\mathbf{f}_i = - \nabla F_i + \mathbf{f}_{ir}$$
$$\mathbf{u}_i = \nabla V_i + \mathbf{u}_{ir} . \tag{5}$$

Then, because of continuity

$$\Delta V_i = 0 \qquad \nabla . \mathbf{u}_{ir} = 0 \tag{6}$$

and the motion equation may be written

$$\nabla p_i^* = - \rho_i(r_i + \kappa)\mathbf{u}_{ir} + \mathbf{f}_{ir} , \tag{7}$$

where $p^*_i = p_i + F_i + \rho_i (r_i + \kappa) V_i$ and $r_i = i (k.U_i - \omega)$. The classical boundary conditions on the electrodes and on the interface are easily expressed in terms of p_i, and it is straightforward to get the distribution of p_i by solving the equation

$$\Delta p_i^* = \nabla . \mathbf{f}_{ir} . \tag{8}$$

2.3 Characteristic Equation and Neutral Curve

Substituting expressions for $p_1(0)$ and $p_2(0)$, obtained from Eq. (8), into the condition of continuity of pressure at the interface, where r stands for surface tension, we obtain

$$p_1(0) - p_2(0) = - \eta \, [\, g \, (\rho_2 - \rho_1) + \Gamma \, k^2] \tag{9}$$

which gives the characteristic equation. The latter may be written as follows when $\sigma_1 \ll \sigma_2$ (in actual cells $\sigma_1 / \sigma_2 \approx 10^{-4}$) :

$$\rho_1 r_1 \, (r_1 + \kappa) + \rho_2 r_2 \, (r_2 + \kappa) + g^* kt \, (\rho_2 - \rho_1) + \Gamma k^3 r + J \gamma_{/2} \, (1 + M + iN) = 0 \tag{10}$$

with

$$\alpha = kH \qquad\qquad t = \tanh kH$$
$$M = 2a \, \alpha t \, (t^2 - t^{-2} - \alpha^{-2})$$
$$N = 2a \, x_{/H} \, \alpha t \, (2t - 2t^{-1} + t^{-1}) \, .$$

The first four terms in Eq. (10) represent the usual characteristic equation, except that g^* is now the effective gravity taking into account the irrotational part of the electromagnetic forces, i.e.,

$$g^* = g - 2a \, \frac{x}{H} \cdot \frac{J \gamma x}{\rho_2 - \rho_1} \, . \tag{12}$$

This effective gravity differs from g only when a is non-zero in the aluminium. This means that the horizontal component of the current density might produce Rayleigh-Taylor instabilities when $a\gamma > 0$, or enhance the stabilizing gravity effect when $a \gamma < 0$.

The last term of Eq. (10) represents the influence of the rotational part of the electromagnetic forces. The reader should notice that it has an imaginary part. This means that it is analogous to the velocity difference $U_1 - U_2$, and that this effect is destabilizing. Consequently, when $U_1 = U_2 = 0$, friction appears necessary to counterbalance the imaginary part of this term. Clearly, even without any shear of the interface some instability may develop, driven only by electromagnetic mechanisms.

The marginal condition such that ω be real follows from (10). With our simplifying assumptions it may be written

$$F^2 = t_{/\alpha} + T \, \alpha t + P(1 + M - QN^2) \, / \alpha^2 \tag{13}$$

with the following definition of the non-dimensional parameters

$$F = \frac{\rho_1 \, \rho_2}{\rho_2^2 - \rho_1^2} \cdot \frac{(U_1 - U_2)^2}{g^* H} \qquad\qquad P = \frac{J \gamma H}{2g^* (\rho_2 - \rho_1)} \tag{14}$$

$$T = \frac{\Gamma}{g^* H(\rho_2 - \rho_1)} \qquad\qquad Q = \frac{J \gamma}{2k^2 (\rho_1 + \rho_2)} \, .$$

For the sake of brevity, we limit the analysis to conditions such that a $\ll 1$ (and M $\ll 1$) and a $x_{/H} = 0(1)$, which seem to correspond fairly well to actual cells, where $x_{/H}$ may be greater than 10. Then Eq. (13) becomes

$$F^2 = t_{/\alpha} + T \, \alpha t + P \, [1/_{\alpha 2} - Q^* G^2(\alpha)] \tag{15}$$

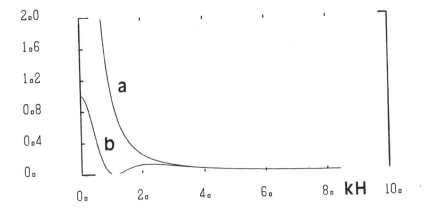

Fig. 1 : Variation of $1/\alpha^2$ (curve a) and $G^2(\alpha)$ (curve b).

with $Q^* = Q.4a^2 x^2/H^2$, and $G(\alpha) = t(2t - 2/_t + 1/_\alpha)$. Figure 1 shows the variations with α of G^2 and $1/_\alpha 2$. One may notice that these two quantities become equal when α is large enough (say $\alpha > 3$), but that $1/\alpha^2$ is of course predominant when $\alpha \ll 1$. The difference $1/_\alpha 2 - Q^* G^2(\alpha)$ may be negative only if $Q^* > 1$. If $1 < Q^* < 9.47$, it is negative for α sufficiently large and it has a minimum for $\alpha \approx$ 2.2 to 2.5. If $Q^* > 9.47$, another band of wave numbers makes the difference negative around $\alpha \approx 0.5$. Of course, since $G(\alpha) = 0$ in the vicinity of $\alpha = 1$, wave numbers of the order of one can never become unstable. Therefore, two different kinds of disturbance may be excited by electromagnetic mechanisms ; those with $\alpha = 2.2$ to 2.5 (corresponding to wave-lengths of the order of 50 cm in cells), and those with $\alpha \approx 0.5$ (corresponding to wave-lengths of the order of 2.5 m in cells). But the second instability requires an electric current density approximately ten times greater than the first.

Figure 2 shows typical neutral curves. Of course, the ordinary minimum of the neutral curve, depending only on gravity and surface tension, may still exist if it concerns a band of wave numbers signficantly larger than those where electromagnetic effects are dominant (right side of Fig. 2) ; then, the neutral curve may have three distinct minima. On the contrary the usual minimum may disappear when it concerns the same wave-numbers as the electromagnetic effects (left side of Fig. 2).

3. AMPLITUDE EVOLUTION EQUATION

In this section, we outline the derivation of an amplitude equation describing the weakly non-linear evolution of a wave-packet. To indicate the nature of the expected result, an amplitude equation derived by Nayfeh and Saric [4] for the classical Kelvin-Helmholtz flow in the absence of surface tension is given below. Provided that we are not near the cut-off wave-number (corresponding to the onset of instability), the disturbance evolves accordingly to :

$$\frac{\partial A}{\partial \tau} - \frac{1}{2} i\omega'' \frac{\partial^2 A}{\partial \xi^2} = \mu |A^2| A , \qquad (16)$$

where τ and ξ are slow scales that will be defined later. For a conservative system, $\omega''(k)$ is real, μ is imaginary, and Eq. (16), often called the cubic Schrödinger equation, can be solved exactly by the inverse scattering method. A particularly interesting solution involves envelop solitons, but this is not the only possibility. Moreover, other amplitude equations are obtained when perturbing away from the stability boundary and in this case inviscid flows (including the Kelvin-Helmholtz) lead to amplitude equations that are second-order in time.

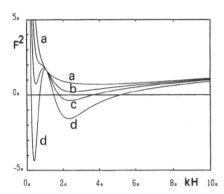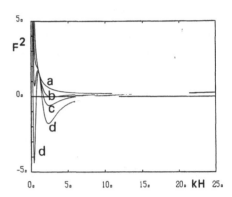

Fig. 2 : Neutral curves for P = 1. Left : T = 0.1, right : T = 0.01 Q* = 1 (curve a), 5(b), 10(c), 20(a).

The generalisation of greatest interest in the present study concerns the treatment of instabilities arising from the Lorentz forces. It develops that a linear term must be added to Eq. (16) and that the constants ω'' and μ will both be complex in general. To obtain an appropriate balance between these terms, as indicated earlier, the Lorentz forces are scaled such that they are of order ε^2. The development will be carried out by employing a version of the multiple-scales method which requires only one slow scale in each of the variables x and t.

We begin by expanding the dependent variables in a series in power of ε and, as usual in such studies, a Taylor series expansion about z = 0 is also employed. The interface displacement, velocity potential and pressure are expanded as follows :

$$\eta = \eta_0 + \varepsilon \eta_1 + \varepsilon^2 \eta_2 + \dots$$
$$\phi = \phi_0 + \varepsilon \phi_1 + \varepsilon^2 \phi_2 + \dots$$
$$\rho = \rho_0 + \varepsilon \rho_1 + \varepsilon^2 \rho_2 + \dots$$

In the $0(\varepsilon^\circ)$ problem, the classical characteristic equation is recovered. However, instead of taking the amplitude constant, as in the usual linear theory for monochromatic waves we write, for example

$$\eta_0 = A(X, T) e^{i\theta} + A^* (X, T) e^{-i\theta}$$

where $\theta = kx - \omega t$ and the slow scales are defined by $X = \varepsilon x$ and $T = \varepsilon t$. The general procedure for deriving the equation satisfied by $A(X, T)$ in hydrodynamic stability problems is described in Benney and Maslowe [5], where a number of illustrative examples are given. If the group velocity ω' is finite, as is usually the case, one obtains

$$A_T + \omega' A_X = \varepsilon (\beta A_{XX} + \delta A + \mu A^2 A^*) \qquad (17)$$

where it can be verified a posteriori that $\beta = i\omega'' /2$.

Equation (17) is the appropriate form in our study, which is interesting, because in the absence of Lorentz forces it was shown in [4] that the amplitude equation at marginal stability is second order in time. A discussion of how the form of the amplitude equation is dictated by the linear dispersion relation can be found in sections 7-3-2 and 7-4-1 of Swinney and Gollub [6] . We have not yet completed the calculation of the "Landau constant" μ, but its value is known in certain limiting cases (e.g. capillary-gravity waves). Therefore the present article concentrates on the evolution of linear wave packets.

Proceeding now to the $0(\varepsilon)$ problem, the variables can be separated by noting that $A_T = -\omega' A_X$ to this order and the time-dependent variables are all taken to be proportional to A_X. Imposing the boundary and interface conditions determines the value of ω'.

At $0(\varepsilon^2)$, the values of the coefficients β and δ in Eq. (17) are determined. This time, to separate variables and solve the resulting nonhomogeneous system, the $0(\varepsilon)$ terms must be included in the amplitude equation. The form of a typical term is illustrated by way of example by writing the pressure as :

$$p_2 = A_{XX} P(z)e^{i\theta} + A P_2(X,z)e^{i\theta} \cdot$$

Because of the way we have scaled the electromagnetic forces, the coefficient β is the same as in a bounded Kelvin-Helmholtz flow. The non-zero value of δ, on the other hand, is a new result due entirely to the Lorentz forces :

$$\varepsilon^2 \frac{\delta}{\omega} = -\frac{J\gamma H}{g(\rho_2 - \rho_1)} a \frac{x}{H} \left(\frac{1}{sc} + \frac{1}{\alpha}\right)$$
$$+ i\left[\frac{J(B_0 - \mu Jx)}{g(\rho_2 - \rho_1)} 2a \frac{x}{H} - \frac{J\gamma H}{g(\rho_2 - \rho_1)} \frac{a}{\alpha}\left(\frac{1}{\alpha^2} + \frac{s^2 + c^2}{s^2 c^2} + 2a \frac{x}{H} kx\right)\right]. \qquad (18)$$

As a final observation, we note that the linearized version of Eq. (17) can be solved exactly by employing a Fourier transform in X if δ is assumed to be constant. It is natural, however to rewrite this equation by employing the coordinates $\tau = \varepsilon T$ and $\xi = x - \omega'T$, corresponding to a frame of reference moving at the group velocity. This leads to the equation

$$\frac{\partial A}{\partial \tau} = \beta \frac{\partial^2 A}{\partial \xi^2} + \delta A \qquad (19)$$

whose solution is given by :

$$A(\xi, \tau) = \frac{e^{\delta\tau}}{\sqrt{4\pi\beta\tau}} \int_{-\infty}^{\infty} A(w, o)\, e^{-[(\xi - w)^2 / 4\beta\tau]} dw \ .$$

A number of special cases, e.g. self-similar solutions, can be obtained easily from the above general solution by an appropriate choice of initial condition. Rather than write these down, we discuss their properties in common. For example, there is a shortening of wave length with distance and a phase that varies slowly in time and space. The dominant effect, of course, is the exponential amplification of A with time which will eventually be modified by nonlinearity.

References

[1] SNEYD, A.D., "Stability of Fluid Layers Carrying a Normal Electric Current", J. Fluid Mech. (1985), vol. 156, p. 223-236.

[2] MOREAU, R. and ZIEGLER, D., "Stability of Aluminum Cells. A New Approach", Light Metals (1986), p. 359-364.

[3] MOREAU, R. and EVANS, J.W., "An Analysis of the Hydrodynamics of Aluminum Reduction Cells", J. Electrochem. Soc. Electrochemical Science and Technology (1984), vol. 131, n° 10, p. 2251-2259.

[4] NAYFEH, A.H. and SARIC, W.S., "Non-linear waves in a Kelvin-Helmholtz flow", J. Fluid Mech. (1972), vol. 55, p. 311-328.

[5] BENNEY, D.J. and MASLOWE, S.A., "The evolution in Space and Time of Non-Linear Waves in Parallel Shear Flows", Studies in Appl. Math. (1975), vol. 54, p. 181-205.

[6] MASLOWE, S.A., "Shear Flow Instabilities and Transition", in Hydrodynamic Instabilities and the Transition to Turbulence (eds. Swinney H.L. and Gollub J.P.) (1982), Springer, p. 181-228.

PHYSICAL AND MATHEMATICAL MODELING OF MHD-PROCESSES
IN ALUMINIUM REDUCTION CELL

V.Bojarevics*,A.I.Chaikovskij*,E.V.Gorbachev*,H.Kalis°,
R.Millere°,Eh.V.Shcherbinin*
*Inst. of Physics,Latvian SSR Acad.Sci.,Salaspils,229021 USSR
°Latvian State University,Riga,226050 USSR

ABSTRACT. A turbulent flow in an aluminium electrolysis cell driven by
the electromagnetic force due to the electric current passing through
the liquid layers of aluminium and electrolyte, and the magnetic field
of this current and the currents in supplying leads, is analysed by
means of numerical and experimental simulation.The experimental cell
models the electric current and magnetic field distributions in dependence of the nonuniform current loads in the anode and cathode elements,
and the associated flow driven in the single fluid layer (mercury). The
mathematical model uses effectively the small relative depth of the
fluid layers and enables to simulate economically the three-dimensional
distributions of electric current and magnetic field, the horizontal
flow and the interface deformation.

Total electric current passing through a modern electrolysis cell
for aluminium production reaches in magnitude 200 kA and higher. The
associated electromagnetic forces mainly determine hydrodynamics of the
adjacent liquid aluminium and electrolyte layers, and, thereby, thermal
balance and physico-chemical processes in the cell [1]. In fact, the
design of modern electrolysis cells has become impossible without
accounting for the MHD-processes within the bath. The vortical electromagnetic forces, first, drive large scale flows in the liquid layers.
Typical Reynolds numbers $Re \approx 10^5 - 10^6$, and the flow is turbulent (this is
promoted by wall roughness, intense gas escape in the electrolyte, a
nonuniform shear layer at the interface). Secondly, the electromagnetic
forces are responsible for a deformation of the interface between the
liquid layers of metal (of density $\rho_1 = 2.3 \times 10^3$ kg/m³) and electrolyte
($\rho_2 = 2.1 \times 10^3$); a wave motion is easily developing there due to the
small difference of densities. The exact knowledge of the electromagnetic force distribution is a key factor in a development of mathematical model. One can explain by the adequate simulation of the electromagnetic forces the success of stationary flow mathematical models [1,2]
based on the two-dimensional Navier-Stokes equations for motion in the
horizontal plane, where a turbulent viscosity is calculated according to
k-ε turbulence model [2], what may be considered also a formal deficien-

205

cy, since this does not account for the vertical momentum transport,i.e.,
the friction of liquid layers at the bottom of bath, anodes, and between
themselves [3,4]. Moreover, the two-dimensional horizontal flow model
cannot account for the effect of channels between the anodes and walls.

The variety of physical effects within the electrolysis cell does
not permit to restrict attention only to the mathematical modeling,
therefore experimental investigations have been initiated with liquid
metal at suitable temperatures [3]-Woods metal,[6]-mercury.

THE EXPERIMENTAL FACILITY permits a separate investigation of
MHD-processes in the liquid electrolyte and metal in a single layer
model (liquid mercury). The facility consists of two working baths containing liquid mercury, placed between imitators of the neighbouring
cells electromagnetic field. All the main elements of the current-carrying external circuit are variable in space and current loads. The model is made from non-ferromagnetic materials what permits additionally
to investigate an effect of different ferromagnetic parts in the industrial cell by indroducing these in the model. The present model repeats
in scale 1:10 the industrial cell of 175 kA of end to end orientation in
the line of cells. The working baths contain liquid mercury layer of
dimensions 1.0x0.42x0.02 m (the depth can be varied).

At the beginning stage the facility included merely one - the graphite - working bath shown in Fig.1. In the process of work electrical
contacts between the graphite anodes and copper rods, and also between
the copper cathode bars and graphite bath bottom, aged (eroded) leading
to a nonuniform distribution of currents along the anodes and cathodes,
what significantly altered the hydrodynamics in the bath.Nevertheless,
this situation with a nonuniform current distribution is quite typical
for industrial cells with disturbances in the technology [6], and we
shall analyse this situation coupled with the corresponding theoretical
model.

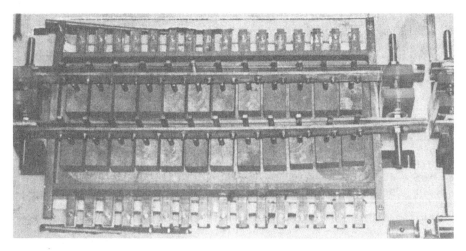

Figure 1. Experimental graphite cell (empty of mercury).

At the second level of the experimental set-up development a working bath has been designed with a controllable current distribution
along the anodes and cathodes: for this purpose each the anode, made in
this model from copper, and cathode (copper or mercury) is connected in
series with a miniature, electrically or mechanically operated mercury
rheostat. The controllable model permits to obtain both the uniform current distribution, and also the most varied horizontal current flow
configurations within the liquid layer, what highly affects the hydrodynamics. The facility of measurements ensures the registration of currents, the measurement of magnetic field by a three-component Hall-ef-
fect probe, the velocity field can be directly visualized and measured
by the use of two-component tensometric probe.

THE THEORETICAL MODEL is conventionally divided in electromagnetic
and hydrodynamic parts. The present model places an emphasis on an adequate simulation of electromagnetic force. The force $f = j \times B$ in the
fluid is generated by an electric current $j = -\sigma \nabla \Phi$ (Φ-electric potential)
and the integral magnetic field B, which is a superposition of the
fields: B_1 due to the internal current j within the cell itself, B_0 due
to the current in the external circuit (leads,bars,neighbour cells), and
B_M due to a magnetization of ferromagnetic parts. The internal current
j is computed dividing the cell into a number of layers of different
electrical conductivity σ (Fig.2), for each of these the equation $\nabla^2 \Phi = 0$
is solved numerically. Boundary conditions are imposed by the entering
and leaving currents through the anodes and cathode elements respectively, at the interfaces of layers both the potential and the normal component of current are continuous. In difference to the preceding models
[1-4], a three-dimensional distribution $\Phi(x,y,z/e)$ within the fluid is
determined employing an expansion by powers of a small parameter $e = H/L$:

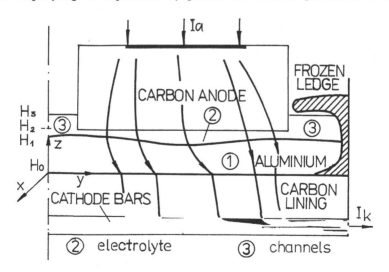

Figure 2. Half-section of the electrolysis cell and the computed
electric current lines.

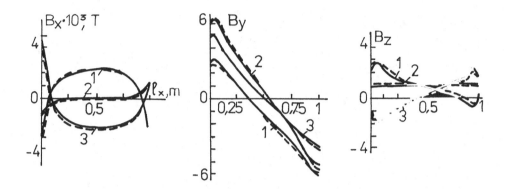

Figure 3. The magnetic field components along the channels of mercury
model (z= 0.02 m): 1 - y=0.185 m, 2 - y=0, 3 - y=-0.185; full line -
predicted, dashed - measured.

$\Phi=\Phi_0+e\Phi_1+e^2\Phi_2+...$, where H is a typical depth of the liquid layer,
L- a horizontal length scale of the layer.

The internal magnetic field B_I is found by integrating Biot-Savart
law over the whole volume V' with the current j:

$$B_I(r)=\mu_0/4\pi \int j(r') \times R \, dV'/|R|^3,$$

where $R=r-r'$, r' is a radius-vector to a current-element, r-to an observation point. It has been found that the accuracy of numerical integration is severely affected by the singularity when the integration point
approaches the observation point, therefore, for a magnitude of $|R|$
close to the numerical mesh element size a linearized j distribution
is integrated analytically over the spatial mesh cell. A similar effect
of the singularity takes place also computing the magnetization M of
ferromagnetic parts by the use of integral equation.

The external magnetic field B_0 is computed integrating Biot-Savart
law over approximately 250 linear current-elements I dl constituting the
external circuit of the single cell, then the structure of the circuit
is periodically transported as a whole to the neighbour cells. The ade-
quacy of the integral magnetic field simulation has been tested for the
physical model with the uniform current distribution over the anodes and
cathodes (Fig.3).

The hydrodynamic model is based on principles of the "shallow
water" theory, which is widely applied in geophysical hydrodynamics. The
analogy of flows could be motivated by the small depth of the liquid layers relative to their horizontal length scale: $H/L=e\approx10^{-2}$. A prediction
method for a depth averaged turbulent flow with free surface is described by Rodi [7]; here we extend it to a two-fluid motion with the deformed interface in the electrolysis cell. Referring to the notation in
Fig.2, an averaging procedure over the i-th layer depth $h_i(x,y,t)=H_i-$
$-H_{i-1}$, e.g., for horizontal velocity components v_{mi}, m=1,2:

$$V_{mi} = h_i{}^{-1} \int v_{mi} \, dz.$$

Integrating the three-dimensional equations of motion according to the procedure we obtain: continuity equations

$$\delta h_i/\delta t + \delta(V_{ni} h_i)/\delta x_n = 0,$$

horizontal momentum equations

$$\delta(V_{mi} h_i)/\delta t + \delta(V_{mi} V_{ni} h_i)/\delta x_n = -\rho_i{}^{-1} \int (\delta P_i/\delta x_m) \, dz + \delta \nu_{Ti} (\delta(V_{mi} h_i)/\delta x_n +$$

$$+ \ \delta(V_{ni} h_i)/\delta x_m)/\delta x_n - k_i V_{mi} - c_i V_{mi} \sqrt{(V_{ni} V_{ni})} + \ \rho_i{}^{-1} \int f_m \, dz, \tag{1}$$

where $i,m,n=1,2$ and the repeated index n implies summation; hydrostatic balance is assumed in the vertical direction: $\delta P_i/\delta z = f_3 - \rho_i g$.
The terms in equations (1) with the coefficients k_i and c_i relate empirically the averaged velocities with tangential stresses at the upper and lower boundaries of the layers, which, according to the estimates [3,4], are the most significant contributions. The term with c_i is the well known friction law [7], yet the linear law with k_i [6] permits to construct a very simple and convenient model suitable for engineering calculations. According to [7], the turbulent viscosity ν_T can be calculated by a modified k-ϵ turbulence model, however the same author notes that ν_T term is often neglected comparing to the tangential stresses. Nevertheless we have tested this assumption performing a flow simulation with $\nu_T = l^2 \sqrt{w^2}$, $w = \delta V_2/\delta x - \delta V_1/\delta y$, $l = 0.1L$ far from the side walls, and $l=0.4n$ if the distance from wall $n \le 0.25L$. The equations of motion are solved by a modified finite-difference method using integro-interpolation approximation of the second order accuracy in space variables (power schema) [8]. Comparing stationary sreamlines in Fig.4 for the mercury model in the case of uniform current distribution along the anodes, it is evident that the result obtained by the use of significantly more elaborate model with the nonlinear term in the left and the term with ν_T (Fig.4a) satisfactorily agrees with the result obtained by retaining from the velocity containing terms merely the linear with k_i (Fig.4b).

Simplicity of the linear model allows to consider also an effect of the channels (region 3 in Fig.2) having the free upper surface. In this case, instead of the unpermeability condition at the side walls,

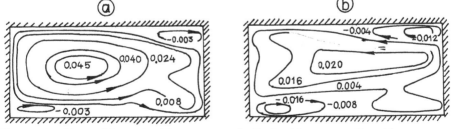

Figure 4. Streamlines in the mercury bath in the case of uniform current $j(y,z)$ along the anodes: a) the nonlinear prediction with ν_T, b) the linear with $k = 0.001$ m/s.

the condition of normal velocity and pressure continuity is set at the
boundary of regions 2 and 3. The linear model is applied to the 175 kA
cell in [9]. Note, when the effect of channels is gradually decreased,
the flow in electrolyte approaches that in aluminium and the maximum
interface deformation decreases up to 2-3 times.

When the electric current is distributed unevenly over the anodes
and cathode bars, the current is forced to flow in the metal layer horizontally [6]. In this case the horizontal current density $\approx \delta I/LH$ appreciably exceeds the vertical current $\delta I/L^2$, and the electromagnetic
force increases in proportion to $L/H=e^{-1}$. As yet mentioned, the situation takes place in practice, and has been investigated in our first
(graphite) model. One of the situations with the three-dimensional current flow in the metal layer is simulated also by the mathematical model.
The prescribed distribution of currents over the anodes leads to j_z distribution over the upper surface of liquid metal shown in Fig.5a, over
the cathodes- to a different j_z distribution over the lower surface (not
shown). The respective horizontal current flow in the liquid metal is
shown in Fig.5b, and it agrees with the experimentally measured by conductive probe. The corresponding mercury flow visualization is drawn in
Fig.6a, and the corresponding linear model simulation is represented by
Fig.6b (without channels).It is evident that the linear model satisfactorily predicts even the flow details directly unseen in the experiment
since the anode blocks cover a larger area of the mercury surface. The
measurements by the use of tensometric probe reveal, that the velocity
in a fixed point grows at a rate I^2 for the total currents $8<I<13$ kA
and a somewhat slower for $I<8$ kA.

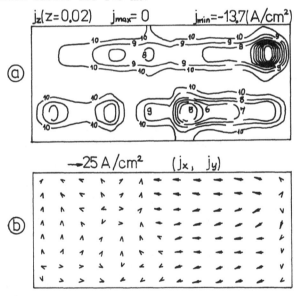

Figure 5. Nonlinear electric current $j(x,y,z)$ simulation in the
graphite model.

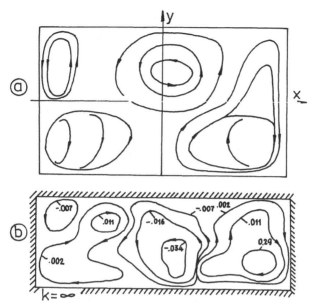

Figure 6. The corresponding streamlines in the graphite model:
a) experimental, b) predicted.

REFERENCES

[1] LYMPANY S.D., EVANS J.W., MOREAU R., 1982, 'Magnetohydrodynamic effects in aluminium reduction cells', Proc. Symp. IUTAM Metall.Appl. MHD,15.
[2] EVANS J.W., ZUNDELEVICH Y., SHARMA D., 1981,'A mathematical model for prediction of currents, magnetic fields, melt velocities, melt topography and current efficiency in Hall- Herault cells',Metall.Trans., Vol. 12B, 353.
[3] MOREAU R., EVANS J.W., 1984, 'An analysis of the hydrodynamics of aluminium reduction cells', J.Electrochem. Soc.,Vol. 131, 2251.
[4] BOJAREVICS V., 1987, 'A mathematical model of MHD processes in aluminium electrolysis cells', Magnitnaya Gidrodinamika, N°1, 107.
[5] BANERJEE S.K., EVANS J.W., 1987, 'Further results from a physical model of a Hall cell',Light Metals,247.
[6] Gorbachev E.V.,et al., 1988, 'Physical modeling of MHD processes in aluminium electrolysis cells', Tsvetniye Metally, N°1, 38.
[7] RODI W., 1980,'Enwironmental turbulence models', Prediction Methods for Turbulent Flows, Hemisphere Publishing Corporation.
[8] PATANKAR S., 1984, Numerical Methods for Problems of Heat Transport and Fluid Dynamics, Energoatomizdat.
[9] BOJAREVICS V., et al., 1988,'A mathematical model for prediction of magnetic field, electric current distribution, liquid medium motion, and interface deformation in aluminium electrolysis cells', Tsvetniye Metally, N°7, 63..

ON THE ANALYSIS BY PERTURBATION METHODS OF THE ANODIC CURRENT
FLUCTUATIONS IN AN ELECTROLYTIC CELL FOR ALUMINIUM PRODUCTION.

J. Descloux & M.V. Romerio
Swiss Federal Institute of Technology
Department of Mathematics
1015 Lausanne, Switzerland

ABSTRACT. The spectrum of anodic current measurements in reduction
electrolytic cells for the production of aluminium is characterized by
the presence of frequencies seemingly related to gravitation forces
only. Assuming that the Lorentz forces are one order of magnitude
smaller than the gravitational ones , perturbation methods are applied
to Navier-Stokes and quasi-stationary Maxwell's equations. An approximation of the spectrum, up to the second order of perturbation, is derived and related stability questions are discussed.

1.   INTRODUCTION

The different studies devoted to electrochemical reactions taking place
in reduction electrolytic cell for the production of aluminium [1],
show that the Faraday efficiency of the global process is strongly depending on the stability of the interface between the electrolytic bath
and the aluminium already produced.
Recordings of the fluctuations of the anodic current density, for sufficiently long period of time constitute an essential source of informations about the motions of this interface. This information is mainly
contained in the frequency spectrum which can be numerically derived
from such recordings.
The analysis of this spectrum shows that some of the observed frequencies are quite close, if not equal, to some of those one would get in
solving Euler's equations for the two considered fluids but under the
action of the gravitational field only. This phenomena has already been
mentioned in the litterature [2].
It may be worthwhile pointing out that no significant damping seems to
affect the corresponding "modes of oscillation".
In the general frame of an analysis on the stability questions for such
electrolytic cells, it seems that a reasonable model should be able to
recover at least part of the striking features which characterize this
spectrum. Our purpose in this paper is to briefly sketch the main ideas
entering the derivation of a model which seemingly satisfies this kind
of requirements. Details can be found in [3].

213

J. Lielpeteris and R. Moreau (eds.), Liquid Metal Magnetohydrodynamics, 213–219.
© 1989 by Kluwer Academic Publishers.

2. THE PHYSICAL MODEL

From a physical point of view our model leans on the following assumptions :
1. The fluids are immiscible, incompressible and Newtonian.
2. In each of the domains in which they are located, the fluids are governed by the Navier-Stokes equations.
3. The electromagnetic fields satisfy quasi-stationary Maxwell's equations in each fluid as well as outside the cell.
4. The Lorentz force field entering the Navier-Stokes equations is one order of magnitude smaller than the gravitational one so that perturbation technics can be applied.
5. The electric current densities, outside the cell, are time independent and supposed to be known.

Being interested in the stability of the system we restrict ourselves to time depending solutions which can be expressed as small "fluctuations" around a stationary state. In agreement with the above mentioned experimental recordings we moreover assume that these fluctuations are not damped by viscous friction.

With these remarks we are thus led to add the following assumptions :
6. There exists a stationary state.
7. The fluctuations around a stationary state are sufficiently small to justify a linearization procedure.
8. The fluctuations of pressure and velocity fields can be described in neglecting all effects related to viscosity.

In order to be more precise, let us consider a parallelepipedic domain Ω of $\mathbb{R}^3$, completely filled by two viscid, immiscible and conducting fluids; each of these fluids occupies a subdomain Ω_j of Ω, of boundary $\partial\Omega_j$, for $j = 1,2$. (figure 1). We denote by Γ the interface between these two liquids. In the frame of reference of figure 1 Γ is described by the equation

$$z = \tilde{h}(x,y,t). \tag{2.1}$$

The system is characterized by its density ρ, its electric conductivity σ and its viscosity η. These parameters have constant values in Ω_1 and Ω_2.

In agreement with the above assumptions the description of the system requires a knowledge of the following fields.
$\underline{u}$: velocity, $\tilde{p}$: pressure, $\underline{\tilde{e}}$: electric, $\underline{\tilde{b}}$: magnetic induction.

These fields satisfy the following equations :

$$\rho\, \partial t\, \underline{\tilde{u}} + \rho(\underline{\tilde{u}},\nabla)\underline{\tilde{u}} = \eta\, \Delta\underline{\tilde{u}} - \underline{\nabla}(\tilde{p}+\rho g z) + \mu_0^{-1}\, \underline{\mathrm{rot}}\, \underline{\tilde{b}} \wedge \underline{\tilde{b}}, \tag{2.2}$$

$$\mathrm{div}\, \underline{\tilde{u}} = 0, \quad \text{in } \Omega, \tag{2.3}$$

$$\partial t\, \underline{\tilde{b}} + \underline{\mathrm{rot}}\, \underline{\tilde{e}} = 0, \quad \text{in } \mathbb{R}^3, \tag{2.4}$$

$$\underline{\mathrm{rot}}\, \underline{\tilde{b}} = \begin{cases} \mu_0\, \sigma(\underline{\tilde{e}} + \underline{\tilde{u}} \wedge \underline{\tilde{b}}), & \text{in } \Omega, \\ \underline{\tilde{j}}_{ext}, & \text{in } \mathbb{R}^3\, /\, \overline{\Omega}, \end{cases} \tag{2.5}$$

$$\text{div } \underline{\tilde{b}} = 0, \qquad \text{in } \mathbb{R}^3, \tag{2.6}$$

where μ_0 is the magnetic susceptibility of the vacuum and g is the gravitation constant.

For any field f, we denote by f_j the restriction of f to $\overline{\Omega}_j$, $j = 1,2$ and by

$$\{f\}_{\Gamma(\tilde{h})} = (f_1 - f_2) \mid_{\Gamma(\tilde{h})} \tag{2.7}$$

the "jump" of f on $\Gamma(\tilde{h})$.

The following conditions are required on the fields :

$$\underline{\tilde{u}}_j = 0, \qquad \text{on } \partial\Omega_j(\tilde{h}) - \Gamma(\tilde{h}), \; j = 1,2, \tag{2.8}$$

$$(\underline{\tilde{j}}_k, \underline{\tilde{n}}_k) = \begin{cases} 0 \text{ on } \partial\Omega_k(\tilde{h}) - \Gamma(\tilde{h}) - \Xi_k, \; k = 1,2, \\ \\ j_{0,k} \text{ on } \Xi_k, \; k = 1,2, \end{cases} \tag{2.9}$$

where $\underline{\tilde{n}}_k$ is the outside unit normal to $\partial\Omega_k$, Ξ_k is the electrode on $\partial\Omega_k$ and $\underline{\tilde{j}}_k = \sigma(\underline{\tilde{e}} + \underline{\tilde{u}} \wedge \underline{\tilde{b}})$, $\underline{\tilde{b}} = 0$ at infinity,

$$\sum_{k=1}^{3} \{(- \tilde{p} \, \delta_{i,k} + \eta(\partial_i\tilde{u}_k + \partial_k\tilde{u}_i))\tilde{n}_k\}_{\Gamma(\tilde{h})} = 0, \; i = 1,2,3, \tag{2.10}$$

(i,k identified with x,y,z)

$$\{\underline{\tilde{b}}\}_{\Gamma(\tilde{h})} = 0, \; \{\underline{\tilde{b}}\}_{\partial\Omega} = 0, \tag{2.11}$$

$$\{(\underline{\tilde{j}},\underline{\tilde{n}})\}_{\Gamma(\tilde{h})} = 0, \; (\underline{\tilde{n}} = \underline{\tilde{n}}_1, \text{on } \Gamma(\tilde{h})), \tag{2.12}$$

$$\{\underline{\tilde{e}} \wedge \underline{\tilde{n}}\}_{\Gamma(\tilde{h})} = 0, \tag{2.13}$$

$$\partial t \, \tilde{h} + (\underline{\tilde{u}}_j, \underline{\nabla} \, \tilde{h}) = \tilde{u}_{j,z}, \; \text{at } z = \tilde{h}(x,y,t), \; j = 1,2, \tag{2.14}$$

$$\int \tilde{h}(x,y,t)dxdy = 0. \tag{2.15}$$

According to our assumptions two steps must now be performed. In the first one the system is linearized around a stationary solution, in the second the system is solved through a perturbation expansion.

In order to carry out this last step we give to the fourth assumption a slightly different form. In fact we replace, in (2.6), $\underline{\tilde{j}}_{ext}$ by $\varepsilon \, \underline{\tilde{j}}_{ext}$ and, in (2.9), $j_{0,k}$ by $\varepsilon \, j_{0,k}$, where ε is a real parameter. It is in this parameter that the perturbation expansion will be performed. It can be shown that, with respect to this parameter, the Lorentz forces are second order in ε.

216

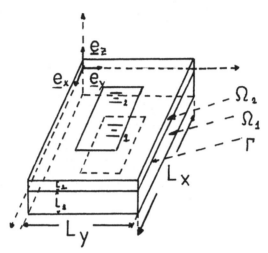

Figure 1. Schematic representation of a parallelepipedic cell. Ω_1 and Ω_2 are the subdomains respectively occupied by the aluminium and the electrolytic bath, Ξ_1 and Ξ_2 are the two electrodes and Γ is the interface between the two fluids (for a static situation).

3.   LINEARIZATION AND ε EXPANSION

Let us introduce some notations. We set :

$$\tilde{\underline{u}} = \underline{u} + \underline{U}, \quad \tilde{p} = p + P, \quad \tilde{\underline{e}} = \underline{e} + \underline{E}, \quad \tilde{\underline{b}} = \underline{b} + \underline{B}, \quad h = \tilde{h} + H, \tag{3.1}$$

where $\underline{u}$, p, $\underline{e}$, $\underline{b}$ and h represent a stationary solution whereas $\underline{U}$, P, $\underline{E}$, $\underline{B}$ and $\overline{H}$ describe the time dependent fluctuations around this same solution,

$$\tilde{w} = (\tilde{\underline{u}}, \tilde{p}, \tilde{\underline{e}}, \tilde{\underline{b}}, \tilde{h}), \quad w = (\underline{u}, p, \underline{e}, \underline{b}, h), \quad W = (\underline{U}, P, \underline{E}, \underline{B}, H). \tag{3.2}$$

Quite formally the system defined by (2.2) to (2.15) can be written in the form

$$T(\tilde{w}) = 0 \quad + \quad \text{boundary conditions}, \tag{3.3}$$

where T is a non linear partial differential operator.
In the same way we have for the stationary solution

$$T(w) = 0 \quad + \quad \text{boundary conditions}. \tag{3.4}$$

The linearization procedure consists in substituting $w + W$ to $\tilde{w}$ in (3.2) and retaining the first order terms in W only. This leads to a <u>linear homogenous equation in W</u> of the form,

$$R(w) \, W = 0, \quad + \underline{\text{linearized}} \text{ conditions,} \tag{3.5}$$

where R is a partial differential operator. (3.5) being linear in W we can look for solutions the time dependence of which is of the form

$$e^{i\omega t}, \text{ where } \omega \in \mathbb{C}. \tag{3.6}$$

In other words all the time derivatives are replaced by $i\omega$ and (3.5) becomes a kind of generalized eigenvalue problem. According to our assumption on the electric current density the solutions w and W, as well as ω, are functions of the parameter ε. We will thus look for solutions in the form of limited power series expansions (up to the second order) in this parameter.

For an arbitrary field f we formally write

$$f(x,y,z,\varepsilon) = f(x,y,z) + \varepsilon \, f'(x,y,z) + \frac{\varepsilon^2}{2} \, f''(x,y,z) + O(\varepsilon^3) \tag{3.7}$$

where

$$f' = \partial_\varepsilon f \Big|_{\varepsilon=0}, \quad f'' = \partial_\varepsilon^2 f \Big|_{\varepsilon=0}. \tag{3.8}$$

We have for instance for $\underline{u}$ and ω the expansions :

$$\underline{u} + \varepsilon \, \underline{u}' + \frac{\varepsilon^2}{2} \, u'' + O(\varepsilon^3), \tag{3.9}$$

$$\omega + \varepsilon \, \omega' + \frac{\varepsilon^2}{2} \, \omega'' + O(\varepsilon^3). \tag{3.10}$$

This type of expansions evidently applies to w and W so that equations (3.4) and (3.5) lead to two new sets of equations and boundary conditions for each power of ε.

It may be worthwhile mentioning that the interface conditions which are derived from the linearized procedure and the ε expansion may become quite intricated.

4. RESULTS AND CONCLUSIONS

The analysis of the different systems associated to the power expansion described above being rather long and tedious we will be compelled restricting ourselves to a discussion of the results only.

4.1. Zero-Order in the Power Expansion

The electric current being vanishing in the zero-order the electromagnetic field is ruled out, implying that the motions of the fluids reduce to oscillations in the gravitational field, around a stationary solution which is a simple hydrostatic solution. The zero-order frequency of these oscillations results from a classical derivation [4]. (Let us remind the reader that according to our assumptions effects related to viscosity are neglected for the fluctuations). In the notations of figure 1 we have

$$\omega^2 = kg\{\rho\} / \sum_{j=1}^{2} \rho_j cthkL_j \quad \text{where} \quad k^2 = \pi[(\frac{m}{L_x})^2 + (\frac{n}{L_y})^2], n, m \in \mathbf{Z} \quad (4.1)$$

where L_1 and L_2 are respectively the distances between bottom and top of the cell to the interface for the hydrostatic solution. In accordance with our notations $\underline{U}$ will represent the hydrostatic velocity mode; in the following we shall assume that $\underline{U}$ is normalised in such a way that it is purely real.

4.2. First-Order in the Power Expansion

The zero-order electromagnetic field being vanishing and the Lorentz force field being formed of products of two magnetic induction terms no contributions of this force appear in the first-order. It is then possible to prove that : $\underline{u}'$, p', h' and ω' are ruled out and that $\underline{U}'$, P' and H' can be considered as vanishing.
The electromagnetic field is different of naught. For the stationary field the contribution to the electric current density within the cell corresponds exactly to the case of "frozen fluids" (absence of motion) with a flat interface and the corresponding magnetic induction field $\underline{b}'$ is given by a Biot-Savart type formula. The fluctuation fields $\underline{E}'$ and $\underline{B}'$ are solutions of a system which depends both on the interface motion and on the induced electric current $\underline{U} \wedge \underline{b}'$.

4.3. Second-Order in the Power Expansion

Our main purpose, for the second-order, is to derive an explicit expression for ω'' in terms of zero and first-order fields. In this paper we will restrict ourselves to Im ω'' (imaginary part of ω'') only.
Let $\underline{C}'$ be a vector field, defined in $\mathbf{R}^3$, and Γ' be a continuous scalar field, defined in $\overline{\Omega}$, satisfying

$$\begin{array}{lll} \text{rot } \underline{C}' = \mu_0\sigma(-\underline{\nabla}\Gamma' + \underline{U} \wedge \underline{b}'), & \text{in } \Omega, & (4.2) \\ \text{rot } \underline{C}' = 0, & \text{in } \mathbf{R}^3/\overline{\Omega}, & (4.3) \\ \text{div } \underline{C}' = 0, & \text{in } \mathbf{R}^3, & (4.4) \end{array}$$

the boundary conditions being such that $\underline{C}'$ is continuous on all the boundaries, $\underline{C}'$ is vanishing at infinity, $(\text{rot } \underline{C}', \underline{n})$ and Γ' are continuous on the interface and $(\text{rot } \underline{C}', \underline{n}) = 0$ on $\partial\Omega$. One can show that this problem is well defined and has, up to a constant for Γ', a unique solution. $\underline{C}'$ is in fact the only contribution of the field $\underline{B}'$ which enters the expression of Im ω''. The same procedure can now be applied to the induced electric current $\underline{U} \wedge \underline{C}'$. This allows us to define the vector field $\underline{A}'$ in $\mathbf{R}^3$ and the scalar field ζ' in $\overline{\Omega}$, satisfying equations (4.2) to (4.4) but with $\underline{b}'$, $\underline{C}'$ and Γ' replaced by respectively $\underline{C}'$, $\underline{A}'$ and ζ'. In other words

$$\text{rot } \underline{A}' = \mu_0\sigma(-\underline{\nabla}\zeta' + \underline{U} \wedge \underline{C}') \quad \text{etc...} \quad (4.5)$$

In terms of these new field we can now write the expression for Im ω". We have :

$$
\text{Im } \omega'' = \frac{\displaystyle\sum_{j=1}^{2} \int_{\Xi_j} \zeta_j' j_{o,j}' \, d\sigma + \sum_{j=1}^{2} \frac{1}{\sigma_j} \int_{\Omega_j(0)} \left| \operatorname{rot} \frac{C_j'}{\mu_0} \right|^2 d\tau}{\displaystyle\sum_{j=1}^{2} \int_{\Omega_j(0)} \frac{1}{2} \rho_j \left| \underline{U}_j \right|^2 d\tau + \int_{\Gamma(0)} g \frac{1}{2}(\rho_1 - \rho_2) |H|^2 d\sigma}. \qquad (4.6)
$$

We recall that in this formula, Ξ_j is the electrode located on the boundary of $\Omega_j(0)$, crossed by a current density $j_{o,j}'$.
One sees that the denominator is the mechanical energy of the two fluids in the zero order. In the numerator the term containing rot $\underline{C}'$ is clearly a Joule dissipative term, for $1/\sigma_j$ is a resistivity and the integral is the square of an electric current density. The last term is more difficult to visualize. It corresponds to the power resulting from the currents entering or leaving the cell with the electric potential ζ_j' defined by (4.5) $(\zeta_j' = \zeta' | \Omega_j(0))$.
What kind of informations can we draw from (4.6) and how can these informations help us answering our initial problem ?
We draw from (4.6) that the system will be stable if the numerator of (4.6) is strictly positive.
In fact relation (4.6) is very tangled for, $j_{o,j}'$ $j = 1,2$ being given, the electric current density can be computed in the cell. In its turn this current density allows, with j_{ext}', to compute the field $\underline{b}'$. The coupling between $\underline{b}'$ and $\underline{U}'$ gives birth to the induced electric current $\underline{U} \wedge \underline{b}'$ and through (4.2) to (4.5) $\underline{U} \wedge \underline{b}'$ generates the fields $\underline{C}'$ and ζ' which are thus both functions of $j_{o,j}'$ and j_{ext}'.
From this fact one easily shows that when rot $\underline{C}' = 0$ then (4.6) is vanishing; the system is thus stable but not asymptotically stable. It however seems rather difficult pursuing any further a search for genuine analytical results.
In order to check the agreement between theory and measurements some numerical computations are presently undertaken.

5.   REFERENCES

[1] GRJOTHEIM K., KROHN C., MALINOVSKY M., MATIAŠOVSK K. & THONSTAD J., 1982, Aluminium Electrolysis, Fundamentals of the Hall-Héroult Process, Aluminium Verlag, Düsseldorf.

[2] URATA N., MORI K. & IKEUCHI H., 1976, 'Behaviour of Bath and Molten Metal in Aluminium Electrolytic Cell', Keikinzoku, vol. 26, n° 11.

[3] DESCLOUX J. & ROMERIO M.V., 'Stability Analysis of an Electrolytic Cell for Aluminium Production by a Perturbation Method', Swiss Federal Institute of Technology, Dept of Math., 1015 Lausanne, Switzerland.

[4] DRAGOS L., 1975, Magnetofluid Dynamics, Abacus Press, Tunbridge Wells, Kent, England, chap. 4.

Session E:
AC Stirring

FLUID FLOWS INDUCED BY ALTERNATING MAGNETIC FIELDS

Yves FAUTRELLE
Laboratoire MADYLAM, ENSHMG, BP 95
38402 ST MARTIN D'HERES Cedex, FRANCE

ABSTRACT : Alternating magnetic fields are commonly used to create or to control liquid metal flows. Various fluid flow patterns may be obtained according to the intensity, the frequency and the phase shift of the applied electric currents. The magnetic field strength governs the amplitude of the velocity whilst the frequency controls both the amplitude and the configuration of the flow. Due to the skin effect, the electromagnetic forces may create strong velocity shears which are responsible for high turbulent intensity. Therefore, electromagnetic stirring is an efficient means for mixing.

I. Industrial context

Alternating magnetic fields are used in a wide variety of practical situations to control the motion of liquid metals and when necessary the free surface.

The applications mainly concern metallurgical devices where it is needed to heat, to stir, to shape or to levitate a liquid metal. Alternating magnetic fields are often preferred to other devices (e.g. D.C. fields) because of the absence of any contacts between the inductor and the bath. Moreover, such a device possesses a large number of degrees of freedom. By acting on both the frequency, the intensity, the phase-shift of the applied electric currents, the geometry of the inductor and the pool, it is possible to obtain a large variety of fluid flows. The main drawback is the rapid decay of the magnetic field in the air which must be compensated by the use of large electrical power.

Owing to the large number of degrees of freedom, is is not surprising to find many industrial configurations. Those may be roughly classified as follows :
- single-phase coil devices,
- multi-phase coil devices.

Single-phase inductor are mainly used for melting or heating. However, it may also be efficient for stirring or levitation. The main examples are [1] :
- coreless or channel induction furnaces,
- induction ladle for treatment,
- levitation,
- continuous casting of aluminium,
- cold crucible furnaces.
Multi-phase inductors are preferred when only stirring is needed. The advantage of such a device is a better control of the liquid metal flow by a suitable choice of the phase distribution of the coil.

J. Lielpeteris and R. Moreau (eds.), Liquid Metal Magnetohydrodynamics, 223–232.

2. Generality

2.1. General equations

Let us consider a closed container of electrically conducting liquid. This container is located inside a coil supplied with alternating electric currents possibly with various phases. Note that this class of problem includes continuous casting, the withdrawal velocity being generally negligible as compared with the electromagnetically driven flow.

The mathematical formulation is generally expressed in terms of the vector potential A and the velocity field u which are respectively governed by Maxwell and Navier-Stokes equations.

In the case of low conductivity materials (e.g., oxides, glasses or plasmas) temperature gradients due to Joule heating may become non negligible. Their main effects are to create non-negligible buoyancy forces or/and large variations of the electrical conductivity [2].

The complete solution of the whole problem is a formidable task, but fortunately in most practical applications many simplifications may occur.

2.2. Non dimensional parameters and orders of magnitude

Many simplifications are possible according to the values of the dimensionless parameters involved in the phenomenon. The main parameters are the following :
- the shield parameter R_ω, namely

$$R_\omega = \mu\sigma\omega L^2 = 2\ (L/\delta)^2, \tag{1}$$

μ, σ, L respectively being the magnetic permeability , the electrical conductivity and a typical length scale of the pool. The length scale δ is the electromagnetic skin depth which quantifies the penetration of the magnetic field into the liquid bath ;
- the Alfvén speed u_a or a typical velocity V such that :

$$u_a = B_0/(\mu\rho)^{1/2}, \quad V = 0((LF_m/\rho)^{1/2}, \tag{2}$$

B_0, F_m, ρ being typical values of the magnetic field and the electromagnetic forces and the density of the liquid metal ;
- the interaction parameter which quantifies the modifications of the electromagnetic forces due to the electric current induced by the motion (hence, the back reaction of the motion), namely

$$N = \frac{\sigma B_0^2 \tau}{\rho}, \tag{3}$$

where τ is a characteristic time scale of the problem. When τ is identified with the viscous diffusion time L^2/ν, $N^{1/2}$ is simply the Hartmann number. In multiphase fields, $1/\tau$ is generally chosen as the angular velocity ω (or U_s/λ, U_s and λ respectively being the travelling velocity and the pole pitch).
- The magnetic Reynolds number R_m such that :

$$R_m = \mu\sigma UL, \tag{4}$$

U being a characteristic velocity of the bath. The magnetic Reynolds number quantifies the modifications of the primary magnetic field due to the electric currents induced by the motion.

A.C. field devices are often aimed at heating a liquid metal, and buoyancy may in some cases be important. An estimate of the ratio between buoyancy and electromagnetic forces is of order of the non dimensional parameter G such that

$$G = \frac{\rho g \beta L}{\mu \sigma k},\tag{5}$$

where g, β, k respectively are the gravity, the coefficient of thermal expansion and the thermal diffusion coefficient. Note that the expression of G is independant of both the magnetic field strength and its frequency.

It is of interest to give the values of the previous parameters in some industrial applications. The results are presented in table 1.

Table 1 : Values of the nondimensional parameters in various practical applications

	σ $\Omega^{-1}m^{-1}$	f Hz	R_ω	R_m	$N = \frac{\sigma B_0^2}{\rho \omega}$	G
20kg - cold crucible induction furnace for metals	10^6	10^4	10^2	10^{-2}	10^{-4}	5.10^{-2}
20kg - induction furnace for glass	10^2	10^5	1	10^{-7}	10^{-8}	10^2
5T - induction furnace for steel	10^6	50	10^2	0,3	10^{-2}	10^{-1}
mold stirring of 200 mm billets in the continuous casting of steel	10^6	2	0,1	0,2	0,1	5.10^{-2}
Secondary stirring of 200 mm - billets in the continuous casting of steel	10^6	50	10	0,2	10^{-3}	5.10^{-2}

Table 1 shows that :

(i) electromagnetic forces are usually predominant except for glasses or oxides,
(ii) the effect of the motion on the primary magnetic field is usually negligible,
(iii) the back reaction of the motion on the electromagnetic forces is negligible except in very low frequency stirring devices, and the electromagnetic and hydrodynamical problems are decoupled,
(iv) the skin effect is important in many applications where it is needed to heat a metal.

3. The force field

Let us first assume that the geometry is 2-dimensional or axisymetric. If the electromagnetic problem and the hydrodynamical one are decoupled (i.e., N << 1), Maxwell equations may be solved first. Solutions are usually expressed in term of a single component vector potential A along the azimuthal unit vector i_θ (i_z in the 2-dimensional x, y, plane), namely [3][4]

$$A = A' i_\theta.$$

For a sinusoidal coil current, A' may be conveniently split into a phase $\varphi(x)$ and a modulus $A(x)$ as follows :

$$A' = Ae^{i(\omega t + \varphi)} + c.c... \tag{6}$$

Calculations of the vector potential and the corresponding electromagnetic forces in specific cases may be found in various works.

Without applied electric potential in the bath, the electromagnetic forces may be expressed in terms of A and φ, namely

$$F_m = J \times B = <F> + F , \tag{7}$$

where the mean part $<F>$ is :

$$<F> = - 1/2 \, \sigma\omega A^2 \, \nabla\varphi, \tag{8}$$

and the pulsating part F is :

$$F = 1/2 \, \sigma\omega A^2 \nabla \, \varphi \cos 2 \, (\omega t + \varphi) + \frac{1}{2}\sigma \, \omega \sin 2 \, (\omega t + \varphi) \left\{ \nabla \left(\frac{A^2}{2}\right) + \frac{A^2}{r} i_r \right\}, \tag{9}$$

r, i_r respectively denoting the radial coordinate and the radial unit vector. The expressions (8), (9) are not convenient for practical calculations because of the multidetermination of the phase. However, it appears interesting for the purpose of interpreting the various phenomena.

The force field depends on the coil geometry but also on the frequency, and we shall consider the two asymptotic limits $R_\omega \gg 1$ and $R_\omega \ll 1$.

3.1. $R_\omega \gg 1$

When the skin depth δ is much smaller than the typical dimension of the pool, solutions of the Maxwell equations exhibit the exponential decay of the vector potential along the normal of the boundary. In the local coordinates of Fig. 1, the expressions of the phase, the vector potential and the electromagnetic forces are [4] :

$$\varphi = - y/\delta + \varphi_0 , \qquad A = B(x) \, \delta e^{-y/\delta} . \tag{10}$$

The function $B(x)$ denotes the magnetic field strength along the boundary of the pool. It is clear from (8) and (10) that variations of B along x, i.e. and end effects are necessary to produce rotational forces. End effects can be strongly enhanced by the presence of corners which produce a local increase of the magnetic field strength [5].

As for travelling or rotating magnetic field, solutions are in the following form :

$$\varphi = - y/\delta + kx + \varphi_0 , \qquad A = B \, (x) \, e^{-\alpha y} / | \, r \, | , \qquad \alpha = Re \, \{r\}, \tag{11}$$

$$r^2 = k^2 + i\mu\sigma\omega , \, k = \pi/\lambda ,$$

where λ is the pole pitch ($\lambda = \pi R$ for a rotating magnetic field generated by a coil of radius R). In that case, variations of the phase along x are created to produce rotational forces, and end effects are no longer necessary. Note that there appears a double skin effect according to both the frequency and the pole pitch which has to be chosen properly.

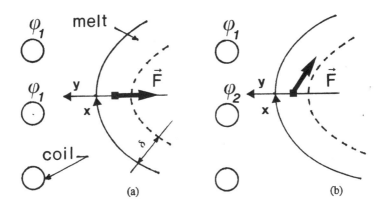

Fig. 1 - Schematic distribution of the electromagnetic forces :
(a) in the single phase case, (b) in the multiphase case.

3.2. $R_\omega \ll 1$

At lowest order, the magnetic field in the pool behaves like in the vacuum. Two cases must be considered according to the phase distribution of the coil.

For single phase coils, the phase is constant in the whole space at lowest order, namely

$$\varphi = \varphi_0 + O(R_\omega) . \qquad (12)$$

From (8) it may be noticed that the mean part of the electromagnetic forces is of order of R_ω^2, and the Lorentz forces are purely oscillating.

That phenomenon does not occur for travelling magnetic field since the spatial variations of the phase are imposed by the phase arrangment of the inductor. The mean part of the electromagnetic forces still remains non zero and rotational.

4. Motion in the single phase case

We shall focus first on the case where the coil electric currents have a single phase distribution.

4.1. Mean motion

The mean motion is characterized by both its magnitude and its configuration.

The power of the electromagnetic forces must be dissipated either by viscous forces or turbulent stresses. This constraint, which is expressed by Batchelor theorem, fixes the amplitude of the velocity [4][6].

4.1.1. Influence of the coil current intensity

The dependance of the characteristic mean velocity U with respect to the magnetic field strength depends on the nature of the flow regime. The magnetic field has a weak effect on the flow pattern, but mainly acts on the amplitude of the velocities. Generally, U may be related to the Alfven speed in the following way [7] :

$$U \approx u_a^p, \tag{13}$$

where $p = 2$ for viscous regimes,
         $p = 1$ for turbulent regimes.

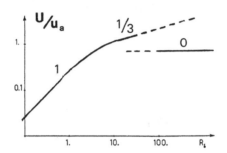

Fig. 2 - Evolution of the mean velocity magnitude with respect to the magnetic field represented by the parameter R_δ.

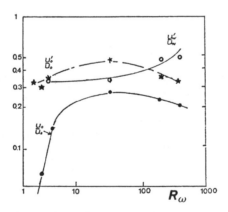

Fig. 3 - Evolution of the mean velocity magnitude with respect to the frequency represented by the parameter R_ω :

• mean characteristic velocity
* turbulent intensity on the axis of the pool
o turbulent intensity near the wall

In the limit of small skin depth $\delta/L \ll 1$, a thin film approach is possible, and the ratio between Lorentz forces and viscous one is bestly represented by the non-dimensional parameter R_δ such that

$$R_\delta = \frac{u_a \delta^2}{\nu L} .$$ (14)

In that case, the various flow regimes have been determined more precisely. Three main regimes may be found according to the value of R_d as shown in Figure 2, namely

(i) $R_\delta \lesssim 1$: viscous forces are predominant with respect to inertia in the skin depth and balance the Lorentz forces :

$$\frac{U}{u_a} = O(R_\delta) ,$$ (15)

(ii) $R_\delta > 1$: the flow regime may be still laminar in the skin depth, but inertia overcomes viscosity whose effects are restricted in a viscous sublayer d_v ; the order of magnitude are [8] :

$$\frac{U}{u_a} = O(R_\delta^{1/3}) , \quad \frac{\delta_v}{\delta} = O(R_\delta^{-2/3}),$$ (16)

(iii) $R_\delta \gg 1$: the flow regime becomes turbulent, and from (13)

$$\frac{U}{u_a} = g(R_\omega, ...) ,$$ (17)

where g is a function of the frequency and the geometry.

4.1.2. Influence of the frequency

The frequency influences both the amplitude and the spatial distribution of the Lorentz forces [7]. In the turbulent regime and for a given geometry, it is of interest to consider the evolution of g from (17) as a function of R_ω. This evolution is illustrated in Fig. 3 in the case of a coreless mercury induction furnace [9]. It may be seen that g decays when $R_\omega \gg 1$ and $\ll 1$ and reaches a maximum value near $R_\omega \approx 30$. When R_ω is large, skin effect is very important and the efficiency of the electromagnetic forces decreases. Note that the decay of U/u_a as R_ω increases is quite slow. Without consideration of corner effects, theoretical predictions yield a $-1/4$ decay law [10].

In the asymptotic limit $R_\omega \ll 1$, the mean electromagnetic force decays rapidly, and from (8) and (12) the order of magnitude of $<F>$ is at least $O(R_\omega^2)$. The decay law of the ratio U/u_a is

$$\frac{U}{u_a} \approx R_\omega^q \text{ with } q \geq 1 .$$ (18)

Note that the decay of the mean velocity is already important as soon as $R_\omega \lesssim 1$. It is not surprising

230

that, in low conducting material, induction melting thermal convection seems generally predominent although R_ω is of order of unity (this value is needed to have a significant electrical coupling between the coil and the melt).

As for the flow configuration, its pattern is mainly linked to the magnetic field distribution in the melt [11]. In the usual induction furnace geometry and for moderate values of R_ω, the magnetic field amplitude is maximum in the medium plane of the pool (cf Fig. 4). The electromagnetic forces behave in a similar way, and the corresponding motion consists in two vortices in a meridian plane. However, when the value of R_ω becomes large, the maxima of B are displaced towards the corners of the pool as well as the maxima of the Lorentz forces. Vortex separation points are located near the corners, and the flow pattern may consist in two or four vortices according to the geometry (cf. Fig. 4).

4.2. Turbulence characteristics

The magnetic field has no significant damping effects on the turbulence. Indeed, the interaction parameter of the turbulence (cf. (3) with τ being identified with the eddy turn-over time) is generally small.

However, one may point out three main features of the turbulence in such configurations :
- the large turbulent fluctuations near the wall for weak skin depth,
- the existence of low frequency fluctuations which, nevertheless, do not contain much energy [12][13].

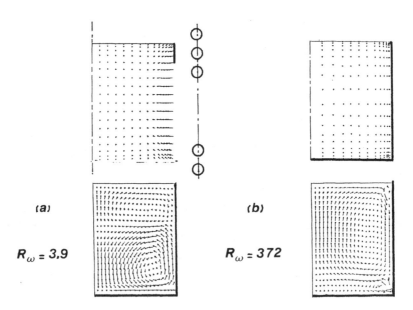

(a)

$R_\omega = 3.9$

(b)

$R_\omega = 372$

Fig. 4 - Evolution of the electromagnetic force distribution and the corresponding flow pattern with respect to the frequency :
(a) $R_\omega \geq 1$, (b) $R_\omega \gg 1$.

Except in the high frequency limit where one observes a shearing effect of the body forces, electromagnetic stirring does not yield very strong shear stresses (as opposed to mechanical stirring). However, the turbulent level is high enough to ensure good mixing properties.

4.3. Effects of the alternating part

The influence of the alternating part of the Lorentz forces is significant only for low frequency magnetic field (as for the bulk motion). Some considerations based on R.D.T. indicate that the alternating body force does not excite significant turbulent structures[10].

However, the fluctuating electromagnetic force may excite the free oscillation modes of the free surface and generate strong amplitude surface waves [14].

5. Motion in the poly-phase case

Many features of the single phase case still hold. Nevertherless, the first significant difference comes from the Lorentz forces which is no longer "repulsive" (cf. § 3). By a suitable choice of the phase distribution, stirring is generally more efficient for a given magnetic field strength. Many studies mainly concern the rotary electromagnetic stirring problem (see for example [15]).

For low values of the interaction parameter (3) and in the turbulent regime, the velocity is usually scaled with V such that [16]

$$V = \left(\frac{\sigma \omega B_o^2 k}{\rho |r|^2} L \right)^{1/2},$$

where k, r are defined in (11). In the turbulent regime, the typical velocity U (or angular velocity Ω in the rotating field case) is proportional to V (or V/R) : $U = \alpha V$. As for the determination of the value of α, there appears a very large discrepancy, especially for the centrifugation problem [16][17][18]. According to the various authors, α ranges from 1 to 50 for quite similar geometries. This discrepancy may come from subtle phenomena linked to the effect of the rotation. Some authors have indeed pointed out the very low turbulence intensity in the experiments. Furthermore, the rotational motion may be subjected to Taylor Couette instability.

As soon as the interaction parameter is of order of unity, the electric currents induced by the motion become non negligible. By Hartmann effect, the total induced electric currents are expelled from the bulk towards the boundary of the domain. This case corresponds to low frequency mold stirring. The practical consequence is a decrease of the efficiency of the stirrer [19].

6. Conclusions

Many works have been achieved so far on liquid metal flows driven by alternating magnetic field. However, the subject is so rich that many works remain to do both from a practical and a fondamental point of view. An example of practical problem is the well-known channel furnace one, whose hydrodynamics is not still well understood. From a fundamental point of view, the very low frequency limit, which has been hardly investigated, could concern new interesting phenomena.

References :

[1] GARNIER M., 1985, "Metallurgical Applications of MHD", Single and multiphase Flows in an Electromagnetic Field, Branover, Lykoudis and Mond eds., vol 100, Progress in Astron. and Aeron. AIAA, Inc. New-York, 589.

[2] CAILLAULT B., 1987, "Fusion d'oxydes par induction - Application à la cristallogénèse", Thèse de Doctorat, INPG (Grenoble, France).

[3] MOREAU R., 1980, "MHD Flows Driven by Alternating Magnetic Field", MHD Flows and Turbulence II, Branover and Yakhot eds., Israel Univ. Press, Jerusalem, 65.

[4] SNEYD A., 1979, "Generation of Fluid Motion in a Circular Cylinder by an Unsteady Applied Magnetic Field", J. of Fluid Mech. vol. 49, n° 4, 817.

[5] MOFFATT H.K., 1984, "High Frequency Excitation of Liquid Metal System", Metallurgical Application of Magnetohydrodynamics", Moffatt and Proctor eds., The Metals Society, London, 180.

[6] TIR L.L., 1976, "Features of Mechanical Energy Transfer in a Closed Metal Circuit in Electromagentic Systems with Azimuthal Currents", Magnit. Gidrodyn. vol. 4, n° 1, p. 51.

[7] FAUTRELLE Y., 1981, "Analytical and numerical aspects of the electromagnetic stirring induced by alternating magnetic fields", J. Fluid Mech., 102, 405.

[8] MESTEL A.J., 1984, "On the flow in a channel induction furnace", J. of Fluid Mech., vol. 147, 431.

[9] TABERLET E., and FAUTRELLE Y., 1985, "Turbulent stirring in an experimental induction furnace", J. of Fluid Mech., vol. 159, 409.

[10] HUNT J.C.R. and MAXEY M.R., 1980, "Estimating Velocities and Shear Stresses in Turbulent Flows of Liquid Metals Driven by Low Frequency Electromagnetic Fields", MHD-Flows and Turbulence II, Branover and Yakhot eds., Israel Univ. Press, Jerusalem, 249.

[11] MIKELSON Y.Y., PAVLOV S., and YAKOVICH A., 1980, "Numerical simulation of MHD Flow in a Region of Arbitrary Axisymetrical Configuration", Magnit. Gidrodin., n° 3, 73.

[12] TRAKAS C., TABELING P. and CHABRERIE J.P., 1985, "Etude Experimentale du Brassage Turbulent dans le Four à Induction", J. de Mec. Théor. App., vol. 3, n° 3, 345.

[13] MOORE D.J., and HUNT J.C.R., 1984, "Flow, Turbulence and Unsteadiness in Coreless Induction Furnaces", Metallurgical Applications of Magnetohydrodynamics, Moffatt and Proctor eds., The Metals Society, London, 93.

[14] GALPIN J.M., 1988, "Ecoulements Engendrés par des Champs Magnétiques Alternatifs Basse Fréquence", Thèse de Doctorat, INP Grenoble.

[15] MOFFATT H.K., 1980, "Rotation of a liquid Metal under the Action of a Rotating Magnetic Field", MHD-flows and Turbulence II, Branover and Yakhot eds., Israel Univ. Press, 45.

[16] DAVIDSON P.A. and BOYSAN F., 1987, "The Importance of Secondary Flow in the Rotary Electromagnetic Stirring of Steel during Continuous Casting", App. Scientific Research, vol. 44, 241.

[17] ROBINSON T. and LARSSON K., 1973, "An Experimental Investigation of a Magnetically Driven Rotating Liquid Metal Flow", J. of Fluid Mech., vol. 60, 641.

[18] SPITZER K.H., DUBKE M. and SCHWERTDFEGER K., 1985, "Rotational Electromagnetic Stirring in Continuous Casting of Round Strands", Met. Trans. B, vol. 17 B, 119.

[19] GLIERE A., 1988, "Contribution à la Modélisation du Brassage Electromagnétique de l'Acier", Thèse de Doctorat, INP-Grenoble (FRANCE).

FLUID METAL FLOW STUDY IN AN INDUCTION FURNACE BASED ON NUMERICAL SIMULATION

Yu.Miķelsons, L.L.Bulygin, A.T.Jakovičs
Latvian State University
226050 Riga
USSR

ABSTRACT. An averaged turbulent flow in an induction furnace is investigated. Firstly, simple energy balance in boundary layer and numerical results of laminar viscosity model are analysed. The area of application of the laminar viscosity model is shown. Numerical results of two-parameter $\kappa-\varepsilon$ turbulence model are presented. A comparison with available experimental data is made. In this type of flows according $\kappa-e$ model the turbulence is not in the state of local equilibrium. Free surface dynamics in potential force field is investigated by using an analytical solution in cases of slight perturbations of fluid surface in cylindrical vessels and by SOLA-VOF technique. The effect of varying several parameters (e.g. viscosity) on free surface dynamics is examined.

1. INTRODUCTION

The interaction between electromagnetic (EM) field and the fluid metal in the coreless induction furaces (CIF), crystallizators causes: 1) turbulent (Re$\simeq 10^4 - 10^5$) recirculating flow due to rotational component of the EM force, 2) formation of a convex surface meniscus due to potential component of the EM force. Both these features strongly influence mass- and heat- transfer processes. To investigate average flow fields in an CIF, a number of approaches are used, namely the laminar models [1, 2], two parameters turbulence ($\kappa-\varepsilon$, $\kappa-w$) models [3] and others. A number of formulas for approximate velocity calculation are presented, balancing different terms in an equation of motion (force and convective term [4], force and diffusive term [5], etc.). In the present work recirculating flow investigation is carried out using two models: 1) one-dimensional laminar model, based on energy conservation, 2) $\kappa-e$ turbulence model [9]; a comparison with experimental data is made. On the basis of the previous results some features of free surface and mass transfer dynamics are investigated.

233

J. Lielpeteris and R. Moreau (eds.), Liquid Metal Magnetohydrodynamics, 233–239.
© 1989 by Kluwer Academic Publishers.

2. LIQUID METAL FLOW IN CIF

2.1. Laminar model

According to the model of "quasi-solid" rotation of liquid metal in CIF
[6] the motion in each quarter of cross section consists of rotating
turbulent core with properties nearly the same as for the solid rotation
motion.

For stationary motion the law of energy conservation holds:

$$\int_V \vec{F} \vec{v} \, dV + \int_{V_{in}} N \, dV = 0, \tag{1}$$

where

$$N_{in} = - N_{dis} = - 2 \varrho \nu \dot{S},$$

V - fluid metal volume.

In the "quasi-solid" flow model the main part of dissipation
occurs in the boundary layer with thickness $\hat{\delta} \sim Re^{-1/2}$ (Fig.1). By assuming linear distribution of the velocity in the boundary layer, we get
(v_o - maximal velocity):

$$\int_{V_{dis}} N \, dV \simeq 2 \pi r_o \varrho \nu v_o^2 / \delta. \tag{2}$$

To evaluate the second integral in (1) the approximate velocity and

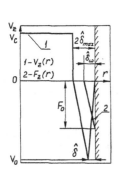

Figure 1. Approximate axial
velocity and force profiles.

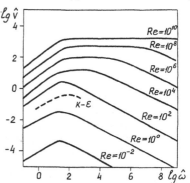

Figure 2. The dependence of the
nondimensional velocity from
nondimemsional frequency.

force distributions are used (Fig.1). The dependence of the nondimensional velocity $\hat{v} = v \cdot (\mu_o \, 6)^{+1/2} / B_o$ from nondimensional frequency
$\hat{w} = w \mu_o \, 6 \, r_o^2$ and Reynolds number $Re = B_o r_o (\mu_o \, 6)^{-1/2} / \nu$ is shown in

Fig.2, 3, the generalisation of them in Fig.4. From these results the existence of four flow regimes follows. A transition from viscous regime ($\hat{v} \sim Re$) to nonviscous ($\hat{v} \sim Re^{1/3}$, in [1] $\hat{v} \sim Re^{0.3}$) regime occures at Re=40 (in [2] Re=100) at low frequencies and Re $\sim \hat{w}$ at high; a transition from low frequency regime ($\hat{v} \sim \hat{w}^{1/2}$) to high frequency regime

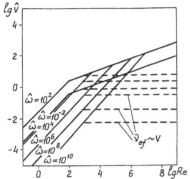

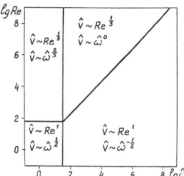

Figure 3. The dependence of the nondimensional velocity from Reynolds number.

Figure 4. Flow regimes.

($\hat{v} \sim \hat{w}^{-1/2}$) occures at $\hat{w}_{opt}$ =40 (in [7] $\hat{w}_{opt}$ =40, [8] $-$ $\hat{w}_{opt}$ =70) at low Reynolds numbers (Re<40). It should be noted, that there exists a flow regime, where velocity does not depend on frequency.

For the turbulent motion it is convenient to introduce an effective viscosity $y_{ef} \sim v$. Assuming, that a turbulent flow regime takes place when Reynolds number exceeds 10^3, a linear relation between velocity and coil current results (a dotted line in Fig.2, 3), which agrees with numerous experimental data (e.g.[11, 12]).

2.2. к-Ɛ model of turbulence.

EM force was computed with programm for arbitrary geometries based on the finite element method [10]. Because of various symmetries only a quarter of qross section is shown in Fig.5 for the case of a long inductor (Hi/Hm=1.6) for two frequencies ($\hat{w}$=10, $\hat{w}$=80). The maximums of effective viscosity, turbulence energy and dissipation rate are located near the center of the vortex; the maximum of generation rate − in the center of metal volume for the case of low frequencies and near the wall for the case of high frequencies. The distribution of the generation-dissipation ratio indicates, that the turbulence according to the к-e model is not in the state of local equilibrium; near the core of the vortex the rate of dissipation is higher than the rate of generation of turbulent energy. This results from the fact, that in the core the mean velo-

236

city gradients are mainly associated with the rotation and the generation of turbulence is weak, and the dissipation, governed by both diffusion and convection, dominates. Effective Reynolds number (based on
maximal values of velocity and effective viscosity) does not depend on

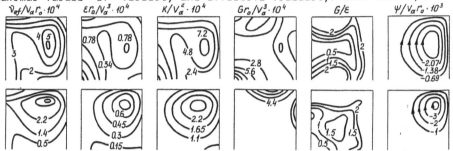

Figure 5. Effective viscosity, rate of dissipation, turbulent energy,
rate of generation, ratio of generation and dissipation rates and
streamlines in a quarter of cross section of CIF for nondimensional
frequencies $\hat{w}=10$ (upper row) and $\hat{w}=80$ (lower row).

current density in the inductor, because of the linear dependence of
both the velocity and effective viscosity on current density. From this
follows the invariance of the structure turbulent flow (local convective
and diffusive momentum transfer ratio, pulsating and mean velocity ratio) of linear current density. This conclusion agrees with experimental

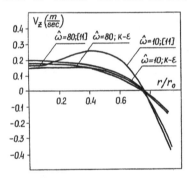

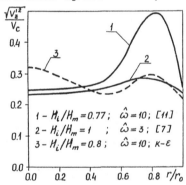

Figure 6. Comparison with
experiment [11] – mean axial
velocity.

Figure 7. Comparison with experiment [7,11] – axial mean
square pulsations normalized
by centerline mean velocity.

data [11, 12]. A comparison of mean axial velocity and mean square
fluctuating axial velocity distributions with experiment [11] is shown
in Fig.6, 7.

3. HOMOGENISATION OF PASSIVE SCALAR.

The computation of the homogenisation of the passive scalar, based on к-е model indicates to the governing role of the effective diffusion on the boundaries of vortices [13] - Fig.8.

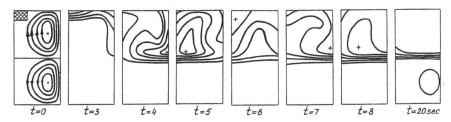

Figure 8. The time evolution of concentration field.

4. FREE SURFACE DYNAMICS IN POTENTIAL FORCE FIELD.

Investigation is made solving Navier-Stokes equations using SOLA-VOF [14] technique for fluid with constant effective viscosity in potential force field. A comparison is made with analytical solutions: 1) for ca-

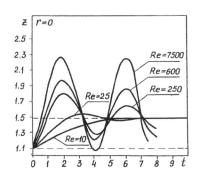

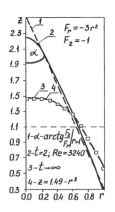

Figure 9. The dependence of the axial coordinate of the free surface on the axis from time.

Figure 10. Comparison of free surface states with analytical solutions.

ses of slight perturbations of free surface in force field $Fr=0$, $Fz=-1$, 2) stationary surface in force field $Fr=-C.r^2$, $Fz=-1$ (Fig.10). It is shown (Fig.9), that depending on the value of viscosity, the free surface dynamics may be periodical or aperiodical. In Fig.11 the development of free surface in the aperiodic case is shown. Because of the

slight dependence of the free surface oscilation frequency in periodic
case and Reynolds number for the transition from one regime to other
from meniscus height, it is possible to determine them with the analytical solutions mentioned above. Dispersion equation for this case is
equal to $w_k^2 = \Lambda_k \cdot th(\Lambda_k \cdot H_m)$, Λ_k - roots of equation $J_1(\Lambda)=0$, J_1 - Bessel
function. In the case of vanishing viscosity one of free surface states
depends only on force ratio on cylindrical surface of the fluid (dotted
line in Fig.10).

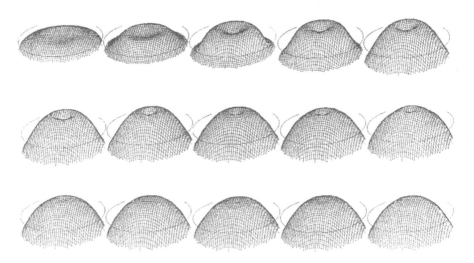

Figure 11. The development of the free surface in aperiodic
case.

5. DISCUSSION.

The laminar model may be useful only for flow pattern investigation. The
dependence of the velocity on current density for great Reynolds numbers
is not in accord with the experimental data. $\kappa-\varepsilon$ model gives results,
that are closer to the experimental. The same dependence follows from
laminar model under assumption $\nu_{ef} \sim v$. The value of the effective viscosity governs the regime of free surface dynamics, its value on the
boundary of vortex strongly influences the time of homogenisation of
passive scalar.

6. REFERENCES.

[1] Fautrelle Y., 1981, 'Analytical and numerical aspects of the electromagnetic stirring induced by alternating magnetic fields', J. Fluid Mech., Vol. 102., 405.

[2] Taniguchi S., Kikuchi A., 1984, 'Flow of Liquid Metal in High Frequency Induction Furnace', J. of the Iron and Steel Inst.of Japan', Vol. 70., N.8., 846.

[3] Szekely J., Chang C., 1977, 'Turbulent electromagnetically driven flow in metals processing: Part 1 formulation', Ironmaking and Steelmaking, N.3., 190.

[4] Khaletzky D., 1975, Etude theoretique du brassage electromagnetique dans les fours a induction. These de Specialite, Grenoble, 102.

[5] Hunt J., Maxey M., 1980, 'Estimating Velocities and Shear Stresses in Turbulent Flows of Liquid Metals Driven by Low Frequency Electromagnetic Field', Proc. of Second Bat Sheva Seminar on MHD Flows and Turbulence, Jerusalem, 249.

[6] Тир Л.Л., 1979, 'Методы расчета движения металла в индукционных тигельных печах', Исследования в области промышленного электронагрева: Труды ВНИИЭТО , М.: Энергия, Вып.9., 83.

[7] Taberlet E., Fautrelle Y., 1985, 'Turbulent Stirring in an Experimental Induction Furnace', J. Fluid Mech., Vol.159, 409.

[8] Тир Л.Л., Столов М.Я., 1975, Электромагнитные устройства для управления циркуляцией расплава в электропечах, М.: Металлургия, 224.

[9] Launder B.E., Spalding D.B., 1974, 'The numerical computation of turbulent flows', Comp. Meth. in Appl. Mech. and Eng., N.3., 269.

[10] Булыгин Л.Л., 1985, 'Расчет электромагнитных полей в аксиально-симметричных МГД-установках методом конечных элементов', Электродинамика и механика сплошных сред. Методы комплексного исследования моделей электродинамических устройств., Рига: ЛГУ им. П.Стучки, 3.

[11] Moore D., Hunt J., 1984, 'Flow, turbulence and unsteadiness in coreless induction furnaces', Metallurgical Applications of Magnetohydrodynamics, 93.

[12] Trakas C., Tabeling P., Chabrerie J., 1984, 'Etude experimentale du brassage turbulent dans le four a induction', Journal de Mecanique Theorique et Apliquee, Vol.3., N3., 349.

[13] Булыгин Л.Л., 1987, 'Гомогенизация расплава в индукционной электропечи в поле однофазного индуктора', 12 рижское совещание по МГД., Саласпилс: ИФ АН ЛатвССР, т.2, 223.

[14] Hirt C., Nicols B., 1981, 'Volume of Fluid (VOF) Method for the Dynamics of Free Boundaries', Journal of computational physics, N.39., 201.

A HIGH FREQUENCY INDUCTION FURNACE FOR OXIDE MELTING

B. CAILLAULT*, Y. FAUTRELLE**, R. PERRIER*, and J.J. AUBERT*

* D. LETI - IRDI - Commissariat à l'Energie Atomique
   D. LETI - CENG, BP 85 X , 38041 GRENOBLE Cedex FRANCE
** MADYLAM Laboratory, BP 95, 38402 ST MARTIN D'HERES Cedex, FRANCE

ABSTRACT. The studied high frequency induction furnace is a skull-melting device similar to the cold crucible technique. Its main advantage lies in its greater electrical efficiency. It allows oxide melting in large melts and may be suitable for crystal-pulling by the Czochralski process. The thermal conditions and the convections required for this purpose are studied both experimentally and theoretically. The electrical and thermal parameters are measured in baths of Sodium Chloride and Lithium Niobate. They are compared to numerical results obtained from the computed magnetic, temperature and velocity fields. The fairly good agreement allows us to understand the thermal convections. Single crystal pulling is possible only if the surface temperature is lowered by a cooling device.

Introduction

The skull melting device is quite similar to the cold crucible technique, except that the cold container acts also as the induction coil [1] [2] [3].

The material to be melted is placed in a water cooled cylindrical coil generally made of copper (see Fig. 1). The inductor is used both to heat the material and to maintain a solidified crust between the melt and the internal wall of the coil.

Such a technique is well suited for oxide melting. The electrical resistivity of such materials decreases sharply at the melting temperature to a value which allows induction [4]. However, they remain at solid state good electrical insulants and ensure the electrical insulation between the bath and the coil. Due to the quite low values of the electrical conductivity of such materials (e.g., $\sigma \approx 50$ $\Omega^{-1}m^{-1}$ for Lithium Niobate), the induction of eddy currents in the melt requires high frequency magnetic field (f $\approx$ 300 kHz).

The advantages of such a device are analogous to those of the cold crucible technique, namely
(i) the melting of high melting point materials,
(ii) the elaboration of high purity materials.

Moreover, the suppression of the air-core transformer in the cold crucible technique allows a much better electrical efficiency. Conversely, the main drawback of this device with comparison with the cold crucible furnace lies in the ability of the material to form the internal solidified crust. The control of the thickness of that crust is essential to avoid short-circuits between the turns of the induction coil.

This kind of process is well suited for single crystal pulling depending on whether the thermal convection can be controlled properly. The aim of the present work is to analyse both experimentally and theorically the flow and the temperature field in the bath.

J. Lielpeteris and R. Moreau (eds.), Liquid Metal Magnetohydrodynamics, 241–246.
© 1989 by Kluwer Academic Publishers.

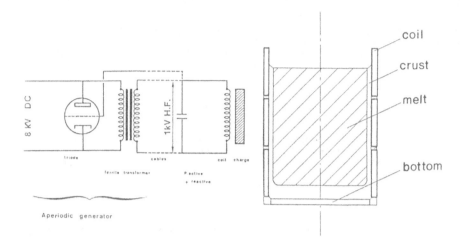

Fig. 1 - Scheme of the skull melting device

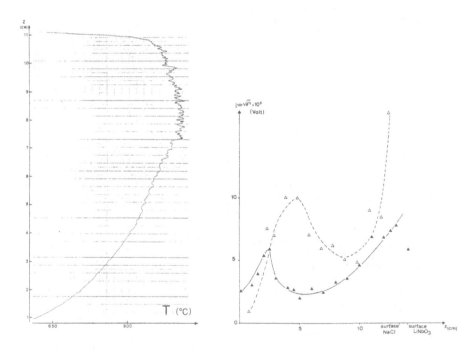

Fig. 2 - Vertical temperature profiles along the axis of the pool in NaCl, (a) instantaneous temperature, (b) fluctuating temperature.

The experiments are achieved on Sodium Chloride (NaCl) and molten Lithium Niobate (LiNbO$_3$) whose melting temperatures respectively are 800° C and 1250° C. The measurements consist in overall power budgets and determination of the local mean and fluctuating temperature. To interpret the experimental observations, we have developed a numerical model of the thermal convection generated by the induced electrical currents in the bath.

1. The experimental device

The apparatus consists in :
(i) an electrical power source,
(ii) a furnace made of a single turn coil and a cooled bottom.
The internal diameter of the coil is 200 mm, and the height of the melt is approximately 120 mm. The consumption of the generator for melting is less than 20 kw, whilst the frequency of the inductor current is about 300 kHz.

Temperature measurements are achieved by means of shielded thermocouples, whose voltage is filtered and kept isolated from the ground [4].

2. Experimental results

3.1. Power budget

By means of measurements and numerical modelling, the contributions of the various parts of the device in the power budget are estimated. They are gathered in Table 1. The global efficiency of the process, which is defined by the ratio between the Joule power dissipated in the melt and the total generator consumption, is about 50 %.

	Joule power dissipated in the bath	Losses at the free surface	Joule power dissipated in the inductor	Losses at the bottom	Maximum overheat
NaCl	6200 w	600 w	600 w	400 w	150° C
LiNbO$_3$	25000 w	7700 w	1200 w	500 w	60° C

Table 1 - Contribution of the various parts of the device in the global power budget

3.2. Temperature measurements

The temperature field is turbulent. Fig. 2 shows vertical profiles of the mean and fluctuating temperature for a sodium chloride bath. As for the mean temperature field, the pool comprises :
(i) an upper unstably stratified zone,
(ii) a lower thermally stable zone.
The turbulent temperature fluctuations are important in the upper part of the pool, but decrease rapidly in the bottom of the bath. Note that they only represent 6 % of the maximum temperature difference in the pool.

It is noticeable that the temperature distribution is quite analogous in LiNbO$_3$ pools, apart from the fact that the stable region is much smaller. In both cases, the overheat remains important (150° C and 50° C respectively for NaCl and LiNbO$_3$).

Fig. 3 - Computed velocity and temperature fields in a half meridian plane in NaCl.

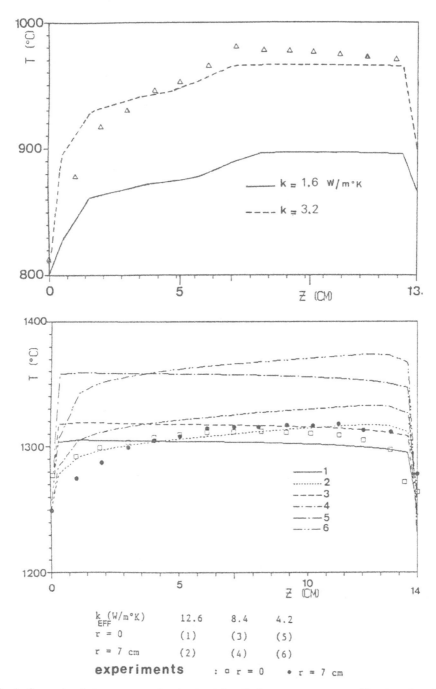

Fig. 4 - Comparison between measured and computed vertical mean temperature profiles along the axis of the pool.

3. Interpretation

The present phenomena may be considered as a peculiar kind of electromagnetically driven flow. The liquid motion is driven more by buoyancy forces generated by Joule dissipation rather than by Lorentz forces. This is a consequence of the low conductivity of the material [5].

As for the thermal convection, one may define a Rayleigh number R_J based on the Joule dissipation such that [4] :

$$R_J = P_r \left(\frac{g\beta R^5}{v^2 k}\frac{j_o^2}{\sigma}\right)^{2/3},$$

where P_r, g, β, v, k, j_o, σ, R respectively denote the Prandtl number, the gravity, the coefficient of thermal expansion, the viscosity, the thermal conductivity, the maximum value of the induced current density, the electrical conductivity and the radius of the pool. Owing to the large Joule dissipation in the bath, the value of R_J is large ($\sim 10^7 - 10^8$), and the convection regime is turbulent.

A numerical model has been developed to predict the induced electrical currents, the liquid mean velocity field and the mean temperature distribution in the melt. The model uses a turbulent viscosity calculated from two additional equations for the turbulent kinetic energy and the turbulent dissipation rate [4]. Fig. 3 shows the computed velocity and temperature fields in NaCl.

From a physical point of view, both the experiments and the predictions show that the flow is a penetrating convection type motion. The strong radiative cooling at the free surface implies a central downward motion. Turbulence is mainly produced near the free surface in the unstable zone and convected in the bulk of the liquid by the upper vortex. The stable thermal gradient down below has a double effect :
- it prevents a deep penetration of the upper vortex,
- it efficiently damps the turbulent fluctuations.

As for Lithium Niobate, the strong radiant surface losses generate high turbulence fluctuations near the free surface. Moreover, it is noticeable in Fig. 4 that the agreement between the computed and measured temperatures is not longer good in LiNbO3. The agreement is improved when a greater thermal conductivity is used in the model. That result indicates that the radiative heat transfers are no longer negligible in LiNbO3.

From a practical point of view, such device is not well fitted for single crystal pulling without any improvements. The reasons come from two correlated factors :
(i) the strong overheat of the bath and the corresponding high free surface temperature,
(ii) the high turbulence level near the free surface.

References

[1] PERRIER R. et TERRIER J., 1983, "Procédé de préparation de céramiques par fusion par induction à haute fréquence", French patent EN n° 83 02328, February 14 th.
[2] WENCKUS J.F., 1975, "Study, Design and Fabricate a cold crucible system", Air Force Cambridge Research Laboratories, 31 march, report n° 1FCRL, TR-75-0213.
[3] ALEKSANDROV V.I., OSIKO V.V., PROKHOROV A.M. and TATARINTSEV V.M., 1978, "Current Topics in Materials Science", North Holland Publishing Co. ,Vol 1, 421.
[4] CAILLAULT B., 1987, "Fusion d'oxydes par induction haute fréquence", Thèse de Doctorat, INP-Grenoble (France).
[5] FAUTRELLE Y., 1988, "Fluid flows induced by alternating magnetic fields", Proc. of IUTAM Symposium, Riga 1988 (See elsewhere in this volume).

DIRECTIONAL MELT-FLOW IN CHANNEL INDUCTION FURNACES

A. Mühlbauer, R. Fricke and A. Walther
Institute of Electroheat, University of Hannover
Wilhelm-Busch-Str. 4
D-3000 Hannover 1
Federal Republic of Germany

ABSTRACT. The electromagnetic Lorentz-forces which are effective in channel induction furnaces are discussed and also the conditions under which a desired directional transit flow of the melt can be obtained. The total action of the electromagnetic driving forces will be divided into three effects: The extension effect, the turn effect and the phase angle effect. For a conical transition from a vertical channel to the furnace bath the extension effect is investigated numerically. As a result, the distribution of the rotational force field and the corresponding fluid-flow pattern are shown in dependence of the extension angle and the frequency, respectively. A physical quantity which characterizes the directional melt flow is presented.

1. INTRODUCTION

Regarding the economy, environmental pollution, specific power rating, energy yield, and safety of handling, channel induction furnaces (CIF) have basic advantages over competing techniques. Large aluminium furnaces with detachable induction units, having capacities up to 50 tons, power rating up to 4,800 kW and high productivity, have been designed and successfully operated [1]. The economical and trouble-free use of CIF and their further technological development are closely related to the melt flow in the channel. A directional transit flow through the channel to the bath is necessary for an effective heat transfer without local overheating. However, flow velocities which are too high can lead to erosion of the refractory lining just as locally strong flow vortices can. Therefore, a smooth melt flow without energy ridge flow vortices permits the required heat transfer from the channel to the bath and a long life of the refractory lining. Many investigations concerning these flow aspects have been undertaken. This paper presents basic relationships between channel geometry, electromagnetic forces and flow in the melt and some results of numerical calculations.

J. Lielpeteris and R. Moreau (eds.), Liquid Metal Magnetohydrodynamics, 247–253.

2. FORCES AND THEIR EFFECTS

Considering a point of a body which moves along a line, the integral of the force density f along this line

$$w_{1.2} = \int_1^2 \vec{f} \; d\vec{s} \quad \text{in} \quad N/m^2 = J/m^3 \tag{1}$$

gives the energy density transferred to this point. If its path is a closed loop, it follows from Stokes' theorem

$$w_0 = \oint \vec{f} \; d\vec{s} = \int_A \text{rot} \; \vec{f} \; d\vec{A} \quad \text{in} \quad J/m^3 \quad . \tag{2}$$

Therefore, the change of energy of this body with the volume V_b is given by

$$W_b = \int_{V_b} w_0 \; dV = \int_{V_b} \int_A \text{rot} \; \vec{f} \; d\vec{A} \; dV \quad \text{in} \quad J \quad . \tag{3}$$

Consequently, the change of the total energy of this body, which consists of both kinetic and potential parts, depends upon the force density field whether it is rotational or not. This can be derived from

$$\vec{t} = \text{rot} \; \vec{f} \quad \text{in} \quad N/m^4 \quad , \tag{4}$$

where t is the rotational force density. If t is zero everywhere, the force density field is purely potential. If t is not zero, a rotational force field exists in which the moved body can change its total energy. This is of importance for the molten metal moving in a channel of a CIF. The mathematical connection between the force density fields and the resulting motions of the melt particles is given by the time-dependent Navier-Stokes equation. For the present case, one obtains

$$\frac{D \; (\varrho \; \vec{v})}{Dt} = -\text{grad} \; p + \vec{S} \times \vec{B} + \varrho \; \vec{g} - \eta \; \text{rot} \; \text{rot} \; \vec{v} \quad , \tag{5}$$

where $D(\varrho\vec{v})/Dt$ is the density of the inertia force, $\varrho\vec{v}$ the momentum density , $-\text{grad} \; p$ the pressure force density, p the pressure, S x B the electromagnetic Lorentz force density, S the electric current density, B the magnetic induction, $\varrho \; \vec{g}$ the gravitational force density , $-\eta \; \text{rot} \; \text{rot} \; \vec{v}$ the friction force density, η the dynamic viscosity, and $\vec{v}$ the velocity.

Thermal buoyancy is not considered here, because it has only a minor influence upon the desired channel flow. Whereas the frictional force density decelerates, the Lorentz force density can accelerate under particular conditions. The necessary condition is

$$\vec{t} = \text{rot} \; \vec{f} = \text{rot} \; (\vec{S} \times \vec{B}) \neq 0. \tag{6}$$

The Lorentz force density can be regarded as the sum of a potential and a rotational part. This relationship is often expressed by

$$\vec{S} \times \vec{B} = -\text{grad}\left(\mu \; \frac{\vec{H}^2}{2} \right) + (\vec{H} \; \text{grad}) \; \mu \; \vec{H} \quad . \tag{7}$$

The first term has a pure potential character. It is generally known as electromagnetic pressure. It is mostly responsible for the melt meniscus in crucible induction furnaces and for pinch effects which can occur in the vertical branches of a channel inductor. The meaning of the second term is not quite so simple. If the Lorentz force density is rotational, the relevant parts are included in this term. This, however, does not imply that it is generally a rotational part that exists, if this term is not zero.

Here the question arises about the conditions under which the melt elements obtain a drive from the Lorentz force density, i.e., under which circumstances Eq. (6) is fulfilled. As one has to deal with alternating electromagnetic fields, variations in both time and space must be considered. This is also the case for time-average force effects, the reason being the limited magnitude of the spreading velocity of an electromagnetic wave which leads to phase differences between the components of the magnetic field; these influences the Lorentz force density and cannot be averaged out.

Effects which results from the basic pattern of current density field lines and magnetic field strength can be estimated more easily. However, one has to take into account that the different phase angles are not considered if the force field is only calculated from the numerical magnitude of the relevant field values; large introduced errors can be the consequence. By means of tensor analysis a more correct interpretation of the phenomena is possible. However, the theoretical effort required is too extensive to be dealt with in detail in this paper. It may suffice to conclude that the distribution of the electromagnetic rotational force density in an actual channel furnace can be assessed qualitatively.

3. DRIVING EFFECTS IN CHANNEL FURNACES OF USUAL GEOMETRY

The total action of the rotational Lorentz force density will be divided into three effects

- Extension effect
- Turn effect
- Phase angle effect,

which it is useful for an analytical evaluation and a computer calculation.

For axisymmetric structures with variations of the cross-section along their axis the extension effect for current conductors can be treated separately (Fig. 1a). The force density and the rotational force density lead here to a driving action in the direction of the larger cross section [2-4]. The example of Fig. 1a is also valid in actual channel furnaces if it is regarded as the transition area from the central channel to the furnace bath. However, an actual channel furnace cannot only be judged by the extension effect, because it is not axisymmetric. The turn effect also has to be considered. It causes a change of direction of the current density lines which are not axisymmetric. Examples for the turn effect are also given in Fig. 1. Fig. 1b shows the usual transition from the lower horizontal channel branch into a vertical section, while Fig. 1c depicts the mouth of the channel into the furnace in an idealized way.

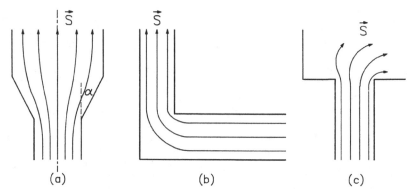

Figure 1. Axisymmetric array (a) showing the extension effect and idealized arrays (b) and (c) showing the turn effect in actual channel furnaces

A numerical treatment of the turn effect is much more difficult than for the extension effect, because it requires a three-dimensional computer model. With the available computers, problems arise with the memory capacity and the calculation time. At present no numerical simulation of the turn effect is possible. Investigations have to be made with experimental models.

The occurence of phase angle differences in the components of the magnetic field and the reason for their formation have already been mentioned. The relevant effects upon the rotational force density distribution and the melt flow are called phase angle effect. This effect can be observed, e.g., if an alternating electric current flows in a structure of infinite length with an axially non-symmetric but constant cross-section. For the directional transit flow in a channel, however, this effect is of no interest, as its driving action always occurs in the cross-sections itself. The resulting flow in the melt is only important for the convective heat transport from the channel to the furnace.

For a qualitative assessment of actual channel furnaces there are 4 basic factors:
- The driving effects which cause channel flow occur at such areas where an extension and/or turn of the electrical current density lines exists. They are mostly found at the junctions of channel and furnace and at the transitions between horizontal and vertical channel sections.
- The resulting flow drive at such points is directed from smaller to larger cross-sections.
- The driving effect increases with the degree of the turn.
- A flow drive in a direction perpendicular to the channel axis leads to flow vortices in the plane of the cross-sections.

In actual cases, however, not only the channel geometry is decisive but also the position of the coil, the yoke and the type of the enclosing case construction. They all influence the resulting electromagnetic field and consequently the distribution of the rotational force density. These effects cannot be treated in a two-dimensional mathematical model, as they generally have no axial symmetry.

4. CALCULATED MODELS FOR THE EXTENSION EFFECT

For the numerical simulation of the flow driving behaviour in channel induction furnaces a computer program, based on the finite difference method, has been developed which can deal with axisymmetric goemetries, i.e., the extension effect. Maxwell's electrodynamic equations and the Navier-Stokes formula (5) furnish the theoretical base. The electromagnetic field, the rotational Lorentz force density and the resulting flow will be calculated. Although a transfer of the results to actual CIFs is limited because of the restriction to axisymmetric arrays, one can obtain much of interesting and important information. With regard to a planned extension of the program to general three-dimensional geometries important deductions can already be made.

As an example of the calculation of flow fields, the case of a conically extending cylinder is presented. In order to solve the corresponding finite-difference equation, the boundary conditions have to be set properly. It is assumed that at the inlet and outlet positions the axial alterations of the velocity are zero and the axial changes of the pressure are constant. The turbulence of the melt flow is considered using an approximate viscosity in connection with the upwind-differencing method used. The extension angle α for a constant frequency and the frequency for an angle α of 45° have been varied. In all cases, the effective power as well as the the material data have been kept constant. An example of these parameter studies is shown in Fig. 2.

Concerning a uniform directional flow, a reliable judgement of the conical channel area is hardly possible, considering either the velocity field or the force density field. Therefore, one has to look for other factors which can be accepted as a measure for the desired axial drive. The dependence of these factors on the frequency or other parameters can be depicted graphically and be used as criteria.

One possibility is to use the resulting mass flow which is given by

$$\dot{m} = \int_A \varrho \, \vec{v} \, d\vec{A} \quad . \tag{8}$$

where A is the cross-section of the array. The mass flow along the z-axis is constant if the melt is assumed to be incompressible. Integrated over the length l of the device one obtains

$$\int_0^1 \dot{m} \, dz = \dot{m} \, 1 = G \quad . \tag{9}$$

G is equal to the magnitude of an axially directed momentum mv_z, where m is the total mass within the array considered and v_z is the average axial flow rate of the melt. The factor G in Eq. (9) seems to be suitable as a measure for the flow drive. Fig. 3 shows G as a function of aperture and frequency.

The electromagnetic force density field not only causes the desired flow in axial direction, but also local flow vortices, shown in Fig. 2. Therefore, a ratio can be formed which permits calculation of the part of the flow according to Eq. (9) in relation to the total motion of the melt which is caused by the Lorentz force. For that purpose one has to

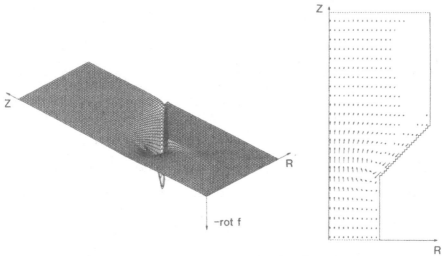

Figure 2. Rotational force density field (left) and fluid-flow pattern (right) for a constant frequency of 50 Hz and an extension angle of 45° (v_{max} = 1.8 m/s, (rot f)$_{max}$ = $-5 \cdot 10^8$ N/m⁴).

integrate the amount of the momentum density over the volume

$$\int_V |\, \varrho\, \vec{v}\, |\, dV = N \quad .$$

(10)

Eq. (10) summates all differential momentums. The resulting value is a measure for the total state of motion of the melt. From Eqs. (9) and (10) the ratio

$$\eta_V \;=\; G/N \;=\; \frac{1}{A}\int_A \varrho\, \vec{v}\, d\vec{A} \;/ \int_V |\varrho\, \vec{v}|\; dV$$

(11)

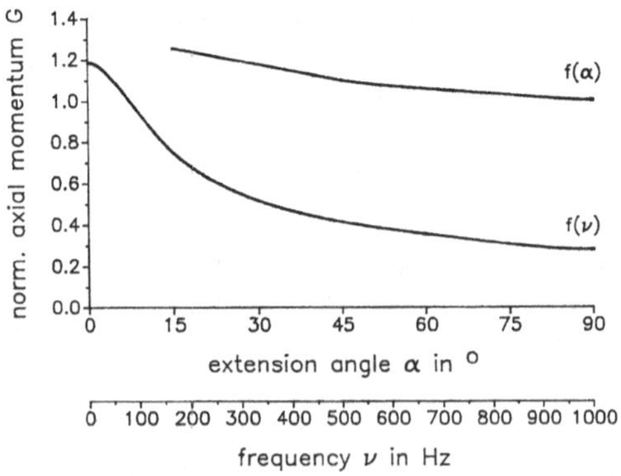

Figure 3. Axial momentum G as a function of extension angle α for a constant frequency of 50 Hz and as a function of the frequency ν for a constant angle α = 45°.

is formed, where η_v is always smaller than 1. η_v is the vortex efficiency, being a measure of the ability of the rotational force density field to form a unidirectional flow in the channel but it is additionally influenced by the geometry and the nature of the fluid. For a large η_v the rotational force density mostly causes axial flow. Local flow vortices then only play a minor role. For a small η_v, on the contrary, the axial drive is small and the flow vortices large. To avoid erosion of the refractory lining by flow vortices a large η_v is desirable. The influence upon η_v of different extension angles α for a constant frequency of 50 Hz and of different frequencies for $\alpha = 45°$ is presented in Fig. 4.

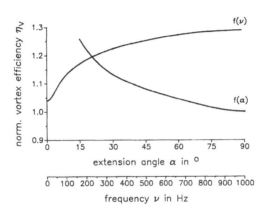

Figure 4. Vortex efficiency η_v versus extension angle α for a constant frequency of 50 Hz and versus frequency ν for a constant extension angle $\alpha = 45°$.

ACKNOWLEDGEMENT

This project has been supported by the Federal Department of Research and Technology of the FRG.

REFERENCES

[1] H. Wicker, In *11. Int. BBC-Fachtagung Induktionsofenanlagen*, 24.1–24.22, Dortmund 1986.
[2] M.Ya. Stolov and M.Ya. Levina, *Elektrotermiya (in Russian)* 151, 12–14 (1975).
[3] I.E. Bucenieks, M.Ya. Levina, M.Ya. Stolov and E.V. Shcherbinin, *Preprint, Inst. Fiz. Akad. Nauk Latv. SSR (in Russian)*, LAFI–021, Salaspils 1980.
[4] A. Mühlbauer, R. Fricke, H. Wicker and F. Feldmann, *Mater. Sci. Technol.*, to be published.

CHARACTERISTIC PROPERTIES OF MHD FLOW IN MAGNETODYNAMIC PUMPS

V.Polischuk, R.Horn, V.Dubodelov, V.Pogorsky,
V.Trefnyak, M.Tsin
Institute of Casting Problems of the Ukrainian SSR
Academy of Sciences, Vernadsky St. 34/1,
252142 Kiev, USSR

ABSTRACT. Magnetodynamic pumps (MDN) are a class of electromagnetic
devices for foundry and metallurgy which combine induction heating of
liquid metal and creation of electromagnetic pressure in the working
zone. MHD flow structure in the latter (both in cross-shaped and T-
piece ones) was studied theoretically by means of numerical computation
of the equations of hydrodynamics and experimentally by photographing
of the visualised stream of the modelling conductive liquid. The investigation showed that there was complicated vortical flow which
strongly influenced the pressure parameters of MDN, and at the same
time essentially increased the possibilities for molten metal treatment
in metallurgical technology.

The essential distinction of magnetodynamic pumps (MDN) [1]
from other types of electromagnetic pumps is the presence of liquid
conductors either apart from the metal transport path or coinciding
with it. MDN-4 channel is a typical example of the former case [2]
(Fig.1,a). The side channels connected to the straight-flow transport
channel form two cross-shaped working zones and, at the same time,
serve as a secondary liquid metal turn in which the electric current of
density j is induced by means of an inductor. By applying the transverse magnetic field B of a corresponding direction to each of the
working zones electromagnetic forces $\overline{F}_1 = \overline{j}_1 \times \overline{B}$ and $\overline{F}_2 = \overline{j}_2 \times \overline{B}$ are produced.
Summed up, these forces pump the liquid metal along the transport
channel. In MDN-6 type magnetodynamic pump [3] the side channels
together with the central one form a T-piece working zone (Fig.1,b).
The difference between the cross-shaped working zone and the T-piece
one is that the side channels of the latter serve both as the liquid
current conductor and the transport channels through which liquid
metal is taken out of the crucible and directed into the central
channel. Electromagnetic force reversal causes the liquid metal to
flow from the central channel into the side ones.

MHD-flow structure in MDN w.z. was studied both theoretically
and experimentally. Theoretical analysis was carried out by means of
numerical computation of the equations of hydrodynamics for the

J. Lielpeteris and R. Moreau (eds.), Liquid Metal Magnetohydrodynamics, 255–261.
© 1989 by Kluwer Academic Publishers.

256

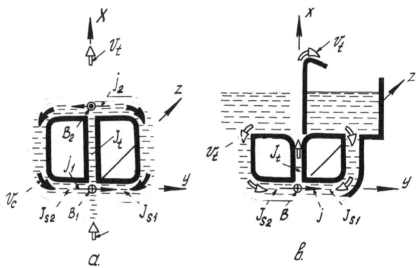

Figure 1. Typical MDN channels with cross-shaped working zone (a) and
T-piece one (b): V_t - transit flow velocity, V_c - circulation flow
velocity, (J_{s1}, J_{s2}, J_t) - current in channels, (B_1, B_2, B) - magnetic
field in working zone, (j_1, j_2, j) - current density.

stationary flow of viscous incompressible liquids [4]

$$\left(\bar{v}\,\nabla\right)\bar{v} = -\nabla P + \frac{1}{Re}\nabla^2\bar{v} + \frac{M}{Re^2}F(x,y); \qquad div\,\bar{v} = 0 \qquad (1)$$

where $Re = v_0 a \rho /2\eta$ - Reynolds number, $M = F_0 a^3 \rho /8\eta^2$ - EMF parameter,
F_0, v_0, a, ρ, η - characteristic values of EMF density, liquid velocity, length, liquid density and viscosity, respectively, Fx(x,y) - EMF
distribution function in w.z. The system of equations (1) with boundary
conditions generally accepted in hydrodynamics was solved by the
method of finite differences in Helmgoltz equation form for current
function ψ

$$\frac{1}{Re}\nabla^2\Omega - \frac{\partial\psi}{\partial y}\frac{\partial\Omega}{\partial x} - \frac{\partial\psi}{\partial x}\frac{\partial\Omega}{\partial y} = -M\frac{\partial F}{\partial y} \qquad (2)$$

where

$$v_x = \frac{\partial\psi}{\partial y}; \qquad v_y = -\frac{\partial\psi}{\partial x}; \qquad \Omega = \frac{\partial v_y}{\partial x} - \frac{\partial v_x}{\partial y} = -\nabla^2\psi.$$

The results of the computation are presented as schemes of current
lines both for halves of T-piece and cross-shaped w.z. The metal flow
structures are shown in Fig.2 for MDN-4 type channel having two w.z.
without the flow through the side channel ($V_c = 0$ in Fig.1,a). EMF
effect boundary is shown by dotted lines - Fx (21, a', Y_F). EMF is of
constant value in the (21 x a') area and diminishing to zero along the
Y-axis in the (21 x Y_F) area.

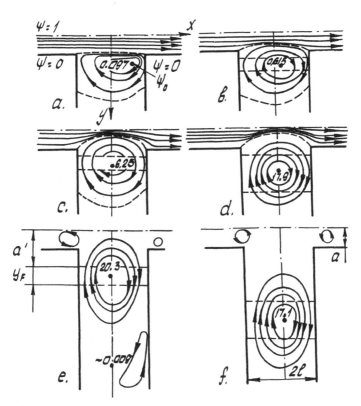

Figure 2. Scheme of the current function lines in half of cross-shaped
working zone at different values of parameters M and Re: a) M = 0;
Re = 2.10² ; M/Re² = 0; b) M = 10⁵; Re = 10³ ; M/Re² = 0.1; c) M = 10⁵;
Re = 10² ; M/Re² = 10; d) M = 10⁶; Re = 10² ; M/Re² = 10² ; e) M = 10⁶;
Re = 0; a' = 2a; Y_F = a; f) M = 10⁶; Re = 0; a' = 4a; Y_F = 2a;

        Fig.2e,f shows the flow structure in w.z. in the case of a
closed transport channel (Re = 0). There is intensive vortical flow in
the side channel region where EMF is of maximum turbulance (rotF_x=max).
The centre of stationary vortex of Ψ_o intensity (i.e. the discharge
between the vortex centre and the wall) is on the Y-axis in the middle
of the EMF field failing region. Both the form and the intensity of
the vortex depend on the value of M, the force gradient along Y-axis
($\partial F_x/ \partial y$) and the side channel width 21. As can be seen from the
comparison, the greater the said gradient and the value of 21, the
larger is   _o and the more rounded shape the vortex acquires. If the
liquid moves through the transport channel (Re ≠ 0) the flow structure
in w.z. is due to the interaction between the transit flow and the
stationary vortex and is defined by the ratio M/Re² . The greater this
ratio (Fig.2,b,c,d), the more compressed by the vortex is the transit

258

flow and for $M/Re^2 > 10^2$ the flow structure in the side channel is defined by a stationary vortex. At $M/Re^2 < 1$ the transit flow replaces the vortex centre in the direction of the flow and defines the vortex. When there is no EMF (Fig.2,a: $F_x = 0$, $M/Re^2 = 0$) the transit flow defines the flow structure in w.z. producing a vortex in the side channel, the intensity of which is several orders less than that of the vortex excited by EMF: for example, at $M/Re^2 = 0$ $\psi_0 = 0.097$ (Fig.2,a), at $M/Re^2 = 10^2$ $\psi_0 = 17.9$ (Fig.2,d).

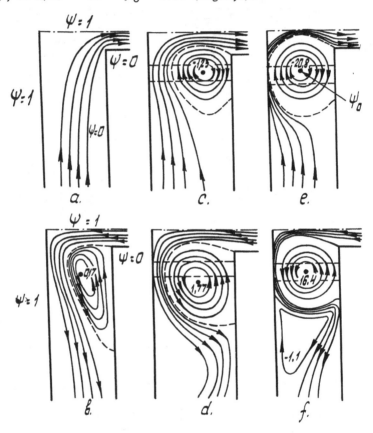

Figure 3. Scheme of the current function lines in half of T-piece working zone at different values of parameters M and Re: a,b) M = 0; Re = 10^2; $M/Re^2 = 0$; c,d) M = 10^5; Re = 3.10^2; $M/Re^2 = 1$; e,f) M = 10^6; Re = 10^2; $M/Re^2 = 10^2$.

Fig.3,c,e shows typical flow structures in T-piece w.z. under the stream confluence conditions (i.e. when the streams move from the two side channels into the central one - Fig.1,b) due to the effect of the force field Fx. Fig.3,d,f gives the structures for the opposite direction of the force field ($-F_x$), i.e. for the stream separation conditions. Stationary vortexes in the side channels are formed both under confluence and separation conditions in the region of the magnetic field reduction (shown by dotted lines) similar to the way it happens in the cross-shaped w.z. Both the form and the intensity of the vortex are defined by the M/Re^2 ratio too. The flow structure at $M/Re^2 = 1$ is shown in Fig.3,c,d. It can be seen that the greater the M/Re^2 ratio, the more intensively compressed by the vortexes is the transit flow. The interaction between the vortex and the transit flow in the T-piece w.z. is stronger than that in cross-shaped w.z. and due to this fact the hydraulic resistance of the former increases. For the sake of comparison the results of the computation at usual hydraulic flow in T-pieces are shown in Fig.3,a,b. Under confluence conditions (Fig.3,a) no vortex appears while under separation conditions, as it was true for the hydrodynamic flow in the cross-shaped w.z., the transit flow excites a vortex of small intensity.

To estimate the degree of certainty and the scope of application for the numerical computation results of the plane problem of laminar flow due to the EMF field effect which is characteristic for the MDN w.z., an investigation of the flow structure in w.z. was carried out by means of photographing the visualized streams of transparent conductive liquid on the XY-plane. Acid solutions (30% H_2SO_4 and 15% HCl) were used as liquids and polystyrene balls (0.5 - 1. mm across) were added for visualization. Flow conditions were varied within the values of Re = (1-5) x 10^3. The magnetic field was excited by real magnetic systems of MDN. The flow structures for characteristic technological conditions of MDN were thus defined.

The photographs of the T-piece w.z. flow are presented in Fig.4. Comparison with similar computed flows (see Fig.4a-3a, 4b-3b, 4c-3c, 4d-3d, 4e-2e) shows that they are well consistent. For instance, under confluence conditions (Fig.4,c) two symmetric vortexes roll along the side channels in the region of the magnetic field reduction. The transit flow is compressed by the vortex towards the lower wall. The transit flow replaces the vortex in the direction of the liquid flow: under confluence conditions the vortex centres are nearer to the central channel (Fig.4,c) and under separation conditions they are farther from it (Fig.4,d). If there is no transit flow (the central channel is closed – Fig.4,e) there are stationary vortices in the side channels, their width being equal to that of the channel and the length being dependent on the force gradient along the Y-axis: the more gentle the force reduction curve is, the more stretched shape the vortex acquires. This structure correlates with the computed vortex structure in the cross-shaped w.z. under no flow conditions - Re = 0 (Fig.2,e,f).

Fig.4,f,g,h presents even more complicated flows in the w.z., those flows have not been computated, but they are observed under technological conditions of MDN application. These flow structures appear when the current in the w.z. is conducted asymmetrically.

The flow structure in the T-piece w.z. at

$$\bar{J} = \bar{J}_{s1} = -\bar{J}_{s2}, \quad |\bar{J}_t| = |\bar{J}_{s1} + \bar{J}_{s2}| = 2J$$

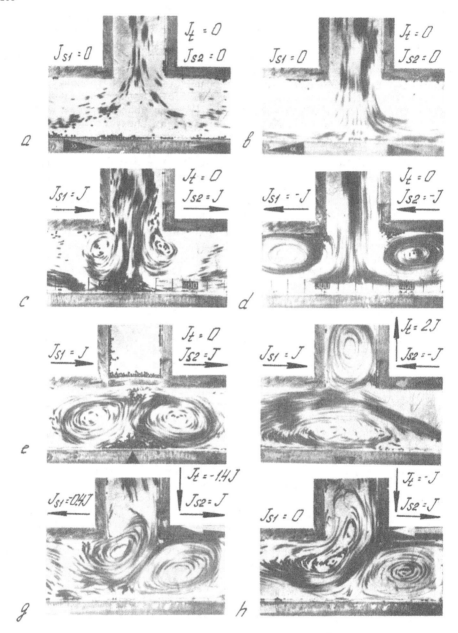

Figure 4. Photographs of flow of transparent conductive liquid and non-conductive balls (0.5 - 1 mm across) within it in T-piece working zone at constant value of magnetic field and different values of current (J_t, J_{s1}, J_{s2}) in channels.

is shown in Fig.4,f. A stationary vortex rolls in the central channel completely occupying its section, and there is no flow through the latter. A transit stream flows from the right side channel to the left one, and a complicated vortical cluster merging two one-direction vortexes rolls in the lower part of the w.z. opposite the transport channel. At $\bar{J}_{s1} \neq \bar{J}_{s2}$, $|\bar{J}_t| = |\bar{J}_{s1} + \bar{J}_{s2}| = 2J$ there is a transit flow in all three channels. Fig.4,g shows the flow structure at $\bar{J}_{s2} = \bar{J}$, $\bar{J}_{s1} = -0.4 \bar{J}$, $|\bar{J}_t| = 1.4$ J. The transit streams run both from the central and the right side channels, flow together into the w.z. and run away into the left channel. There are two stationary vortexes of a complicated shape. The flow structure at $\bar{J}_{s2} = \bar{J}$, $\bar{J}_{s1} = 0$, $|\bar{J}_t| = -J$ is presented in Fig.4,h. The main transit flow circulates through the central and the left side channels.

The last three examples show that by varying the ratio of the electric currents in the side channels one can get an infinite amount of transit flow combinations of an extremely complicated vortical structure in the w.z.

The described characteristic properties of the MHD-flows in MDN essentially determine the pressure parameters and should be taken into account while designing MDN. Moreover, these flows are very promising for molten metal treatment in metallurgical technology.

REFERENCES

[1] POLISCHUK V., 1971, 'Application of Magnetodynamic Devices in Industry', *The Magnetohydrodynamics*, No.1, 118.

[2] USSR Patent 176184, Cl. 59a, II. Electromagnetic Induction Pump/ V.Polischuk, M.Tsin. - Publ. 26.10.65, Bul. No.21.

[3] USSR Patent 288183, Cl. 21c, 18/02. Induction Channel Furnace with a Magnetic Core / V.Polischuk, M.Tsin. - Publ. 03.12.70, Bul. No.36.

[4] BURDE G., HORN R., YAKUSHIN V., 1972, 'On the Liquid Motion in the MHD-pump with Cross-shaped Active Zone', *The Magnetohydrodynamics*, No.3, 93.

Session F:
Electromagnetic Shaping

DEFLECTION OF A STREAM OF LIQUID METAL BY MEANS OF AN ALTERNATING MAGNETIC FIELD

J. ETAY
MADYLAM
Institut de Méchanique de Grenoble
B.P. 95
38400 St Martin d'Hères Cédex
France

A. J. MESTEL & H. K. MOFFATT
D.A.M.T.P
Cambridge University
Silver Street
Cambridge CB3 9EW
United Kingdom

ABSTRACT. The deflection of a thin, two-dimensional stream of liquid metal due to external high-frequency currents is investigated. A simple theoretical model assuming uni-directional flow is presented. The relationship between the angle of deflection of the stream and the power supplied to the perturbing currents is determined. Experiments are performed in which a free-falling mercury sheet is deflected by two anti-parallel line currents. The agreement between theory and experiment is reasonable, despite a tendency towards three-dimensionality in the latter.

1. Introduction

For many purposes involving the processing of liquid metals, it may be desirable to be able to deflect a stream through a given angle. In this paper we describe how such deflection can be achieved by the action of the magnetic pressure p_M associated with a high-frequency magnetic field produced by current sources outside the liquid stream. We give a simple one-dimensional analysis of the phenomenon and compare the theoretical predictions with the results of some experiments performed with mercury. A fuller description of both the theory and experiments may be found in Etay, Mestel & Moffatt [1].

We consider a two-dimensional configuration, in which a sheet of metal, of initial thickness d_0 and uniform velocity u_0, moves under the influence of alternating line currents. Such sheets are used in industrial processes such as the manufacture of metallic ribbon. In order to analyse the effect, we make certain simplifying assumptions, as follows:

(a) We assume that the field frequency, $\omega/2\pi$ is sufficiently high for it to be reasonable to treat the effect of the fields entirely in terms of the magnetic pressure on the liquid surface. This requires that the magnetic skin-depth $\delta = (2/\mu_0\sigma\omega)^{1/2}$ (where σ is the electrical conductivity of the liquid and μ_0 its permeability) be small compared with the undisturbed thickness of the stream, d_0.

(b) We assume that d_0 is small compared with the scale L on which the magnetic pressure varies over the surface, as determined by the current source distribution and the surface curvature.

(c) Finally, we assume that gravity may be neglected at least over the scale L on which the deflection takes place. In terms of u_0, the length-scale on which gravity acts is $l_g = u_0^2/g$.

These three assumptions imply a hierarchy of length-scales as follows:

$$\tfrac{1}{2}\delta \ll d_0 \ll L \ll l_g \tag{1}$$

265

The factor of 1/2 is included in (1) because the magnetic pressure is quadratic in the magnetic field and thus decays twice as quickly. Assumptions (a), (b) and (c) are obviously restrictive, but they permit significant progress to be made. The limitations of the analysis are considered in [1].

2. Quasi-one-dimensional analysis

The assumption (b) above allows the use of a quasi-one-dimensional analysis, in which the liquid stream is in effect located by the position of its left-hand boundary. This is a curve C parameterised by arc length s from some fixed point O, as in figure 1. The stream thickness, $d(s)$, is (weakly) non-uniform when deflection occurs. We suppose that, upstream of the region of magnetic influence, the stream has uniform velocity u_0 and thickness d_0 so that the volume flux is $Q = u_0 d_0$. As viscous effects are negligible, the flow is irrotational by virtue of the assumption (a) which ensures that the sole effect of the magnetic field is to provide a magnetic pressure distribution over the liquid surface.

Let (s, n) be taken as coordinates tangent and normal to C, as in figure 1, and let $u(s, n)$ be the velocity within the stream, effectively parallel to C. To leading order in the small parameter d_0/L, the velocity is uniform, $u \approx u_0$, and the n-component of the equation of motion is

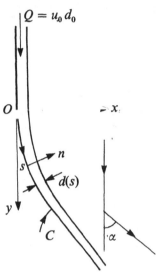

Figure 1: The coordinate system.

$$\frac{\partial p}{\partial n} = -\rho u^2 K(s), \tag{2}$$

where ρ is the liquid density, $p(s, n)$ the pressure, and $K(s)$ is the curvature of C at position s. To the same approximation, the appropriate boundary conditions are

$$p(s, d_0) = p_0 - \gamma K \qquad \text{on the right-hand boundary,}$$
$$p(s, 0) = p_0 + \gamma K + p_M \quad \text{on the left-hand boundary,} \tag{3}$$

where p_0 is atmospheric pressure, γ is the surface tension and $p_M(s)$ is the magnetic pressure. In writing (3), we are assuming that all the current sources occur on the left-hand side of the metal stream. Integrating (2) using (3) we get

$$p(s, n) = p_0 - \gamma K + \rho u_0^2 K(d_0 - n) \tag{4}$$

and hence the required magnetic pressure $p_M(s)$ is given by

$$p_M = \rho u_0^2 d_0 K(s) \lambda \qquad \text{where} \qquad \lambda = 1 - \frac{2\gamma}{\rho u_0^2 d_0} . \tag{5}$$

The perturbation to the velocity $u(s, n)$ and width $d(s)$ may now be found from Bernoulli's theorem,

$$p + \tfrac{1}{2}\rho u^2 = p_0 + \tfrac{1}{2}\rho u_0^2 , \tag{6}$$

and mass conservation. From (4) and (6), we have

$$u = u_0 + K \left[\frac{\gamma}{\rho u_0} - u_0(d_0 - n) \right], \qquad \text{and so} \tag{7}$$

$$Q = \int_0^{d(s)} u \, dn = u_0 d - \tfrac{1}{2} K u_0 d_0^2 \lambda \; . \qquad (8)$$

As also $Q = u_0 d_0$, the stream thickness, d, is given by

$$d(s) = d_0 + \tfrac{1}{2} K d_0^2 \lambda \; . \qquad (9)$$

It is clear from the form of (5) that the inertia of the uniform stream acts somewhat like a negative surface tension. As the magnetic pressure is positive, the direction in which the stream is deflected (as determined by the sign of K) may be seen from (5) to depend on the sign of λ. We are mainly concerned here with parameter values such that λ is positive, i.e. when the magnetic pressure generates a momentum flux away from the currents. When $\lambda < 0$, the magnetic pressure is strongly resisted by surface tension, with the momentum flux being less important. In both these cases the jet thickness, $d \geq d_0$ everywhere. It is of interest to note that when both λ and p_M are zero, any shape C gives rise to an admissable steady state, to lowest order in the jet thickness. Somewhat curiously, the path of the stream is maintained by its own surface tension.

Equation (5) defines a non-linear relation between the curve C and the external currents. In theory, any desired stream path C (for which $K > 0$) can be obtained by choosing the distribution of current sources to give the required magnetic pressure in (5). One such distribution (and there will be many others) is provided by placing coils so as (in effect) to provide a current sheet $J(s) \cos \omega t$ near to the deflected stream. This current sheet produces a magnetic field $B(s) \cos \omega t$ in the gap between the coils and the stream, where $B(s) = \mu_0 J(s)$, and an associated time-averaged magnetic pressure $p_M = B^2 / 4\mu_0 = \tfrac{1}{4} \mu_0 J^2$. Hence the required magnetic pressure (5) is achieved provided

$$\mu_0 J^2 = B^2 / \mu_0 = 4 \rho u_0^2 \lambda K d_0 \; . \qquad (10)$$

The required deflection of the stream can then be maintained.

If we use the expression $K = d\psi / ds$, where ψ is the angle that the tangent to C makes with the horizontal, then by integrating (10) we obtain the result

$$4 \rho u_0^2 d_0 \lambda \, \alpha = \int_C \frac{B^2}{\mu_0} \, ds \qquad (11)$$

where α is the total angle of deflection of the stream. We may express α in terms of the power supplied to the coils (per unit length in the z-direction). This power, W, is balanced by the Joule heating in the metal stream

$$W = \int \frac{|\overline{\nabla \wedge \underline{B}}|^2}{\mu_0^2 \sigma} \, dV = \frac{\sigma \delta^2}{\mu_0^2} \int_0^\infty e^{-2n/\delta} \, dn \int_C B^2(s) \, ds = \frac{\delta \omega}{4} \int_C \frac{B^2}{\mu_0} \, ds \qquad (12)$$

in terms of the skin-depth approximation. Together (11) and (12) give a simple relationship between the angle of deflection and the power supplied to the external coils,

$$\alpha = \frac{W/\omega \delta}{\rho u_0^2 d_0 \lambda} \; . \qquad (13)$$

Recalling the definition of δ, we see that the power needed to obtain a given deflection α behaves as $\sqrt{\omega}$, provided ω is sufficiently large for the skin-depth approximation to apply. Thus, in practice there will be an optimal frequency at which the deflection effect is still pronounced, but for which the power dissipated in the metal is relatively low. This optimum value will occur when δ / d_0 is $O(1)$.

3. Deflecting action of weak line currents.

It is difficult to find the stream path C analytically for a given current distribution, as the magnetic field is strongly coupled with the shape of C. A problem of this type was solved numerically by Shercliff [2] in his work on the shaping of liquid metal columns, when the balance is between surface tension and magnetic pressure. If the deflection is weak, however, then progress is possible by perturbation analysis. We illustrate this with reference to the action of a concentrated line current $I\cos\omega t$ placed at a distance L from a stream whose undisturbed position is $0 < x < d_0$. When $I \neq 0$, we suppose that the stream is symmetrically perturbed as indicated in figure 2. To leading order, however, the magnetic field distribution may still be calculated as if the stream were in the undisturbed position. There is then an image current $-I\cos\omega t$ at $x = L$, $y = 0$, and the net magnetic field on $x = 0$, is $\underline{B} = (0, B_y, 0)$ where

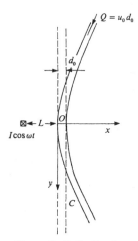

Figure 2: Deflection by a weak current source.

$$B_y = \frac{\mu_0 I}{\pi}\frac{L}{L^2 + y^2}\cos\omega t, \qquad \text{so that} \qquad p_M = \frac{\mu_0 L^2 I^2}{4\pi^2}\frac{1}{(L^2 + y^2)^2} \ . \tag{14}$$

The deflection of the surface, from (5) with $K \approx \mathrm{d}^2 x/\mathrm{d}y^2$, is given by

$$\frac{\mathrm{d}^2 x}{\mathrm{d}y^2} = \frac{N}{4\pi^2}\frac{L^3}{(L^2 + y^2)^2} \qquad \text{where} \qquad N = \frac{\mu_0 I^2}{\rho u_0^2 d_0 L\lambda} \ . \tag{15}$$

We integrate this with 'initial' conditions $x(0) = \frac{\mathrm{d}x}{\mathrm{d}y}(0) = 0$ with the result

$$x(y) = \frac{N}{8\pi^2}\, y\, \tan^{-1}\left(\frac{y}{L}\right) \ . \tag{16}$$

When $|y| \gg L$, we have $x \sim \frac{N}{16\pi}|y|$, so that the net angle of deflection is given by

$$\alpha = \frac{N}{8\pi} \ . \tag{17}$$

Equation (17) is strictly valid only provided the deflection is weak, i.e. $\alpha \ll 1$; however the results of section 2 suggest that an arbitrarily large deflection of the stream may be achieved if the dimensionless parameter N is increased to a sufficiently large value. This parameter may be regarded as the magnetic interaction parameter, giving a measure of the transverse flux of momentum generated by the magnetic forces relative to the flux of momentum $\rho u_0^2 d_0$ in the incoming stream.

The magnetic field lines of a line current and its image consist of a family of co-axial circles. Thus, the analysis of this section and the result (17) in particular may be extended to the more realistic case when the line current is replaced by a circular wire, of radius r, whose axis lies on $x = -b$, $y = 0$. At high frequency, the current in the wire flows near the surface and is equivalent to a line current at $x = -L$, $y = 0$, where $L^2 = b^2 - r^2$. For this value of L, the deflection obtained is given by (17) with (15). It should be noted that no such exact result holds for low frequencies (or D.C.) for which the surface of the wire is not a magnetic field line.

In the experiments that follow we use for electrical convenience a device which may be modelled by two opposite line currents a fixed distance, a, apart. It is a simple matter to extend the above theory to include this case. The magnetic field may be represented as a sum of two terms similar to

(14) which can then be squared and integrated to calculate the deflection angle α. When the two line currents lie in a horizontal plane a distance a apart, the result corresponding to (17) is

$$\alpha = \frac{N_h}{8\pi} \qquad \text{where} \qquad N_h = N\frac{a^2}{(L+a)(2L+a)} \tag{18}$$

whereas when they lie in a vertical plane we find

$$\alpha = \frac{N_v}{8\pi}\left(1 + \frac{1}{1 + \frac{a}{L}\frac{N_v}{8\pi}}\right) \qquad \text{where} \qquad N_v = N\frac{a^2}{a^2 + 4L^2}\,. \tag{19}$$

In (19), α varies between a value due to deflection by both branches of the inductor in turn, and one where only the uppermost branch is important, according to the relative size of a/L and N.

So far, we have supposed that the current sources are all placed to one side of the stream. More complex stream deformation may clearly be achieved if current sources on both sides are used. Because of the two-dimensionality of the problem and the high frequency approximation, the magnetic fields on either side of C do not interact. The stream path, C, will still be determined by the differential equation (5) with the appropriate value of p_M being the difference between the magnetic pressures on either side of C. In [1], it is described how two line currents, one on each side of the metal stream, can act as a sensitive control over the ultimate direction of the stream.

4. Experiments

Experiments were performed in order to measure the deflection of a nearly two-dimensional stream by line currents for various values of the interaction parameter N. The experimental set-up used is described in full by Etay & Garnier [3]. Essentially, the facility consists of a mercury-filled hydraulic circuit containing a freely falling column. The initial cross-section of the column is determined by a gently converging nozzle ending in a slit 1mm. by 39mm. The alternating magnetic field is supplied by insulated copper inductors connected to adjustable capacitors, and powered by a 100kWatt generator. The inductors are made from hollow copper tubing of 3-4mm. diameter through which cooling water is passed. A single line current would give rise to a large self-inductance with resultant inconvenience for the tuning capacitor. Instead, hairpin-shaped inductors were built, with the current flowing along one branch of the hairpin and returning along the other. In the middle of the inductor the field is approximately that due to two line currents. The distance between the branch axes is 15mm. The current flowing in the inductor varies from 0 to 1520 Amps. at a frequency of 350kHz. The skin depth, δ, is therefore 0.8mm. and the magnetic field near the sheet varies from 0 to 2000 Gauss. A higher current could not be obtained with this kind of copper tube inductor. In Riga we were taught the phrase "The experiment was a success: the apparatus burned out!" Indeed, one of the inductors used in our experiments proved very "successful!"

A truly two-dimensional geometry cannot be achieved experimentally, even in the absence of electromagnetic effects. Surface tension causes rolls to form at both ends of the initally flat sheet which grow as the metal falls, as in figure 3a. Of course, in any industrial application, the sheet would be longer and thus end effects less important than in these experiments. A further departure from two-dimensionality occurs due to the slight twisting of the sheet when it leaves the nozzle. When an alternating, high frequency current flows in the inductor the departures from two-dimensionality increase. The ends of the sheet are repelled further from the inductor than is the middle, due to the three-dimensionality of the magnetic field there. The resulting curved cross-section aggravates the tendency of the sheet to contract and its thickness increases. The measurement of the deflection angle can thus become difficult. This effect can be seen in the photograph of figure 3b which shows the "worst" kind of deflection that can occur.

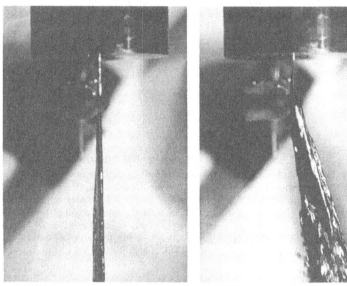

(a) No current, $I = 0$.   (b) With current, $I = 1520$Amps.

Figure 3: Evolution of a two-dimensional mercury sheet.

The greatest error in calculating the experimental value of N derives from the estimate of L, defined in terms of the distance between the centre of the coil and the liquid stream. The value of L was measured on a photograph taken with a $45°$ mirror placed below the inductor. For maximal effect, L should be small, when the accuracy of this measurement is low. The deflection angle, α, was measured on photographs taken by a camera mounted adjustably in the plane perpendicular to the metal sheet. The three-dimensionality of the deflected jet ensures that the measurement of α is not easy. It was decided to define α as the average of the maximum and minimum deflections attained over the sheet. A curious feature of the observed deflection was that the stream did not appear to bend until below the position of maximum pressure, whereas theoretically this should occur very slightly above that position. This was probably an observational effect.

Two sets of experiments were performed, one with the inductor horizontal (so that only one branch of the inductor has a large effect on the stream) and the other with it vertical. The deflection was noticably greater when the inductor was vertical. The measured values of α (in degrees) are plotted in figure 4 as functions of N_v and N_h. The theoretical line (18) and curve (19) are drawn on the figure for comparison. The agreement is satisfactory for α small, but not surprisingly, the theory over-estimates the obtained deflection angle for N (and α) large. The manner in which this occurs appears to be fairly systematic. There are a number of reasons why this might be expected. First, equations (18) and (19) apply only when α is small, whereas the experiments cover a wide range of α. The "weak deflection" approximation, whereby the magnetic field may be easily calculated, thus breaks down. Secondly, assumptions (a) and (b) are only weakly satisfied in the experiments. As a result, the unidirectional irrotational flow assumed in the theory may not be wholly accurate. Thirdly, it is clear that the three-dimensionality of the experiments will lead to a reduced deflection. It also hinders observation of α as we have already discussed. Finally, we should recall the problems inherent in the measurement of L. It may well be that there is a systematic under-estimation of L when α is appreciably large. Bearing in mind the practical difficulties, it was felt that the agreement between theory and experiment was satisfactory.

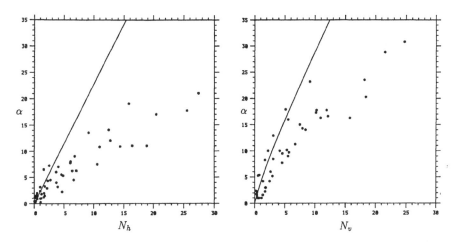

Inductor horizontal   Inductor vertical

Figure 4: Deflection angle α in degrees plotted against N_h and N_v.

6. Concluding remarks

The use of electromagnetic fields as controlling devices in the metallurgical industry is growing. Liquid metal may be stirred, heated shaped and transported without resort to mechanical means. Usually the fluid flow that results is complex in nature, and is often turbulent. In this paper, by contrast, we have investigated a process for which the flow is particularly simple, in which the ultimate position of a stream of liquid metal is controlled. The analysis is fairly general and may be extended to cover particular geometries of industrial interest. Although the fluid dynamical techniques we have used are elementary, they do seem to give a reasonable description of the real behaviour, as witnessed by experiment.

The main limitations of the process derive from its reliance on two-dimensionality and a very high-frequency. In [1] a second configuration is considered which avoids these difficulties, relying instead upon the repulsive force that exists between electric currents. If a thin, cylindrical metal jet is made to carry a current, then it can be deflected by other suitably positioned currents. The equilibrium position for such a jet is calculated, and the question of its stability is briefly discussed.

In conclusion, the authors would like to offer their congratulations to the organisers of this very successful and enjoyable symposium, and to express their gratitude for the lavish hospitality shown to them in Riga.

References

[1] ETAY J., MESTEL A.J., & MOFFATT H.K., 1988 'Deflection of a stream of liquid metal by means of an alternating magnetic field.' J. Fluid Mech Vol. 194, 309.

[2] SHERCLIFF J.A., 1981 'Magnetic shaping of molten metal columns.' Proc. R. Soc. Lond. A Vol. 375, 455.

[3] ETAY J., & GARNIER M., 1982 'Le contrôle electromagnétique des surfaces métalliques liquides et ses applications.' J. Méch. Theor. Appl. Vol. 1, 911.

THE SHAPE OF LIQUID METAL JETS UNDER A UNIFORM TRANSVERSE MAGNETIC FIELD

S. Oshima and R. Yamane
Tokyo Institute of Technology
O-okayama 2-12-1 Meguro-ku
Tokyo 152
Japan

ABSTRACT. The shape of a horizontal jet of a liquid metal issuing from an electrically insulated nozzle under a uniform transverse magnetic field is discussed. Our primary interests are to determine the shape of the free surface of the jet, and to demonstrate the possibility of the magnetic shape-control of liquid metal jets by using a D.C. magnetic field. In case of a jet issuing from a nearly-circular nozzle, the basic equations of MHD flows are simplified under the assumption of a small deformation and solutions of the approximate analysis are presented. In the case of a jet issuing from a nozzle with a large aspect ratio, an approximate analysis is carried out by dividing the jet into two region. Whether the jet is nearly-circular or not, the shape change of the jet due to surface tension is suppressed under a uniform transverse magnetic field. The shapes obtained by the approximate analysis are shown to be in good agreement with those of the experimental results using mercury.

1. INTRODUCTION

In recent years, studies on the magnetic shape-control of liquid metals by using a A.C. magnetic field is being actively conducted. For example, interesting results were given from the theoretical study of Shercliff [1] and the experimental investigation of Etay-Garnier[2].

On the other hand with the development of superconductors, it has become relatively easy, from a technological viewpoint, to generate a strong static magnetic field in a relatively large space. Previously, we have elucidated the effects of a magnetic field gradient on the shape of a liquid metal jet[3] by an approximate analysis and experiments with mercury, and have demonstrated the possibility of the magnetic shape-control of liquid metal flows by using a D.C. magnetic field.

Many studies on a liquid jet without a magnetic field have been conducted since Rayleigh[4]; however, the theoretical studies on the shape change of a liquid metal jet under a magnetic field are rare. In this paper, we elucidate the effects of a uniform transverse magnetic field on a liquid metal jet, for the development of a noncontacting shape-control method using a D.C. magnetic field.

J. Lielpeteris and R. Moreau (eds.), Liquid Metal Magnetohydrodynamics, 273–279.
© 1989 by Kluwer Academic Publishers.

2. ANALYSIS

2.1. The Shape of a Jet Issuing from a Nearly-Circular Nozzle

In this section, we give an outline of our analysis for the shape change of the jet issuing from a nearly-circular nozzle. We choose the coordinate system as shown in Fig. 1., and assume that the following conditions are satisfied during the shape change.

1) Though the jet is not circular, the deflection from a circle is small ($\epsilon = |S-a_e|_{max}/a_e \ll 1$).
2) Jet is slender, and the derivatives of variables with respect to z is much smaller than those with respect to r and to θ ($\delta = a_e/1 \ll 1$).
3) Initial velocity of the jet is uniform.
4) Induced magnetic field is much smaller than the applied magnetic field ($Rm = \sigma\mu w_o a_e \ll 1$).
5) The effects of the gravity and the viscosity on the shape change are negligible small.
6) Shape of the jet is symmetric with respect to both the x- and the y-axis.

In these assumptions, S is the distance measured from the center of the jet, l the longitudinal characteristic length of the jet, a_e the equivalent radius ($a_e = \sqrt{b_n h_n/\pi}$), w_o the cross sectional mean velocity, σ the electric conductivity, and μ magnetic permeability.

We express the velocity and the electric potential as follows

$$\mathbf{v} = \mathbf{v}_o + \mathbf{v'}, \quad \Phi = \Phi_o + \Phi', \quad \Phi_o = (\mathbf{v}_o \times \mathbf{B}_o) \cdot \mathbf{r}_p \qquad (1a\text{-}c)$$

where, $\mathbf{v}_o$ is the uniform velocity ($\mathbf{v}_o = w_o\mathbf{k}$), $\mathbf{v'}$ the additional velocity due to the shape change, $\mathbf{B}_o$ the applied magnetic flux density, Φ_o the electric potential induced by the uniform velocity, Φ' the additional electric potential due to the shape change, and $\mathbf{r}_p$ the position vector. We define the following nondimensional variables.

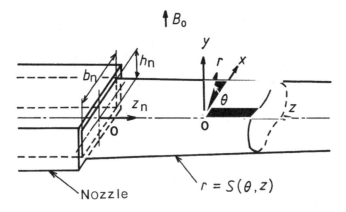

Figure 1.　Coordinate system.

$$(v_r^*, v_\theta^*) = (v_r', v_\theta')/\epsilon\delta w_o, \quad v_z^* = v_z'/\epsilon w_o$$

$$(J_r^*, J_\theta^*) = (J_r, J_\theta)/\epsilon\sigma w_o B_o, \quad J_z^* = J_z/\epsilon\delta\sigma w_o B_o$$

$$p'^* = p/\epsilon\delta^2\rho w_o^2, \quad \Phi^* = \Phi'/\epsilon w_o B_o a_e$$

$$(B_r^*, B_\theta^*) = (B_r, B_\theta)/B_o, \quad r^* = r/a_e, \quad z^* = z/l \tag{2}$$

On the assumptions 1)-5), we can simplify the basic equations as follows

$$\frac{\partial \mathbf{v}^*}{\partial z^*} = -\nabla_2^* p'^* + N\mathbf{J}^* \times \mathbf{Bo}^* \tag{3}$$

$$\mathbf{J}^* = -\nabla^*\Phi^* + \mathbf{v}^* \times \mathbf{B}_o^* \tag{4}$$

$$\nabla^* \cdot \mathbf{v}^* = 0, \quad \nabla_2^* \cdot \mathbf{J}^* = 0 \tag{5a-b}$$

where $\nabla_2^* = e_r(\partial/\partial r^*) + e_\theta/r^* (\partial/\partial\theta)$ is the two-dimensional nabla, and N= $\sigma B_o^2 l/\rho w_o$ is the Stuart number.

On the assumptions 1)-4), we can write the condition of zero normal flow through the free surface as follows

$$v_r^* = \partial S^*/\partial z^*, \quad J_r^* = 0 \quad (r^* = 1) \tag{6a-b}$$

where $S^*=(S-a_e)/\epsilon a_e$.

The jump in pressure at the free surface Δp, which is due to the surface tension, is given as follows

$$\Delta p = p|_{r=S} - p_a = \gamma(1/r_1 + 1/r_2) \tag{7}$$

where $1/r_1+1/r_2$ is twice the mean curvature of the surface, and p_a is the pressure of the surrounding atmosphere. Equation (7) can be written in the nondimensional form as follows

$$p^* = (1/W) \cdot (S^* + \partial^2 S^*/\partial\theta^2) \tag{8}$$

where, $W=\rho w_o^2 a_e^3/\gamma l^2$ is the Weber number, and $p^*=p'^*+p_a^*-1/\epsilon W$ the nondimensional pressure. Even though p'^* in Eq. (3) is replaced by p^*, equation (3) still holds. The initial conditions we used are as follows

$$\mathbf{v}^*|_{z^*=0} = \mathbf{k}, \quad S^*|_{z^*=0} = S_o(\theta), \quad \partial S^*/\partial z^*|_{z^*=0} = 0 \tag{9a-c}$$

We solved the differential equations (3)-(5) under the assumption 6) and the conditions (6a-b) and (9a-c), and obtained the following equations.

$$J_r^* = J_\theta^* = v_z^* = \Phi^* = 0 \tag{10}$$

The rest of quantities are dependent on the initial shape disturbance $S_o(\theta)$, the Stuart number N and the Weber number W. We found the following three kinds of solutions for the lowest mode ($S_o = \cos 2\theta$).

276

i) In the case that $N^2W < 96$ $(\alpha^2 < \omega_o^2)$

$$S = - \omega_o \omega^{-1} e^{-\alpha z} \sin(\omega z - \psi_1) + (N/2) \omega^{-1} e^{-\alpha z} \sin \omega z \qquad (11)$$

where, $2\alpha = N/2$, $\omega_o^2 = 6/W$, $\omega^2 = \omega_o^2 - \alpha^2$, and $\tan\psi_1 = \omega\alpha^{-1}$.
ii) In the case that $N^2W = 96$ $(\alpha^2 = \omega_o^2)$

$$S = (1 + Nz/4)e^{-Nz/4} \qquad (12)$$

iii) In the case that $N^2W = 96$ $(\alpha^2 = \omega_o^2)$

$$S = (s - t)^{-1} [(s - N/2)e^{-sz} - (t - N/2)e^{-tz}] \qquad (13)$$

where, $-s$ and $-t$ are the solutions of the quadratic equation $x^2 + 2 x + \omega_o^2 = 0$. Equations (11)–(13) correspond to the three solutions for a simple harmonic oscillation with a damper; namely, a damped, a critical damping and an overdamping oscillations. We also obtained the solutions for the shape change of the higher mode ($\cos4\theta$ mode). Analytical results are shown in the next chapter, with the experimental results.

2.2. The Shape of a Jet Issuing from a Nozzle with a Large Aspect Ratio

In case of the jet issuing from a nozzle with a large aspect ratio, the effect of the surface tension is restricted nearby both edges of the jet. Kozuka et al.[5] analyzed the film jet under an alternating magnetic field by separating the jet into two region; one is the central region where the thickness is almost uniform, the other is the side region where the jet is rounded off by the surface tension (see Fig. 2). In such a case, the motion of both the sides is governed by the following equations.

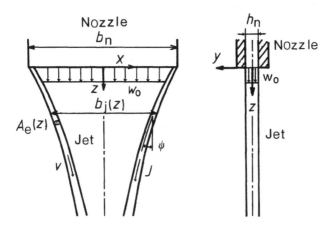

Figure 2. Analytical model of the jet with a large aspect ratio.

$$\frac{d}{dz}(A_e v^2 \cos\psi) - w_o^2 h_n \tan\psi = 0 \tag{14}$$

$$\frac{d}{dz}(A_e v^2 \cos\psi) - w_o^2 h_n \tan\psi + \frac{2\gamma}{\rho}\tan\psi - \frac{1}{\rho}JB_oA_e\tan\psi = 0 \tag{15}$$

$$\frac{d}{dz}(A_e v^2 \sin\psi) - \frac{2\gamma}{\rho} + \frac{1}{\rho}JB_oA_e = 0 \tag{16}$$

where, A_e is the cross-sectional area of the side, v the velocity, ψ the angle clockwise angle measured from the z-axis to the center line of the side region. Equation (14) is the equation of conservation of mass, and Eqs. (15) and (16) are the equation of motion.

When an electrically conducting media having a width bj moves at a uniform velocity w_o in a uniform magnetic field, the electric-field potential at the sides Φ_o can be expressed as follows

$$\Phi_o = -w_oB_ox \tag{17}$$

Assuming that Eq. (17) holds also in the jet, we can express the electric current density at the side region as follows

$$J = -\sigma\frac{d\Phi e}{ds} = \sigma(woBo/2)\frac{dbj}{ds} \tag{18}$$

We assume that the initial area and the initial velocity are given by following equations.

$$A_e = A_o, \quad v = w_o/\cos\psi_o \quad (z = 0) \tag{19a-b}$$

where ψ_o is the initial value of ψ. Assuming that $d\psi/dz|_{z=0} = 0$, we obtain the following equation.

$$\psi_o = \sin^{-1}\left[\frac{\sqrt{N_h^2 + 8/W_h} - N_h}{2}\right] \tag{20}$$

where, $N_h = \sigma B_o^2 h_n/\rho w_o$ is the Stuart number based on the nozzle height h_n, and $W_h = \rho w_o^2 h_n/\gamma$ the Weber number. After all, the streamwise distribution of the jet width can be obtained by solving Eqs. (14)-(16) under the initial conditions given by Eqs.(19) and (20).

3. RESULTS AND DISCUSSION

The shapes of the mercury jet measured by a spot electrode probe are shown in Fig. 3. The solid lines in the figures show the theoretical results. The origin of the z_n-axis is located at the nozzle exit. In the case without a magnetic field, the cross-sectional shape of the jet issuing from a noncircular nozzle oscillates along the streamwise direction. A jet, which is longer than it is wide, is changed to be long to side to side (see Fig. 3a). However, in the case with a uniform vertical

278

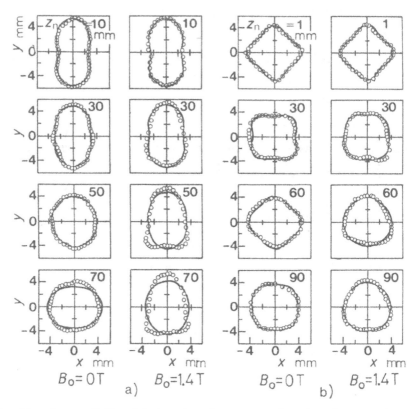

Figure 3. The shape of the jet issuing from a nearly-circular nozzle:
a) rectangular nozzle ($b_n \times h_n$ = 5mm $\times$ 10mm); b) square nozzle ($b_n = h_n$ = 7mm). The exit mean velocity w_o = 2.0m/s.

magnetic field, this oscillation in shape is highly suppressed. In the case that the jet issues from a square nozzle, the shape disturbances of the jet mainly consist of the mode of cos4 . The solid lines in the figure show the theoretical results. In the case without a magnetic field, the jet changes in shape just as if it were made a quarter revolution around its barycenter, and then it is restored to the original shape. The corners of the jet become obscure at the downstream. In the case with a magnetic field, the shape oscillation is suppressed by the electromagnetic force, and the shape gradually approaches a circle.

Figure 4 shows the change in half width $(b_{i_o} - b_i)/2$ of the jet issuing from a nozzle with a large aspect ratio, where b_{i_o} is the initial width of the jet. The width change is suppressed by the electromagnetic force; however, it is not affected by the initial width. The width change is attributed to the movements of the both sides caused by the surface tension, therefore, is not affected by the initial width until the both sides meet. The analytical results shown by solids lines

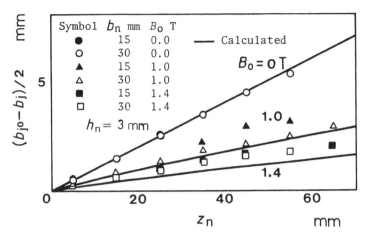

Figure 4. Change in half width of the jet issuing from a nozzle with a large aspect ratio. The exit mean velocity w_o = 1.25m/s.

well agree with experimental results.

Provided that the initial angle $\psi_0 = \pi/2$, we obtain the following equation from Eq. (20).

$$W_h \text{crit} = 2/(1+N_h) \tag{21}$$

When the Weber number of the film jet W_h is smaller than $2/(1+N_h)$, the surface tension becomes greater than the sum of the kinematic pressure and the electromagnetic force, so that the jet becomes unstable and breaks down. Equation (21) shows that applying a uniform transverse magnetic field will make a film jet stable.

REFERENCES

[1] SHERCLIFF, J. A., 1981, 'Magnetic Shaping of Molten Columns' Proc. R. Soc. London, Vol. 375, Ser. A, 455.

[2] ETAY, J. and GARNIER, M., 1982, 'Some Applications of High Frequency Magnetic Field', Proc. Symp. IUTAM, 190.

[3] OSHIMA, S., YAMANE, R., MOCHIMARU, Y., & MATSUOKA T., 1987, 'The Shape of a Liquid Metal Jet under a Nonuniform Magnetic Field', JSME Int. Journal, Vol. 30, No. 261, 437.

[4] RAYLEIGH, L., 1879, 'On the Capillary Phenomena of Jets', Proc. R. Soc., Vol. 29, No. 196, 71.

[5] KOZUKA, T, ASAI, S, & MUCHI, I, 1987, 'Effect of Applying Electromagnetic Force on Falling Behaviour of Molten Film', J. of the Iron and Steel Inst. of Japan, Vol. 73, No. 7, 828.

ELECTROMAGNETIC CONTROL OF LIQUID METAL FREE SURFACES

M.Garnier , J.Etay , A.Gagnoud , F.Garnier.
MADYLAM
BP95
38402 ST Martin d'Hères
France

ABSTRACT. High frequency magnetic fields induced in liquid metal within a small skin depth result in pressure forces which may be equivalent to forces exerted by walls. Electromagnetic control of liquid metal free surface in the scope of suppressing the walls is facing with specific problems : stability problems of free surfaces in the presence of A.C. fields with non uniform distributions in the skin depth ; free boundary problems for the prediction of equilibrium shape of a liquid metal volume submitted to the effect of a given inductor or, conversely, the determination of the inductor able to impose a given shape to a liquid metal volume. Both analytical and numerical methods to solve these problems will be presented in connection with real situations such as levitation melting in conical inductors , and in cold crucible , or electromagnetic shaping in the application of mouldless strip casting.

1.Introduction

When an alternating magnetic field **B** faces an electrically conducting material, induced currents **J** appear in the material surface bed. This is the *electromagnetic skin effect*. When the frequency of the magnetic field $\omega/2\pi$ or the electrical conductivity of the liquid metal σ is large enough then , the magnetic field lines become quite tangential to the surface and the magnetic field intensity $|B|$ decreases exponentially in the electromagnetic skin depth δ_m. Indeed the Lorenz force **F** due to the combination of the inducting magnetic field and the induced current is normal to the surface and can be written as

$$F = \frac{|B_s^2|}{2\mu\,\delta_m} \exp(2n/\delta_m)\ n$$

It is directed normal to the surface. Then, when the surface is free, the jump of normal stress is modified. The behaviour of the surface is the result of a competition between magnetic pressure, surface tension and inertia. This property of the magnetic field can be used for shape, guide and stabilize liquid metal domain without having resort to any wall. This phenomena are well illustrated in the electromagnetic levitation melting process [1].

2. Numerical modelling of electromagnetic levitation

Classical levitation melting device consists in a conical helical inductor with several turns , and one or two counterspires. When the load of metal is molten the local balance between magnetic pressure, surface tension, gravity and hydrodynamical pressure governs its shape. This is a typical example of coupled

J. Lielpeteris and R. Moreau (eds.), Liquid Metal Magnetohydrodynamics, 281–286.
© 1989 by Kluwer Academic Publishers.

phenomena : a free boundary problem is to be solved to determine the equilibrium shape and the stability conditions of a given molten metal volume with respect to electrical and geometrical parameters of the system. A non linear coupling arises through boundary conditions which have to be expressed along the unknown free surface. In the following calculations, the magnetostatic hypothesis is used , so the hydrodynamic inside the melt is ignored.

Two methods are tested [2] . They make possible to calculated the equilibrium shape of levitated liquid with respect to both electrical and geometrical parameters of the system.

*The *local method* defines the equilibrium shape of the free surface as resulting from the local balance between pressure , irrotational electromagnetic forces , surface tension and gravity. This is written as

$$\frac{2\,\gamma}{\rho\,g\,L}K^* + z^* + \frac{B_o^2}{2\mu\rho\,g\,L}B_s^{*2} = Cst$$

where γ is the suface tension, ρ the density , L the radius of the corresponding spherical load, B_o a typical intensity of the magnetic field , K^* the non dimensional curvature , z^* the vertical coordinate referred to L . A supposed velocity is introduced such as

$$u = \nabla\phi$$

Though the boundary condition ϕ is known on the boundary $\delta\Omega$ and because the continuity equation ϕ is harmonic in the domain Ω . So its normal derivativ can be found on the surface using , for example , boundary elements techniques. Using the found normal velocity , a new boundary is found. And so on until the displacement velocity vanishes.

*In the *global method* , the equillibrium shape of the free surface corresponds to the minimum of the total energy E of the system.

$$E = E_\gamma + E_g - E_M$$

with

$$E_\gamma = \int_{\delta\Omega} \gamma\, ds \qquad E_g = \int_{\Omega} \rho\, g z\, dV \qquad E_M = \int_{R^3} \frac{B^2}{2\mu}\, dV$$

Iterativ process starting from initial volume Ω_0 , towards equilibrium shape is defined by

$$\delta\Omega_{k+1} = \delta\Omega_k + \tau_{k+1}$$

with τ the displacement such as

$$\tau_{k+1} = -\left(\frac{B^2}{2\mu} + \rho g z + \gamma K + \lambda \right)_k n_k$$

In both modellings each iteration needs to compute magnetic field and local curvature of $\delta\Omega$.

Figures 1 and 2 give typical examples of computes free surfaces with the help of both global and local methods which give similar results. These latter are in good agreement with classical observations. The most $B_0^2L/\mu\gamma$ is increasing the most different from a sphere is the shape. When $\gamma/\mu\rho gL$ decreases liquid metal flows down. A fundamental difficulty which prevents levitated volumes to be very large clearly appears. Near the axis of symmetry the magnetic field is weak, moreover the thiner the skin depth the stronger this trend. On the axis only surface tension can balance gravity. But a spherical load would never drop. Computational methods make possible to define magnetic field distribution suitable for imposing a nearly spherical shape. A cold crucible inductor is able to create such field. Moreover it allows strong strirring motions of the melt which homogenize temperature and composition. This phenomena have to be introduced in the present software to increase efficiency of the levitation process.

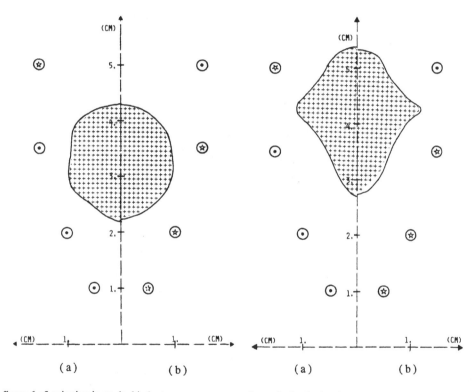

(a)	(b)

figure 1 : Levitation in conical inductor
Generatrix of the equilibrium shape obtained for
I = 1070A ; γ = 10N.m^{-1}
(a) Local method δ/a=0
(b) Global method δ/a=0.07

figure 2 : Levitation in conical inductor
Generatrix of the equilibrium shape obtained for
I = 2100A ; γ = 0.1N.m^{-1}
(a) Local method δ/a=0
(b) Global method δ/a=0.07

3.experiments on liquid metal streams

An inductor able to produce very thin sheet of molten metal will be presented. This study is done within the framework of direct strip casting research.

In such process a sheet of liquid metal is solidified directly on moving cooled substratum. The metal have to be supply shaped as near as possible from the final product which will be manufactured. An approach consists in using a free liquid sheet falling under gravity from a slit made from refractory material. Two main problems arise.

*the slit-shape nozzle is liable to plugging or erosion.

*hydrodynamic instabilities devellop like flapping or surface tension effects.

The use of alternating magnetic field provides an interesting solution to these problems. Using electromagnetic shaping, the geometry of the stream impinging on the cooling surface can be different from the nozzle one. Moreover some works demonstrated both theoretically and experimentally that alternating magnetic fields can stabilize liquid free surfaces of liquid metal. In the system we propose the liquid metal flows from the tundish through several circular nozzles of acceptable diameters to prevent plugging and erosion. Magnetic fields are then used to shape each of the circular jet into a liquid metal sheet and force them to coallesce into a single larger sheet which is then guided where it must be solidified.

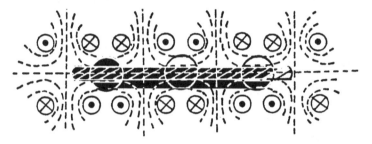

figure 3
Sketch of the inductor used to shape 3 initially circular jets.

Using the inductor sketched on figure 3 each initially circular jet flows in a parallel magnetic field where it stretches itself along magnetic field lines, and coalesce with its neighbours. Because this inductor is built with units it could be exended. Morover its symmetry plane insures a strong guiding of the liquid sheet. The distancy between unit-axis is equal to the distancy between jet-axis. For example to shape three jets a eigth units inductor has to be used. The order of magnitude of the magnetic field required to shape a sheet has been calculated for a 1 mm depth and for mercury steel and aluminium , and lies in the table 1 [3] . The intensity of the magnetic field needed to control steel is 2.5 time greater than the one used on mercury.

A side view of experiments performed with such inductor on mercury and on steel are display on photos 1 and 2. The used nozzle owns 3 holes of 8 mm diameter and 18 mm between axis. The jet velocity is about 2 m.s^{-1} . The inductor owns 8 units. A sheet of mercury of 70 x 2.2 mm^2 cross-section is created needing 25kW.

The main difficulty to increase the electromagnetic shaped ribbon width arrises with the choice of the frequency. This one have to be as high as possible to insure efficient electromagnetic effects. But, because between 20 to 50% of the supply power is used up in the inductor it is also indispensable to reduce its resistivity, so the frequency. This shows the neccesity to modelling the inductor taking account of the alternating nature of the electric current. This presents two main difficulties:

* this kind of inductor is fully three-dimensional and ,

* the scale created by the thin electromagnetic skin depth inside the material of the inductor is much smaller than its total characteristic length which increase the number of mesh-points.

photo 1
coalescence of three mercury jets in a single ribbon

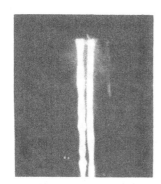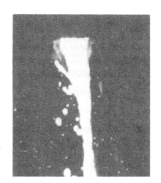

photo 2
coalescence of two steel jets in a single ribbon

	mercury	steel	aluminium
ρ density ($kg.m^{-3}$)	$1.3 \; 10^4$	$6.95 \; 10^3$	$2.32 \; 10^3$
σ electrical conductivity ($\Omega.m$)$^{-1}$	$1.09 \; 10^6$	$8.33 \; 10^5$	$4.12 \; 10^6$
electromagnetic skin depth corresponding to a frequency of 350 kHz (mm)	0.81	0.93	0.3
surface tension ($N.m^{-1}$)	0.485	1.81	0.840
order of magnetic field to supply to shape a sheet of 1 mm depth (Gauss)	914	2360	259

Table 1 : order of magnitude of the magnetic filed required
to shape a sheet of liquid metal

4.Conclusion

Now, our works are devoted to resolve the following problems:
        * It is urgent to introduce, in the numerical modelling, the as yet ignored coupling between the shape of the free part of the boundary of the liquid metal domain electro-magnetically controlled. This must be done to know the effects of the stirring motion in the homogenization of the temperature and the composition.
        * Moreover, an effort must be made to optimize the inductor we built in order to promote their transfer towards an industrial scale, where the width of the shaped ribbon must be more than 1 meter large. This prescribe the three-dimensionnal modelling of the electrical part of the inductor.

references

[1]E.C.Okress, D.M.Wroughton, G.Comenetz, P.H.Brace et J.C.R.Kelly may 1952 "Electromagnetic levitation of solid and molten metals" J.of Applied Physics **vol23** number3 pp545-552

[2] A.Gagnoud, J.Etay et M.Garnier 1986   "Le problème de frontière libre en lévitation électromagnétique" Journal de Mécanique théorique et appliquée **vol5** n°6 pp911-934

[3] J.Etay 7 Janvier 1988 "Le problème de frontière libre en magnétodynamique des liquides avec champ magnétiques alternatifs : aspects fondamentaux et applications" Thèse d'état- INPGrenoble-France

DEFORMATION OF AN ELECTRICALLY CONDUCTING DROP IN A MAGNETIC FIELD

G. GERBETH*, M. KAUDZE**, A. GAILITIS**
* Central Institute for Nuclear Research, Rossendorf
  P.O. Box 19, Dresden 8051 GDR
** Institute of Physics, Latvian SST Academy of Sciences, 229021 Riga
   Salaspils, USSR

ABSTRACT. The shape of a falling drop in an inhomogeneous magnetic field is investigated both theoretically and experimentally. This problem is of actual interest for a lithium droplet divertor in magnetically confined fusion reactors. The first part of the theoretical analysis is based on a one-mode approximation of the drop oscillations assuming an ellipsoidal shape of the drop. In the second part higher modes are included and the possibility of drop division is discussed. The numerical results are in qualitative agreement with first experiments.

Introduction

The shape is investigated of an electrically conducting drop free falling in an inhomogeneous magnetic field $B_0 = B_0(x) e_z$. This problem is of actual interest for a proposed lithium droplet divertor or limiter in magnetically confined fusion reactors /1/. A falling drop experiences the following effects due to entering and leaving the magnetic field :
  (i)   change in velocity
  (ii) transverse displacement,
  (iii) drop distorsion.
Estimations of these effects are given in /2/ assuming the magnetic field entry length to be much larger than the radius R_0 of the spherical drop. For usual parameters the change in velocity is negligible compared to the constant drop velocity u while the transverse displacement can reach the value of the drop radius. The drop distorsion is characterized by the dimensionless quantity

$$c = \frac{R_o^3}{8T} \sigma u B_o \left| \frac{dB_o}{dx} \right|$$ (1)

where T, σ denote surface tension and electrical conductivity, respectively, and $|dB_0/dx|$ is a typical value of the field gradient /2/. For $c \ll 1$ drop distorsion is negligible whereas $c \sim 1$ indicates an appreciable deviation from the spherical shape of the drop. This drop distorsion can more easily be analysed if the action of the inhomogeneous magnetic field on the falling drop is replaced by the action of a time-dependent field $B_0 = B_0(t) e_z$ on a drop at rest.

Drop distorsion was estimated by comparison of various energy contributions /3/ and by a numerical procedure assuming the drop to be two-dimensional /4/. In the present paper the three-dimensional shape of the drop is determined both theorically and experimentally.

J. Lielpeteris and R. Moreau (eds.), Liquid Metal Magnetohydrodynamics, 287–292.

2. Basic equations

The induced velocity **v** within the drop is determined by the MHD equations

$$\rho d\mathbf{v}/dt = -\partial p/\partial r + \sigma(\mathbf{E} + \mathbf{v} \times \mathbf{B_0}) \times \mathbf{B_0}, \tag{2}$$

$$\text{rot } \mathbf{E} = -d\,\mathbf{B_0}/dt \tag{3}$$

and the incompressibility condition

$$\text{div } \mathbf{v} = 0 \tag{4}$$

Here ρ, p, $\mathbf{E}$ denote density, pressure and electric field, respectively. The viscous contribution in (2) is neglected because it is in the order of $(\eta^2/\rho \ TR_0)^{1/2} \ll 1$ compared to the pressure gradient (η-dynamic viscosity). A usual polar coordinate system (r, φ, z) is used in the following. Symmetry relative to the magnetic field direction $\mathbf{e_z}$ requires $v_\varphi = 0$ and

$$\mathbf{E} = (\mathbf{r} \times d\,\mathbf{B_0}\,/\,dt)/2.$$

For further calculations all quantities are made non-dimensional using R_0 as characteristic length and the plateau value B_0 as characteristic field. The time is measured in units of $1/\omega_0 = (\rho \ R_0^3/T)^{1/2}$ where ω_0 is the characteristic frequency of drop oscillations /5/. The velocity is given in units of $R_0\omega_0$ and the pressure in units of $\rho \ (R_0\omega_0)^2$. In this way equation (2) is rewritten in the form

$$d\mathbf{v}/dt = -\partial p/\partial r - Nb(t) \ [0.5 \ rdb(t)/dt + bv_r] \ \mathbf{e_r} \tag{5}$$

where the interaction parameter $N = \sigma B_0^2/\rho\omega_0$ and the notation $B_0(t) = B_0 b(t)$ are introduced. The boundary conditions are that of a constant drop volume and the Laplace condition $p_S = K$ relating the pressure at the surface to the curvature. In the following the unknown shape of the drop is denoted by $r = g(z, t)$.

3. Solution

In a first step the drop is approximated by an ellipsoid $g^2 = (1 - z^2/a^2) \ / \ a$ where $a(t)$ is the large half axis. The velocity is taken in a form corresponding to the basic axisymetric oscillation of a free drop /5/

$$v_z = e(t)z, \ v_r = -e \ r/2 \tag{6}$$

This expression is compatible with the ellipsoidal shape of the drop if $e(t) = sa(t) \ dt/a$. The pressure is then obtained from (5)

$$p = p_0 \ (t) + (e_1 \ r^2 + e_2 \ z^2) \ / \ 4 \tag{7}$$

where

$$e_1 = de/dt - e^2/2 + Nb \ (be - db/dt), \quad e_2 = -2 \ (e^2 + de/dt).$$

The Laplace condition is taken into account in an equivalent variational form

$$\delta \int pd^3r = \delta \int dS$$

which reduces in the present one-mode analysis to

$$\frac{d}{da} \int pd^3 r = dS/da$$

where S denotes the surface of the drop. In this way the time-derivative of e(t) is determined by

$$e_2 a - e_1 a^{-2} = \frac{15a^2}{2\pi} dS/da.$$

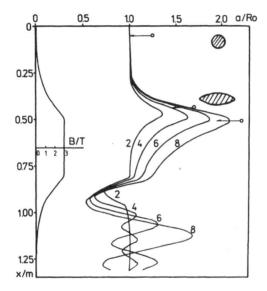

Figure 1. Large half axis of an ellipsoidal drop fallling with various velocities u = 2, 4, 6, 8 m/s in an inhomogeneous magnetic field.

Consideration of higher oscillation modes without restriction to an ellipsoidal shape of the drop is possible taking

$$v_z = A_{mn} z^{2m-1} r^{2n-2}, \tag{6a}$$
$$v_r = B_{mn} z^{2m-2} r^{2n-1}, \tag{6b}$$
$$p = C_{m'n'} z^{2m'-2} r^{2n'-2} \tag{7}$$
$$g^2 = a_p z^{2p-2} \tag{8}$$

where m, n = 1, 2, ... NF, (m+n $\leq$ NF + 1), m', n' = 1, 2, ... NF + 1, (m' + n' $\leq$ NF + 2), p = 1, 2, ..., NG, and the summation convention are used. The coefficients A, B, and C are related to each other by (4) and (5). Compatibility between (8) and (6a, b) requires

$$\partial g/\partial t + v_z \partial g/\partial z = v_r \tag{9}$$

on the surface of the drop. The Laplace condition and in a complementary way the boundary condition (9) are realized by

$$\sum_{i=1}^{M} (p_s(zi) - K(zi))^2 \rightarrow \min. \tag{10}$$

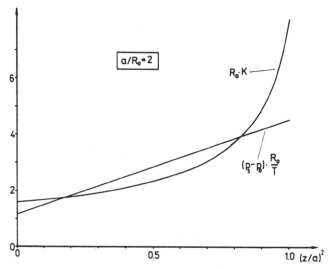

Figure 2. Comparison between surface pressure and curvature for an ellipsoid with $a/R_0 = 2$.

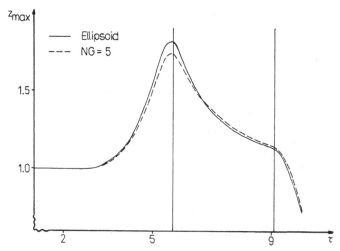

Figure 3. Drop dimension in magnetic field direction. The two vertical lines indicate beginning and end of the constant magnetic field region.

4. Results

The described equations are solved numerically using a usual Runge-Kutta method. Initially a spherical drop is presumed. The parameters of the drop are taken corresponding to the experiment where the eutectic alloy In-Ga-Sn was used : $\rho = 6545$ kgm^{-3}, T = 0.535 N/m, $\sigma = 3.33 * 10^6$ A/Vm, Ro = 2.6mm.The magnetic field (see Fig. 1) has a plateau value Bo = 3T (which leads to N = 67.1 for the interaction parameter) with a length of 0.3 m. Entrance and exit region are modelled by a Lorentz type curve. These parameters lead to an approximate value of c = 0.5 u (u in m/s) for the non-dispersional parameter (1) indicating a significant deformation of the spherical drop.

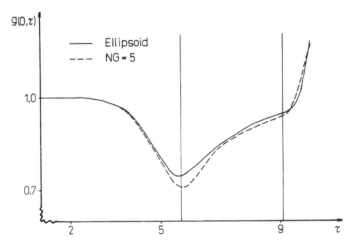

Figure 4. Drop thickness in the middle of the drop (z = 0). The two vertical lines indicate beginning and end of the constant magnetic field region.

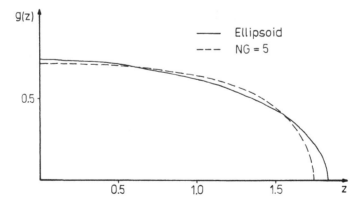

Figure 5. Shape of the drop at magnetic field entrance.

The shape of an initially spherical drop falling with various velocities (u = 2, 4, 6, 8 m/s) has been analysed in the ellipsoidal approximation. The large half axis is shown in Fig. 1. A significant deformation of the drop is found in the entrance and exit region of the magnetic field. The drop is elongated in the magnetic field direction e_z in the entrance region. This entrance distorsion is damped in the homogeneous part of the field whereas the exit deformation continues to exist in the form of undamped oscillations.

The maximum dimension of the drop was measured in a first experiment by means of a high speed camera. The results are shown in Fig. 1 (u = 5.8 m/s). In the experiment the air resistance of the drop must be taken into account which leads to an initial deformation of the drop before entering the magnetic field. This initial deformation is estimated as Δ (a/Ro) = 0.26 /6/ and is shown by arrows in Fig. 1. Notwithstanding the uncertainties connected with this first experiment the predicted drop shape agrees with the measuring values qualitatively.

Fig. 2 shows the correspondance between pressure and curvature for a drop distorsion of a/Ro = 2. These curves emphasize the error connected with the ellipsoidal approximation. In order to overcome this uncertainty higher modes have been included in the numerical procedure as described above. The values NG = 5, NF = 3 proved to be the favourable parameters since higher values lead to numerical problems. The results for the maximum drop dimension, the drop thickness in the middle of the drop and the shape of the drop at magnetic field entrance are presented in Fig. 3, 4, 5, respectively. The difference between the ellipsoidal and the correct shape of the drop in Fig. 5 corresponds to the discrepancy between pressure and curvature of an ellipsoid as shown in Fig. 2.

5. Conclusions

A falling drop in an homogeneous magnetic field experiences a strong deviation from its spherical shape for $c \geq 1$. This deformation is well described assuming an ellipsoidal shape of the drop. The precise form of the drop deviates from an ellipsoid in a way shown in Fig. 5. There is no tendency of drop division up to $c \sim 50$.

References

/1/ DEMJANENKO W.N. et al. 1988, Magnetohydrodynamics, n° 1, 104.
/2/ WALKER J., WELLS W.M., 1979, ORNL/TM-6976.
/3/ MURAWJOW E.W., 1980, Magnetohydrodynamics, n° 3, 43.
/4/ ELKIN A.B., 1984, Magnetohydrodynamics, n° 2, 73.
/5/ GAILITIS A., 1966, Magnetohydrodynamics, n° 2, 79.
/6/ CLIFT R. et al., 1978, Bubbles, Drops, and Particles, Academic Press, New York, 181.

CONTROLLED DECOMPOSITION OF LIQUID METAL JETS AND FILMS IN TECHNOLOGICAL AND POWER DEVICES

V. N. Dem'janenko,
Res. Works "Energy", Moskow
E.V. Murav'ev
Inst. Atomic Energy, Moskow
Yu. M. Gorislavets, I.V. Kazachkov, A.F. Kolesnichenko, N.V. Lysak, V.O. Vodjanjuk
Inst. Electrodynamiks, Ukr. SSR Acad. Sci., Kiev,
252680, Kiev-57, pr. Pobedy, 56
USSR

ABSTRACT. Controlled decomposition of liquid-metal jets and films into drops of equal size is caused by periodic perturbations-electromagnetic forces due to the interaction of current in a jet outflow or a film with internal or external magnetic fields. Since problems of the transient flow of the conducting liquid with free boundaries have been solved. The conditions of the decomposition of jets and films into drops of equal size are formulated. The types of jet and film liquid metal atomizers producing the monodispersed granulometric composition of particles of dispersed metal are described. Examples of the application of liquid metal dispergators to metallurgy and to free surface contact devices planned for future power plants are given.

The problem of the control of the free melt boundary behaviour and in particular of the control of jets and film decomposition [1] is important at this time because of the development of new technological and power devices with a molten-metal working medium [2] and because of the modernization of the existing ones. This paper deals with some problems of the control of jets and films decomposition by alternating electromagnetic fields in granulators- the devices intended for producing the monodisperse metal powders- and in molten-metal drop curtains of tokamaks. The research methods used are mathematical modelling performed with the aid of computers, and checked against experimental result.

The modern technology of physical modelling does not allow us to study the evolution of free boundaries of jets and films effectively enough and to determine the dynamic and kinematic parameters of continuous flow (disperse flow - after decomposition). Therefore mathematical models are being developed, the most complex of which are realized on a high speed computer. Thus, the computation of the jet decomposition which involves considerable displaying of inertia forces, was performed [3] by means of the numerical solution of nonstationary Navier-Stokes equations of the following nondimensional type :

$$\text{Sh} \frac{\partial}{\partial t} \int_V u \, dV - \int_S U u n \, dS = -\text{Eu} \int_s p n \, dS + \frac{1}{\text{Re}} \int_s \pi n \, dS + \frac{1}{\text{Fr}} \int_V A \, dV,$$

$$\frac{\partial}{\partial t} \int_V dV = \int_S U n \, dS, \tag{1}$$

where $\text{Sh} = r_0/u_0 T$, $\text{Eu} = p_0/\rho u_0^2$, $\text{Re} = u_0 r_0/\nu$, $\text{Fr} = u_0^2/a r_0$, - the numbers of Strouhal, Euler, Reynolds, Froude, V - elementary volume, S - limiting its surface, U - the surface moment

293

J. Lielpeteris and R. Moreau (eds.), Liquid Metal Magnetohydrodynamics, 293–298.
© 1989 by Kluwer Academic Publishers.

294

velocity relative to liquid. p, π, A - the voltage tensor components (pressure, viscous friction, surface tension, electromagnetic forces, etc).

The system of equations (1) with corresponding boundary conditions has been solved numerically. This allows determination of the form of the free surface, as it develops with time, for different physical situations. These results are presented in Fig. 1, which shows where the free boundaries of formed drops are marked, the continuous lines in them indicate isotachs, dotted lines - isobars. The formation of constrictions is characterized by the emergence of the local features of pressure and velocity.

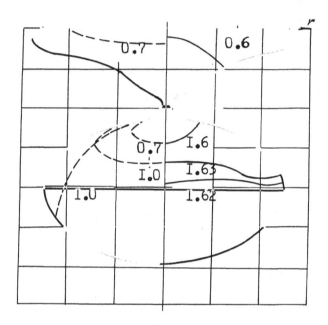

Fig. 1 : Illustration of the result of numerical solution of the Navier-Stokes equations for some variant of free boundary jet shaping. The continuous lines indicate isotachs, dotted lines - isobars.

The mathematical model of liquid-metal film spreading out radially on a horizontal surface, which is surrounded by a non conducting medium is obtained [1] in accordance with the schema shown on Fig. 2. The paper discusses the problems of wave excitation and propagation in the film, which are affected by a progressive electromagnetic wave of the following type :

$$h = h_m \ (z, r) \ exp \ i \ (kr + m\theta - wt) \tag{2}$$

where k, m - the wave numbers on coordinates r, θ, w - frequency, h - vertical component of magnetic field strength $h = \{\theta, \theta, h\}$. The induction system producing this field is considered in [4].

The linearized system of magnetohydrodynamic equations with corresponding boundray conditions , averaged over the film thickness, results in [1, 5] :

$$\frac{\partial h}{\partial t} + \frac{1}{r}\frac{\partial h}{\partial r} = \frac{1}{Re_m}\left(\frac{\partial^2 h}{\partial r^2} + \frac{1}{r^2}\frac{\partial^2 h}{\partial \theta^2} + \frac{1}{r}\frac{\partial h}{\partial r}\right),$$

$$\frac{\partial^3 \xi}{\partial t \partial r^2} + \frac{1}{r}\frac{\partial^3 \xi}{\partial r^3} + \frac{1}{r^2}\frac{\partial^3 \xi}{\partial t \partial \theta^2} + A_1\frac{\partial^2 \xi}{\partial t^2} + \frac{(1+2A_1)}{r}\frac{\partial^2 \xi}{\partial t \partial r} + \left(\frac{A_2}{We} + \frac{1+A_1}{r^2}\right)\frac{\partial^2 \xi}{\partial r^2} \qquad (3)$$

$$+ \frac{A_2}{We^2} \cdot \frac{\partial^2 \xi}{\partial \theta^2} + A_3\left(\frac{\partial \xi}{\partial t} + \frac{1}{r}\frac{\partial \xi}{\partial r}\right) + \frac{(1-\rho_{21})}{Fr^2}A_2\xi = -0.5A_2 h^2,$$

where A_i - Reynolds number function, Re_m - Reynolds magnetic number, h -the film surface disturbance. The system (3) allowing for (2) was integrated numerically on the computer for some variants of the boundary conditions problems. The examples of such calculations are given in [1,5].

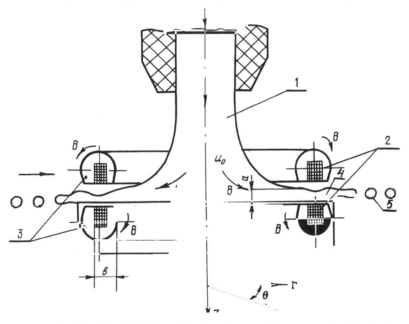

Fig. 2 : 1 - Vertical jet, 2 - coils with different directional currents, 3 - magnets, 4 - film, 5 - drops.

The research revealed some peculiarities of the electromagnetic excitation of jets and films decomposition, to determine the optimal operations of MHD granulators and to develop a number of such devices intended to obtain granular half-finished product of diverse metals.

The other example [2] of the application of the controlled decomposition of molten-metal jets is the tokamak drop diaphragm in which the molten-metal jets decomposition is performed by the introduction of periodic electromagnetic forces with the frequency of $f = 2\sqrt{2}\pi r_0/u_0$. The afore mentioned forces are formed by the interaction of current flowing through the melt and the magnetic azimuthal (or poloidal) field of tokamak. The behaviour of current changing in time, the form and dimension of molten metal drops are interconnected. In order to obtain particles of predetermined dimensions and forms, the current has a preset harmonic composition. The molten metal drop curtain functions as a diaphragm and a thermal recepted "restored" molten-metal wall. The significant MHD resistances, usually encountered with the pumping of molten metal in the blanket, are practically eliminated from this system, because delivery branches forming the jets and supplying the melt to the nozzles completely looped inside the magnetic field. The drops moving in a sufficiently strong magnetic field of tokamak does not encounter as strong a resistance as does a continuous flow.

Let us assume the form of the jet-drop curtain to be two films in the discharge chamber. Let us consider the decomposition point should be connected with the amplitude of disturbing electromagnetic forces at the inputs into the nozzles.

Assuming that the radial component of velocity u_r in the jet source is smaller than the axial one u_z, jet swirl is absent ($u_\varphi = \theta$), the jet radius is much smaller than the length of its nondecayed part, outflow is laminar, and Re << 1, the equation of Navier-Stokes are :

$$\frac{\partial u_z}{\partial t} + u_z \frac{\partial u_z}{\partial z} = -\frac{1}{r}\frac{\partial}{\partial z}(p_s + p^*) + 2\,\nu\,\frac{\partial^2 u_z}{\partial z^2},$$

$$\frac{\partial(ru_r)}{\partial r} + \frac{\partial(ru_z)}{\partial z} = 0 . \tag{4}$$

If $r = x(z, t)$ is the equation for the axially symmetric generatrix of the free boundary, then the boundary conditions are as follows

$$\frac{\partial x}{\partial t} + u_z \frac{\partial x}{\partial z} = u_r, \quad \frac{2u_r}{r} + \frac{\partial u_z}{\partial z} = 0 \tag{5}$$

on $r = \xi(z, t)$, i.e. on the free boundary, and $u_r = 0$ with $r = 0$. The internal jet pressure is

$$P_{zz} = P_\sigma + 3\,\nu r\,\frac{\partial u_z}{\partial z} + p^*, \quad r = \xi(z,t),$$

$$P_{zz} = -p + 2\nu r\,\frac{\partial u_z}{\partial z},$$

$$p \approx -\frac{1}{3}(P_{zz} + P_\pi + P_{\varphi\varphi}) = -\frac{1}{3}(p_{zz} + 2p_\sigma). \tag{6}$$

The pressure $p_\sigma = (R_1^{-1} + R_2^{-1})$ is caused by the surface tension, p^* is the ambient pressure (outside jet), p_{zz}, p_{rr}, $p_{\varphi\varphi}$ are the components of the voltages tensor.

Let's assume the friction condition on the free boundary to be absent $r = \xi(z,t)$, $\partial U_z/\partial r = 0$ and $\partial p/\partial r = 0$. Then, we obtain the equation

$$\frac{\partial^2 \xi}{\partial t^2} + 2u_0 \frac{\partial \xi}{\partial z} + u_0^2 \frac{\partial^2 \xi}{\partial z^2} = \frac{r_0}{2\rho}\frac{\partial^2}{\partial z^2}(p_\sigma + p^* + p) + 2\nu\left(\frac{\partial^2 \xi}{\partial z^2} + u_0 \frac{\partial^3 \xi}{\partial z^3}\right). \tag{7}$$

The periodic solutions (7) are found as $\xi = r_0 + \alpha\,\text{expi}(wt - kz)$, where $\alpha = \alpha(t)$ is the relative radial deviation of the free boundary from its equilibrium position, corresponding to a cylindrical surface.

The equation (7) has the periodic solution, as well as a diverging one, the right part of the equation (7) being absent. The diverging solution is as follows

$$\alpha = \exp(-ht)(\alpha_0\,\text{ch}\sqrt{h^2 + x^2}\,t + \frac{\alpha_0 h + U_0}{\sqrt{h^2 + x^2}}\,\text{sh}\sqrt{h^2 + x^2}\,t), \qquad (8)$$

if :

$$\Delta^2 < 1 + 2(m-1)^2\,We^{-1}, \qquad (9)$$

where $We = \sigma/\rho r_0\,u_0^2$ - Weber number, $m = w/u_0 k$, w/u_0 is a velocity of disturbance wave progation along the free jet.

Condition (8) determines the surface instability for the liquid circular cylinder. The maximum of amplitude takes place if the wave number is $\Delta_0^2 \approx 1/2$, depending on the velocity of propagation of disturbance u_0. If $m=1$ these disturbances in a jet propagate with the outflow velocity u_0, if $m = 0$, then wave velocity is equal to zero.

Both the length of the undisintegrated part of the jet as a function of a given disturbance force, and the given disturbance force sufficient for jet decomposition into drops at a given distance from the source, can be determined from the condition (9).

In accordance with (1) and (3), the density of the disturbed force in the jet source, and the resulting deformation velocity of the generatrix, are connected by the relation

$$Eu_m = \frac{jBa}{\rho u_0^2} = -2\bar{u}_0 .$$

If $z = \pi(0,5 + 2n)/\Delta$, where n are the wave numbers giving the distance of the decomposition point from the jet source, then :

$$Eu_m \geq \frac{2\sqrt{h^2 + x^2}}{\text{sh}\sqrt{h^2 + x^2}\,t} .$$

The argument values being small

$$Eu_m \geq \frac{\sqrt{2}}{\pi(1 + 5n)}$$

the jet decomposes at the distance λ_n from the jet source. On assuming the demanded value n, in accordance with condition (10), we can determine the current in molten metal, resulting in sufficient disturbance for jet decomposition to occur.

The form of drop flow (see fig. 3) is completely achieved in the discharge chamber and enters the inductive canal. The canal consists of two coaxial pipes, inside the other. In its underside the external pipe has holes of 2 mm, which form the free metal jets decomposed into drops. The inductor solenoid is put into the internal pipe. The ferromagnetic metal core is absent. The inductor forms an alternating electromagnetic field at the frequency of $f = u_0/2\pi d_0$, where u_0 - a consumption outflow velocity, d_0 - diameter of the hole. The inductor field induces currents in molten metal in the space between the pipes. Interacting with the poloidal field in the discharge chamber, these currents form variable electromagnetic forces, which assist in the modulation of outflow velocity. The vectors of these forces and velocity coincide in the nozzle. The value of current density $j = \rho\, u_0^2\, Eu_m/Ba \approx 0,5.10^6$ A/m^2 is supported easily in the real installation. This fact allows us guarantee the jet decomposition already at the distance of one wave-length $z = \sqrt{2}\,\pi d_0$.

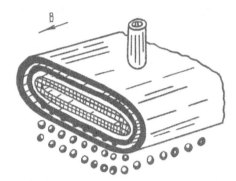

Fig. 3 : Induction device forming liquid-metal droplet flow, using toroidal field of tokamak.

The results of the estimations given in the paper, demonstrate that the curtain of free drops falling appears to be a device more economic than a sheet is.

In similar way - at the first wave-length of melt disturbed jets - the working process occurs in MHD granulators for technological purposes. The appearance of oxide films on the drop surface of real metals in metallurgy (such as aluminium, copper) results, however, in the removal of the jet decomposition point from its source. In this case the form of the drops is considerably non-spherical, and the process of decomposition appears to be unstable. This circumstance has been avoided either by putting the sources of decomposing jets in a neutral gaseous atmosphere or by passing the current of polyharmonic composition through the melt. In the latter case, the particle form correction has been carried out due to the sharp decay of outflow velocity at the end of the drop formation period. That is why the MHD melt metal granulation devices developed in Institute of Electrodynamics of the Acad. of Sci. of the Ukr. SSR are equipped either by specialized current sources or by the system of general or local protection from atmospheric effects.

References

[1] KOLESNICHENKO A.F., KAZACHKOV I.V., VODJANJUK V.O., LYSAK N.V., 1988, Capillary MHD flow with free bounds, Kiev, Nauk.dumka, 176.
[2] KOLESNICHENKO A.F., 1980, Technological MHD devices and processes, Kiev, Nauk.dumka, 192.
[3] KOLESNICHENKO A.F., LYSAK N.V., 1985, Numerical modelling of free bounds of conducting liquids, Kiev, 36. (Preprint - 426/Institute of Electrodynamics of the Academy of Sciences of the Ukr. SSR).
[4] KAZACHKOV I.V., KOLESNICHENKO A.F., 1984, 'Desintegration of melt metal films effecting of by electromagnetic field'., Tech. electrodinamika, vol. 3, 16.
[5] KAZACHKOV I.V., 1986, Electromagnetic excitation of parametric vibrations in melt metal films, spreading over nonconducting mediums, Kiev, 52. (Preprint - 454/Institute of Electrodynamics of the Academy of Sciences of the Ukr. SSR).

Poster Session 2:
AC Fields

MORE ACCURATE SKIN-DEPTH APPROXIMATIONS

A. J. MESTEL
D.A.M.T.P
Cambridge University
Silver Street
Cambridge CB3 9EW
United Kingdom

ABSTRACT. The penetration of a high frequency alternating or rotating magnetic field into a conductor is considered. The magnetic Reynolds number is assumed to be small. The standard skin-depth approximation is shown to be incorrect in the interior of the conductor, leading to large relative errors, especially as regards the curl of the Lorentz force. Two terms of the correct expansion are calculated. It is found that a "focussing" effect occurs, and that at some interior points the field is several times larger than expected. The behaviour near such singular points is found. The implications for electromagnetically driven fluid flow are considered.

1. Introduction

Time-dependent fields have some advantages over D.C. fields as a means of controlling the behaviour of liquid metals. For steady fields to exert influence they must interact with currents which are either externally supplied or induced by fluid motion. The application of D.C. devices is thus limited, especially as the magnetic Reynolds number, R_m, tends to be small in practice:

$$R_m = UL\mu\sigma \ll 1 \tag{1}$$

where U and L are typical velocity and length scales, while μ and σ are the permeability and conductivity of the metal. When alternating or rotating fields are used, however, eddy currents may be automatically induced in the metal, giving rise both to a large heating rate and to a rotational Lorentz force which drives fluid motion. Metals may thus simultaneously be melted and stirred, as occurs in induction furnaces.

For many applications, a high frequency driving field is desirable. As is well known, in these circumstances the magnetic field tends to be excluded from the interior of conductors, apart from a surface layer of typical thickness

$$\delta = \left(\frac{2}{\mu\omega\sigma}\right)^{\frac{1}{2}} \tag{2}$$

where ω is the rotation rate of a rotating field or the angular frequency for A.C. fields. The exclusion of the magnetic field gives rise to a magnetic pressure distribution over the metal boundary, which can be used to control the behaviour of free surfaces (e.g. [1]).

A simple analytical model of the behaviour described above is widely used, in which the magnetic field drops exponentially with normal distance from the surface. For some purposes this description

J. Lielpeteris and R. Moreau (eds.), Liquid Metal Magnetohydrodynamics, 301–307.
© *1989 by Kluwer Academic Publishers.*

is sufficient. It is accurate to order δ/L near the surface, where L is determined both by the curvature of S and the form of the imposed field. Frequently, however, one is interested in the field in the interior of the metal. During continuous casting of metals, for example, a solid layer forms around the liquid core. Electromagnetic stirring devices must rely on the field penetrating to the liquid core. It is thus important to calculate the field accurately in the interior of the metal. Usually for such applications, the important magnetic feature is the curl of the Lorentz force, $\nabla \wedge (\underline{j} \wedge \underline{B})$, which is even less accurately portrayed by the simple model. For most practical purposes the ratio $\delta/L \geqslant .2$, and the simple estimates are inadequate. In this paper we aim to provide an accurate, uniformly valid asymptotic representation of $\underline{B}$ as $\delta \to 0$.

2. Formulation of the problem

We consider a two-dimensional region V of metal bounded by S and surrounded by vacuum $\overline{V}$. The extension of the theory to three dimensions is not difficult. We represent the magnetic field $\underline{B}$ by a flux function $\psi(x, y)$ so that

$$\underline{B} = \Re e \left[\nabla \wedge (0, 0, \psi) \, e^{i\omega t} \right] \tag{3}$$

where $\Re e$ denotes the real part. The magnetic Reynolds number is assumed small as in (1) so that the magnetic induction equation reduces to

$$\nabla^2 \psi = \frac{2i}{\delta^2} \psi \qquad \text{in } V \tag{4}$$

$$\nabla^2 \psi = -\mu_0 j_E \qquad \text{in } \overline{V} \tag{5}$$

where j_E represents any externally imposed current sources and μ_0 is the vacuum permeability. For simplicity we shall assume $\mu = \mu_0$, but differing permeabilities can be incorporated into the theory without difficulty. The boundary conditions for ψ are then

$$\psi, \ \frac{\partial \psi}{\partial n} \quad \text{continuous on } S, \qquad \psi \to \psi_\infty \quad \text{as } |\underline{x}| \to \infty, \tag{6}$$

where ψ_∞ is externally imposed. Equations (4) and (5) with (6) define a unique solution for ψ for every δ. The field, induced current and Lorentz force then follow from ψ. In particular, fluid motion in V is driven by the time-averaged curl of the Lorentz force

$$\underline{G} = \overline{\nabla \wedge (\underline{j} \wedge \underline{B})} = \frac{1}{\delta^2 \mu} \Im m \left[\nabla \psi \wedge \nabla \psi^* \right] \tag{7}$$

where $\overline{}$, $*$ and $\Im m$ denote respectively a time-average, complex conjugate and imaginary part.

We are interested in the solution to (4), (5) and (6) in the limit as $\delta \to 0$. In the opposite limit, as $\delta \to \infty$, we can expand

$$\psi = \psi^{(0)} + \frac{2i}{\delta^2} \psi^{(1)} + \dots \tag{8}$$

and reduce the problem to a series of potential problems

$$\nabla^2 \psi^{(m)} = j^{(m)} \tag{9}$$

where $j^{(m)}$ is known in terms of $\psi^{(m-1)}$. For each m, (9) can be solved by standard techniques. Our aim is similarly to reduce the high frequency problem to potential theory, performing the expansion as $\delta \to 0$

$$\psi = \psi_0 + \frac{\delta}{1+i} \psi_1 + \left(\frac{\delta}{1+i} \right)^2 \psi_2 + \dots \qquad \text{as } \delta \to 0. \tag{10}$$

3. The naive skin-depth approximation

Local to the surface, we define inward normal and tangential orthogonal coordinates n and s as in figure 1. Inspection of (4) immediately reveals the need for variation on a small length scale within V as $\delta \to 0$. As the solution of (5) in $\overline{V}$ need have no such small scale structure, we expect normal derivatives to dominate:

$$\frac{\partial}{\partial n} \gg \frac{\partial}{\partial s} \, . \tag{11}$$

A necessary condition for (11) to hold is that $K\delta \ll 1$ everywhere on S, where $K(s)$ is the curvature of S. Clearly, this is asymptotically satisfied as $\delta \to 0$ unless S has a sharp corner where K is infinite. The magnetic field is very large near such corners. This important case is considered in [2].

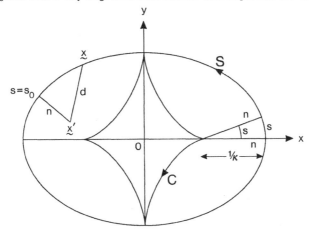

Figure 1: Illustration of n, s, d and C for a given boundary S.

The simplest skin-depth approximation is obtained by substituting into (4) the relation

$$\nabla^2 \approx \frac{\partial^2}{\partial n^2} \, . \tag{12}$$

Together with the assumption that $\psi \to 0$ as $n/\delta \to \infty$, this leads straightforwardly to the solution

$$\psi \approx f(s) e^{-(1+i)n/\delta} \quad \text{in } V \, . \tag{13}$$

The function $f(s)$ may be found from the external potential problem (5) along with the boundary conditions

$$\frac{\partial \psi}{\partial n} = -\frac{1+i}{\delta} \psi \quad \text{on } S, \qquad \psi \to \psi_\infty \quad \text{as } |\underline{x}| \to \infty \, . \tag{14}$$

To lowest order in δ, $\psi = 0$ on S, and $f(s) = \delta B_S/(1+i)$, where $B_S(s)$ is the surface magnetic field which would result were the field totally excluded from V, as for a perfect conductor.

The expression (13) is accurate to $O(\delta)$ near the surface ($n = 0$), but as n increases the percentage error becomes large. The naive approximation (13) has the wrong amplitude for the exponential decay in the interior of V. Even if one is only interested in the surface behaviour, before the next order term may be calculated it is necessary to find an improved approximation.

4. An improved approximation

A better approximation allows for the effects of the surface curvature, $K(s)$. From figure 1 it is clear that when the curvature is positive, the scale factor corresponding to s in an orthogonal coordinate system, h_s, is a decreasing function of n. In fact, when s represents arc-length on S,

$$h_n \approx 1, \qquad h_s \approx 1 - K(s)n . \tag{15}$$

Substituting these values in the expression for the Laplacian in curvilinear coordinates, and assuming that h_s remains $O(1)$ so that s-derivatives may be neglected, we obtain

$$\nabla^2 \psi \approx \frac{\partial^2 \psi}{\partial n^2} - \frac{K}{1 - Kn} \frac{\partial \psi}{\partial n} = \frac{2i}{\delta^2} \psi \tag{16}$$

giving the improved approximation as $\delta \to 0$

$$\psi = \frac{f(s)}{(1 - Kn)^{1/2}} e^{-(1+i)n/\delta} \qquad \text{in } V. \tag{17}$$

Expression (17) is uniformly valid to $O(\delta)$ provided $(1 - Kn)$ is $O(1)$ as $\delta \to 0$. Physically, both (17) and (13) state that the field at any point in the interior of V is determined solely by the field at the geometrically closest point on the surface, to lowest order in δ. Viewed in this way, it is not surprising that the correct expansion (17) breaks down when $n = 1/K(s)$, at the centre of curvature of a point on S. At such a point it is to be expected that the fields due to equidistant points on the surface should interfere with each other. A similar process occurs in wave propagation known as "aberrated focussing."

5. Higher order approximations

To obtain higher order approximations it is best to transform the partial differential equation (4) into an integral equation over the boundary, S. To this end, we introduce the function

$$H(|\underline{x} - \underline{x}'|) = -\frac{1}{2\pi} K_0 \left[\left(\frac{1+i}{\delta} \right) |\underline{x} - \underline{x}'| \right] \tag{18}$$

where K_0 is a modified Bessel function. The function H satisfies

$$\nabla^2 H = \frac{2i}{\delta^2} H + \widehat{\delta}(\underline{x} - \underline{x}') \tag{19}$$

where $\widehat{\delta}$ is the Dirac delta-function. Then from (4) and (19) using Green's theorem,

$$\psi(\underline{x}') = \int_V \left(\psi \nabla^2 H - H \nabla^2 \psi \right) \, \mathrm{d}V$$

$$= \int_S \left(H \frac{\partial \psi}{\partial n} - \psi \frac{\partial H}{\partial n} \right) \, \mathrm{d}S \tag{20}$$

as the normal coordinate n is defined inwards. Let $\underline{x}'$ be any point in V, $\underline{x}(s)$ be any point on S, and let the distance between these two points, $d(s)$, have a minimum value $n(\underline{x}')$ when $s = s_0(\underline{x}')$, as in figure 1. In the vicinity of this minimum, we may expand

$$d^2(s) \equiv |\underline{x} - \underline{x}'|^2 = n^2 + (1 - K(s_0)n)\hat{s}^2 + \beta\hat{s}^3 + \gamma\hat{s}^4 + \ldots \tag{21}$$

where $\hat{s} = s - s_0$, while $\beta(\underline{x}')$ and $\gamma(\underline{x}')$ are geometrical properties of S. We now expand the values of ψ and its normal derivative on S in a manner similar to (10):

$$\left.\begin{array}{l}
\psi = \psi_0 + \dfrac{\delta}{1+i}\ \psi_1 + \left(\dfrac{\delta}{1+i}\right)^2 \psi_2 + \ldots \\[3mm]
\psi\big|_S \equiv f(s) = f_0 + \dfrac{\delta}{1+i}\ f_1 + \left(\dfrac{\delta}{1+i}\right)^2 f_2 + \ldots \\[3mm]
-\dfrac{\partial\psi}{\partial n}\bigg|_S \equiv g(s) = g_0 + \dfrac{\delta}{1+i}\ g_1 + \ldots
\end{array}\right\} \tag{22}$$

Substitution of (21) and (22) into (20) and assuming that $(1 - Kn)$ is $O(1)$ as $\delta \to 0$, leads, after some analysis, to the asymptotic results

$$f_0 = 0, \qquad f_1 = g_0, \qquad f_2 = g_1 + \tfrac{1}{2}Kf_1, \qquad \text{on } S; \tag{23}$$

$$\text{while} \quad \left\{\begin{array}{l}
\psi_0 = 0 \\[2mm]
\psi_1 = \dfrac{e^{-(1+i)n/\delta}}{(1-Kn)^{1/2}}\ f_1(s_0) \\[4mm]
\psi_2 = \dfrac{e^{-(1+i)n/\delta}}{(1-Kn)^{1/2}}\left[f_2 + \dfrac{\frac{1}{2}nf_1''}{1-Kn} - \dfrac{\frac{3}{2}n(\gamma f_1 + \beta f_1')}{(1-Kn)^2} + \dfrac{\frac{15}{8}n\beta^2 f_1}{(1-Kn)^3}\right]
\end{array}\right\} \quad \text{in } V, \tag{24}$$

where $'$ denotes the derivative with respect to s evaluated at the 'nearest point', $s = s_0$. The functions f_1 and f_2 can be found by solving the external potential problems for ψ_1 and ψ_2 with boundary conditions similar to (14) obtained from (23). We have thus achieved our aim of reducing the high frequency diffusion equation to a series of potential problems, at least for points such that $(1 - Kn)$ is $O(1)$. The expansion in (24) clearly breaks down when $(1 - Kn)$ is $O(\delta^{1/3})$. We must now investigate the behaviour near points where $n \approx 1/K$.

6. Behaviour near involute

The function $d(s)$ defined in (21) will have at least one minimum and one maximum over S. Provided such stationary points are separated by $O(1)$-distances, equation (24) will apply. If there are two or more such minima, then expansions of the form (24) corresponding to each minimum in turn may be added to give the total field. Usually, the term corresponding to the global minimum of d over S will be exponentially greater than the rest, and a very small relative error will result from considering only this term. Difficulties occur when two stationary values of $d(s)$ occur very close together on S, so that $\underline{x}'$ lies close to the locus of the centre of curvature of S, and $n \approx 1/K(s_0)$. This locus is known as the *involute* of S. For example, the involute of the elliptical domain bounded by $\frac{x^2}{a^2} + \frac{y^2}{b^2} = 1$, is a characteristically cusped curve, C, given by

$$\left(\frac{ax}{a^2 - b^2}\right)^{\frac{2}{3}} + \left(\frac{by}{a^2 - b^2}\right)^{\frac{2}{3}} = 1, \tag{25}$$

drawn in figure 1 for $a/b = \sqrt{2}$. The function $d(s)$ has one or two minima over S according to whether $\underline{x}'$ is outside or inside C. In this example, the points where C intersects the x-axis are particularly critical because at these points $1/K$ is a *global* minimum of d.

Provided $n \gg \delta$, equations (18), (20) and (22) imply, to leading order,

$$\psi_1 = \left(\frac{1+i}{2\pi n\delta}\right)^{\frac{1}{2}} f_1(s_0) \int_S e^{-(1+i)d/\delta} \, ds \; . \tag{26}$$

The expression for ψ_1 in (24) follows from (26) and (21), provided $1 - Kn$ is $O(1)$. Greater care must be taken over the asymptotics of (26) close to the involute, when, conversely, $1 - Kn$ is small. There does not seem to be a simple representation of the integral in (26) for general values of β and γ. When $\beta = 0$, however, as would occur on an axis of symmetry ($y = 0$, $a - b^2/a < |x| < a$ in figure 1, for example) it is not hard to show that provided γ is $O(1)$ and positive

$$\psi_1 = \frac{e^{-(1+i)n/\delta}}{(1 - Kn)^{1/2}} f_1(s_0) \, \mathrm{P}\left[\frac{1+i}{16n\delta\gamma}(1 - Kn)^2\right], \quad \text{where} \quad \mathrm{P}[\lambda] \equiv \sqrt{\frac{2}{\pi}}\lambda^{1/2}e^{\lambda}\mathrm{K}_{\frac{1}{4}}(\lambda) \; . \tag{27}$$

In (27), $\mathrm{K}_{1/4}$ is a modified Bessel function so that as $\lambda \to \infty$, $\mathrm{P}[\lambda] \to 1$, in agreement with (24), while as $\lambda \to 0$, $\mathrm{P}[\lambda] \propto \lambda^{1/4}$. The value of the flux function ψ is therefore increased by a factor proportional to $\delta^{-1/4}$ due to the geometrical "focussing" of the field. The curl of the Lorentz force, G, is increased as δ^{-1}. Still greater amplification is possible when $\gamma \leqslant 0$. At the centre of a circular boundary, for example, the focussing leads to an increase in $\psi \propto \delta^{-1/2}$, while G increases as δ^{-2}, although in many problems the symmetry leads to cancellation of the largest term. Of course, as $\delta \to 0$ the exponential decay is dominant. It is nevertheless clear that the internal field, though small, can be many times greater than the value predicted by the naive approximation.

7. Implications for fluid flow

Once ψ is known in the interior of V, the curl of the Lorentz force may be calculated from (7). This quantity, which determines the mean fluid motion, is found to be more evenly distributed over the main body of the fluid than the simple skin-depth approximation might lead one to expect. Stirring in the core region is thus enhanced, while the intensity of the wall-jets characteristic of this sort of flow is reduced. As we have already observed, this is especially important when the field must penetrate solid conductors before it can influence the liquid metal.

As an example, for which there exists analytic solutions both for the field and the fluid velocity, we consider the effect of a rotating, uniform field about a circular region of metal, $0 < r < a$. In figure 2a, G is drawn as a function of r/a for $\delta/a = 0.35$. Three curves are drawn, corresponding to the exact solution (E), the naive approximation (N), and the accurate approximation (A). Even for this fairly high value of δ, curve (A) is a very good approximation, except near the centre when the entirety of S is approximately equidistant.

Circular fluid motion is driven by the rotating field. In figure 2b, the non-dimensionalised velocity, u_θ, corresponding to each of the three curves in figure 2a is shown. Clearly, the cumulative effect of the errors in G lead, for curve (N), to considerable distortion of the velocity profile.

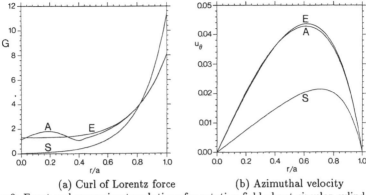

(a) Curl of Lorentz force   (b) Azimuthal velocity

Figure 2: Exact and approximate solutions for rotating field about circular cylinder.

Analytic solutions for non-circular boundaries are uncommon. Figure 3 corresponds to a rotating uniform field around the elliptical boundary of figure 1. The values of G along the x-axis are plotted for the approximate solution (A) and the "exact" (numerically calculated) solution (E), for $\delta = .3\sqrt{ab}$. Once again, the agreement is satisfactory. The fluid motion is more complicated for non-circular shapes, with a structure that depends on the Reynolds number. We shall not discuss the flow pattern here, but clearly an accurate representation of G is essential if the velocity is to be correctly predicted.

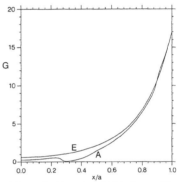

Figure 3: Rotating field about elliptic cylinder for $a/b = \sqrt{2}$. G is drawn as a function of x/a along $y = 0$.

8. Concluding remarks

The motivation behind this paper was to enable the electromagnetic properties within a region of liquid metal to be represented analytically, with reasonable accuracy when the skin-depth is small but not negligible. As we have shown, it is possible to improve considerably on the simple approximation, which is unreliable in the metal interior. The analysis may be extended to three dimensions without difficulty, although the method is most useful when the external potential problem can be solved easily, for which conformal mapping may be useful in two dimensions.

References

[1] ETAY J., MESTEL A.J., & MOFFATT H.K., 1988 'Deflection of a stream of liquid metal by means of an alternating magnetic field.' Proceedings of IUTAM symposium Riga 1988. (See elsewhere this volume.)

[2] MESTEL A.J., 1986 'Diffusion of an alternating magnetic field into a sharply cornered conductive region.' Proc. R. Soc. Lond. A Vol. 405, 49.

OVERALL AND LOCAL THICKNESS MEASUREMENT OF LAYERS WITH DIFFERING ELECTRICAL PROPERTIES

F. R. Block, W. Lesch and P. Palm
Technical University of Aachen
5100 Aachen
Federal Republic of Germany

ABSTRACT. The spreading of electromagnetic fields depends on the distribution of the electrical conductivity and magnetic permeability everywhere in the region of the field. Therefore, the resulting field yields information about the properties of a particular material.

Assuming simple material equations, especially the proportionality of magnetic induction to magnetic field strength, and a simple geometric arrangement of the materials, i.e. a plane or rotationally symmetrical configuration with discrete homogeneous steps, it is possible to determine the thickness and the above mentioned parameters of each layer by measuring the field at one point using different frequencies.

In this paper the solution is expanded to the case in which the measuring coil is not coaxially arranged to the transmitter coil. Also in the case of the cylindrical problem, instead of measuring the overall thickness using coaxial coils, local measurements are discussed which are made using small flat coils the shapes of which are adapted to that of the bar surface. In both cases the number of layers is not restricted. Finally, the thickness of the layers will be calculated from the measured values.

1. INTRODUCTION

There are two important technical applications for measurements using electromagnetic fields:

1. Measuring the wall thickness of containers that are only approachable from one side, such as containers for galvanizing or nuclear processes.

2. Measuring the skin thickness of metals, the conductivity of which changes on solidification, e. g. in continuous casting plants.

The literature on this topic deals with two groups of problems: The first group deals with flat coils of circular cross-sections arranged parallel to the plane surface of large solids. The solids consist of one, two or three layers. The coils are located on one or both sides of the solid or are imbedded between the layers. The receiver coil is always coaxial to the transmitter coil [1]. The second group deals with

J. Lielpeteris and R. Moreau (eds.), Liquid Metal Magnetohydrodynamics, 309–316.
© *1989 by Kluwer Academic Publishers.*

long cylindrical coils arranged coaxially to a rotationally symmetrical, cylindrical solid which itself consists of several layers. The coils surround the solid or are surrounded by it [1-3].

In the theoretical study the layers are assumed to be infinite in extent and homogeneous. Nevertheless, the theory does provide an estimation for localized effects, given that the signal is essentially determined by the material local to the transmitter and measuring coils.

2. SOLUTIONS OF THE FIELD EQUATIONS

2.1 Plane Problem

In the case of plane layers it is assumed that the solid extends far beyond the actual measuring area. Round coils are used as transmitter and receiver. They are placed parallel to the upper surface of the layers. The difference between the emfs of the system under investigation and that of a reference system with a homogeneous solid are determined.

The vector potential is defined by

$$\vec{B} = \text{curl } \vec{A} \ . \tag{1}$$

and the Coulomb gauge

$$\text{div } \vec{A} = 0 \ . \tag{2}$$

The electric field strength can be written as

$$\vec{E} = - \frac{\partial \vec{A}}{\partial t} \ . \tag{3}$$

The vector potential $\vec{A}$ satisfies the partial differential equation

$$\text{curl curl } \vec{A} + \mu\varepsilon \frac{\partial^2}{\partial t^2} \vec{A} + \mu\sigma \frac{\partial}{\partial t} \vec{A} = 0 \ . \tag{4}$$

The second term can be neglected at the low frequencies generally used.

When using a harmonic current in the transmitter coil, it is convenient to use for each harmonic physical variable the representation:

$$X = \text{Re}(\underline{X} \ e^{i\omega t}) \ , \tag{5}$$

where the complex amplitude is characterized by a bold and underlined letter. If the coordinates are chosen in such a manner, that the centre of the transmitter coil is placed on the z-axis, then for symmetry reasons, the vector potential has only an azimuthal component and is independent of the azimuthal coordinate. Using cylindrical coordinates it follows from equation (4) that the amplitude of the complex vector potential in layer l satisfies

$$- \frac{\partial^2}{\partial z^2} {}^l\underline{A} - \frac{\partial^2}{\partial r^2} {}^l\underline{A} - \frac{1}{r} \frac{\partial}{\partial r} {}^l\underline{A} + \frac{1}{r^2} {}^l\underline{A} + {}^lk^2 \ {}^l\underline{A} = 0 \tag{6}$$

with $\quad {}^lk^2 = - \varepsilon^l\mu\omega^2 + i^l\mu^l\sigma\omega; \ l = 0,1,\ldots n. \tag{7}$

The separation parameter is chosen as ${}^l\nu^2$. The general solution for each layer l is obtained by superposition of particular solutions,

$$^1\underline{A}(\nu) = \left[{}^1C_1(\nu)J_1(\nu r) + {}^1C_2(\nu)N_1(\nu r) \right] \left[{}^1C_3(\nu)e^{{}^1\nu z} + {}^1C_4(\nu)e^{-{}^1\nu z} \right] \tag{8}$$

with $\qquad {}^1\nu = \sqrt{\nu^2 + {}^1k^2}$. $\hspace{4cm}$ (9)

In this equation J_1 and N_1 are the Bessel- and Neumann Function of the first order respectively. Since $N_1(\nu r)$ has a singularity at r=0 and the vector potential must be finite there, the coefficient ${}^1C_2(\nu)$ is zero, so that without loss of generality the coefficient ${}^1C_1(\nu)$ can be set equal to one. The parameter ν can only have real values; otherwise the function $J_1(\nu r)$ tends to infinity as r→∞. For symmetry reasons only positive values of ν need be considered. Therefore the general solution is

$$^1\underline{A}(r,z) = \int_0^\infty J_1(\nu r) \left[{}^1C_3(\nu)\,e^{{}^1\nu z} + {}^1C_4(\nu)\,e^{-{}^1\nu z} \right] d\nu \quad \text{for } 1 = 0,..,n. \tag{10}$$

The coefficients are determined from radiation and boundary conditions, namely that the electromagnetic fields must vanish for r→∞ and z→±∞ and that both the tangential components of the electric and the magnetic fields must be continuous, except at the border surface which contains the circular current loop of the transmitter between the areas 0 and 1.
From the radiation conditions in the z direction, it follows that

$$^0C_3(\nu) = {}^nC_4(\nu) = 0. \tag{11}$$

The boundary conditions give:

$$^1\underline{A}(r,z_1) = {}^{1+1}\underline{A}(r,z_1) \qquad\qquad \text{for } 1=0,1,\ldots,n-1, \tag{12}$$

$$\frac{1}{{}^1\mu}\frac{\partial}{\partial z}\,{}^1\underline{A}(r,z)\Big|_{z=z_1} = \frac{1}{{}^{1+1}\mu}\frac{\partial}{\partial z}\,{}^{1+1}\underline{A}(r,z)\Big|_{z=z_1} \qquad \text{for } 1=1..,n-1 \quad \text{and} \tag{13a}$$

$$\frac{1}{{}^0\mu}\frac{\partial}{\partial z}\,{}^0\underline{A}(r,z)\Big|_{z=0} = \frac{1}{{}^1\mu}\frac{\partial}{\partial z}\,{}^1\underline{A}(r,z)\Big|_{z=0} - (I/R_1)\,\delta(r/R_1-1). \tag{13b}$$

The boundary conditions that both the normal components of the magnetic induction and of the electric displacement must be continuous does not give additional restrictions. The Dirac-Distribution is replaced by

$$\frac{1}{R_1}\,\delta(r/R_1-1) = \int_0^\infty J_1(\nu r)J_1(\nu R_1)\nu R_1 d\nu .$$

Assuming that (12) and (13) hold also for the integrands, linear equations result between the coefficients. They are homogeneous for $1 \geq 1$.

$$\begin{bmatrix} {}^1C_3 \\ {}^1C_4 \end{bmatrix} = \begin{bmatrix} \alpha_1 & \beta_1 \\ \gamma_1 & \delta_1 \end{bmatrix} \begin{bmatrix} {}^{1+1}C_3 \\ {}^{1+1}C_4 \end{bmatrix} \qquad \text{for } 1 = 1,2,..n-1 . \tag{14}$$

Using (11) and

312

$$\prod_{l=1}^{n-1} \begin{bmatrix} \alpha_1 & \beta_1 \\ \gamma_1 & \delta_1 \end{bmatrix} = \begin{bmatrix} \overset{*}{\alpha} & \overset{*}{\beta} \\ \overset{*}{\gamma} & \overset{*}{\delta} \end{bmatrix} \tag{15}$$

it follows that

$$^1C_3 = \overset{*}{\alpha}\, {}^nC_3 \text{ and } {}^1C_4 = \overset{*}{\gamma}\, {}^nC_3 . \tag{16}$$

Assuming $^0\sigma = {}^1\sigma = 0$, giving $^0\nu = {}^1\nu = \nu$ with $^0\mu = {}^1\mu$ and $z_0 = 0$ enables the integrands of (12) with l=0 and of (13b) to be written as

$$^0C_3 + {}^0C_4 = {}^1C_3 + {}^1C_4 \text{ and } -{}^0C_4 = {}^1C_3 - {}^1C_4 - \mu_0 I\, R_1\, J_1(\nu R_1) . \tag{17}$$

To determine the emf in the receiver, which is placed at region 0, $^0C_4$ is calculated from (16) and (17)

$$^0C_4 = \left[\frac{\overset{*}{\gamma}}{\overset{*}{\alpha}} + 1 \right] \frac{\mu_0 I R_1}{2}\, J_1(\nu R_1) . \tag{18}$$

The complex amplitude of the emf in the receiver is:

$$\underline{U}^I = - i\omega \int_{\partial S} {}^0\underline{A}\, \vec{e}_\varphi\, d\vec{x} \text{ with } {}^0\underline{A} = \int_0^\infty {}^0C_4 J_1(\nu r)\, e^{-\nu z}\, d\nu. \tag{19}$$

For the numerical calculation of the measured difference between the signal $\underline{U}^I$ and the signal U^{II} of a reference system it is convenient to carry out the subtraction before the integration.

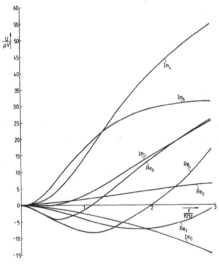

Fig. 1: Real and imaginary parts of the difference signal versus frequency for different horizontal distances D between the coil centres
Parameters: one loop coils $R_1 = R_2 = 25$ mm, $I = 1$ A, air gap $d_1 = 1$ mm,
$d_2 = 2$ mm, $\mu = \mu_0$; $\sigma_0 = \sigma_1 = 0$, $\sigma_2 = 1$ MS/m, $\sigma_3 = 0{,}9$ MS/m, vertical
coil distance $H = 1$ cm, $D = 0$; 2; 4; 6 cm

Real and imaginary parts or amplitude and phase of the difference signal as a function of the frequency supply the required information.

Fig. 1 demonstrates that it may be advantageous to use non-coaxial coils, because there are more sensitive signals.

2.2 Cylindrical case

For investigating multilayered cylinders coils the sides of which run in the axial and azimuthal directions are used, see Fig. (2).

To solve the differential equation (4) for this arrangement, superordinate vector potentials are used [4,5]. Since its divergence is zero, the vector potential can itself be displayed as the curl of a vector, which can be described as the sum of a component in z direction and another component in the x-y plane. It is assumed that the last one can be derived as curl $\underline{P}_2\,\vec{e}_z$ from a scalar function $\underline{P}_2$. The vector potential

$$^1\vec{\underline{A}} = \mathrm{curl}\ \left[\ ^1\underline{P}_1\vec{e}_z + \mathrm{curl}\ ^1\underline{P}_2\vec{e}_z\ \right] \tag{20}$$

obeys equation (4) if the scalar functions $^1\underline{P}_1$ and $^1\underline{P}_2$ obey the scalar differential equations:

$$\Delta\,^1\underline{P}_j = \frac{\partial^2}{\partial r^2}\,^1\underline{P}_j + \frac{1}{r}\frac{\partial}{\partial r}\,^1\underline{P}_j + \frac{\partial^2}{\partial z^2}\,^1\underline{P}_j + \frac{1}{r^2}\frac{\partial^2}{\partial \varphi^2}\,^1\underline{P}_j = k^2(r)\,^1\underline{P}_j,\ j=1,2. \tag{21}$$

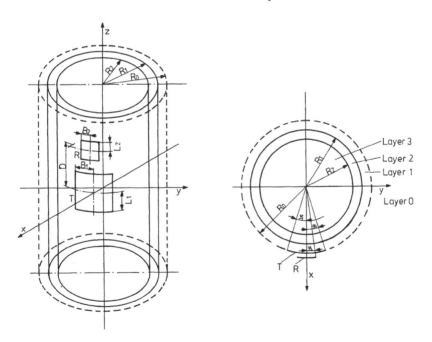

Fig. 2: Sketch of the cylinder- and coil-system and its cross-section

314

The separation parameter for the terms which depend only on the azimuthal coordinate is chosen as m^2, and that for the terms which depend only on the axial coordinate as ν^2. The general solution for each layer 1 is obtained by superposition of particular solutions:

$$^1\underline{P}_j(m,\nu,r,\varphi,z) = \left[\,^1C_{1j}(\nu)\,\sin(\nu z) + \,^1C_{2j}(\nu)\,\cos(\nu z)\,\right]\cdot$$

$$\left[\,^1C_{3j}(m)\,\sin(m\varphi) + \,^1C_{4j}(m)\,\cos(m\varphi)\,\right]\cdot\left[\,^1C_{5j}(\nu)\,I_m(^1\nu r) + \,^1C_{6j}(\nu)\,K_m(^1\nu r)\,\right]$$

with $^1\nu = \sqrt{\nu^2 + \,^1k^2}$ for $j = 1,2$; $1 = 0,1,\ldots,n$ and $m = 0,1,\ldots$, (22)

where I_m is the modified Bessel function of the m-th order and K_m is the modified Neumann function of the m-th order.

In order that the solution be unique, the parameter m is an integer. Since the fields must remain finite for $|z|\to\infty$, ν must be real. For symmetry reasons it can be assumed that $\nu \geq 0$.

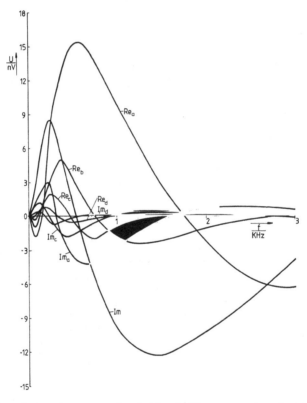

Fig. 3: Real and imaginary part of the difference signal vs. frequency
Parameters: -see Fig. 2 - $L_1 = B_1 = 22$ mm, $L_2 = B_2 = 20$ mm, $D = 120$ mm,
$H = 0$ mm, $\sigma_0 = \sigma_1 = 0$, $\sigma_2 = 1{,}47$ MS/m, $\sigma_3 = 1{,}35$ MS/m, $\mu = \mu_0$,
$R_0 = 130$ mm, $R_1 = 120$ mm, $R_2 = $ a) 110 mm, b) 100 mm, c) 90 mm, d) 80 mm

As above the coefficients are determined from the radiation and the boundary conditions. The transmitter 1 is represented by a current loop between the areas 0 and 1. The current distribution on this surface is

$$\vec{K}(R_1,\varphi,z)/\underline{I} = \frac{1}{L_1}\left[\delta(\frac{z}{L_1}+1)-\delta(\frac{z}{L_1}-1)\right]\left[\varepsilon(\frac{\varphi}{\varphi_1}+1)-\varepsilon(\frac{\varphi}{\varphi_1}-1)\right]\vec{e}_\varphi +$$

$$+\left[\varepsilon(\frac{z}{L_1}+1)-\varepsilon(\frac{z}{L_1}-1)\right]\left[\delta(\frac{\varphi}{\varphi_1}-1)-\delta(\frac{\varphi}{\varphi_1}+1)\right]\frac{1}{R_1\varphi_1}\vec{e}_z . \qquad (23)$$

The factors are represented by Fourier series or Fourier integrals:

$$\vec{K}/\underline{I} = -\frac{2}{\pi}\int_0^\infty \sin(\nu L_1)\sin(\nu z)\,d\nu \cdot \frac{2}{\pi}\left[\frac{\varphi_1}{2}+\sum_{m=1}^\infty \sin(m\varphi_1)\frac{\cos(m\varphi)}{m}\right]\vec{e}_\varphi +$$

$$+\frac{2}{\pi}\int_0^\infty \sin(\nu L_1)\frac{\cos(\nu z)}{\nu}\,d\nu \cdot \frac{2}{\pi}\sum_{m=1}^\infty \sin(m\varphi_1)\sin(m\varphi)\frac{1}{R_1}\vec{e}_z . \qquad (24)$$

Consequently the coefficients of the particular solution with the same ν or m can be determined in the same way as for the plane case [6].

From $\underline{P}_1$ and $\underline{P}_2$ the magnetic induction can be derived:

$$^1\underline{B}_r = \frac{\partial^2}{\partial z\,\partial r}\,^1\underline{P}_1 - {}^1k^2\frac{1}{r}\frac{\partial}{\partial\varphi}\,^1\underline{P}_2 ,$$

$$^1\underline{B}_\varphi = \frac{1}{r}\frac{\partial^2}{\partial\varphi\,\partial z}\,^1\underline{P}_1 + {}^1k^2\frac{\partial}{\partial r}\,^1\underline{P}_2 ,$$

$$^1\underline{B}_z = \frac{\partial^2}{\partial z^2}\,^1\underline{P}_1 - {}^1k^2\,^1\underline{P}_1 \qquad , \qquad 1 = 0,1,..n. \qquad (25)$$

In the areas zero and one, the electrical conductivity is zero. Therefore, the radial component of the magnetic induction, which determines the voltage induced in the receiver, depends only on $\underline{P}_1$. This voltage has been calculated numerically, using (19).

Fig. 3 demonstrates how the signals depend on the thickness of the solidified skin $(R_1 - R_2)$.

3. CALCULATING THE LAYER THICKNESS FROM THE MEASURED VALUES

The amplitude and phase shift of the emf relative to the applied current in the transmitter coil are obtained as functions of frequency.

In order to detect n parameters, e. g. layer thicknesses, at least n amplitudes or phase shifts at different frequencies must be measured.

It is expedient to apply additional information so that the range of expected solutions is reduced. For instance, the layer thicknesses can only be positive and the value of the product $\mu\sigma$ changes monotonically from layer to layer during cooling processes. With the aid of such restrictions, the inverse calculation can usually be carried out by known methods. For example in the four layer model investigated the solution was found in a few steps using the Newtonian gradient procedure. When the coil distances from the surface as well as the conductivities and permeabilities were given, the unknown thicknesses of the layers 2 and 3 could be determined.

4. ACKNOWLEDGEMENT

This research was supported by Bundesminister für Forschung und Technologie, Bonn, under grant 03 S 393 5. The authors would like to assume responsibility for this paper.

5. REFERENCES

[1] Dodd C.V. & Deeds W.B.,1968,'Analytical solutions to eddy-current probe-coil-problems', Journ. Applied Physics, Vol. 39, 2829
[2] Witting G.,1970,'Theoretische und experimentelle Unters. z. Wirbelstromprüfung des zweischichtigen Zylinders', Materialprüf., Vol. 12, 81
[3] Hansen W.,1983, Grundlagen und erste Versuche zur e. m. Messung der Schalendicke bei stranggegossenen Knüppeln, Diss. RWTH Aachen
[4] Hannakam L.,1973,'Wirbelströme in einem massiven Zylinder bei beliebig geformter erregender Leiterschleife', Arch.f.Elektrot., Vol. 55, 207
[5] Moffat H.K.,1978, Magnetic field generation in electrically conducting fluids, Cambridge University Press, 13
[6] Palm P.,1988/89, Zur Theorie der Detektion elektrischer Leitfähigkeitsverteilungen mit Hilfe von induzierten Strömen und ihre Anwendung auf die lokale Strangschalendickenmessung, Diss. RWTH Aachen

DETERMINATION OF MHD MACHINE PARAMETERS USING THE ID MODEL OF A NONUNIFORM FLOW

I.Gavare, R.Krisbergs, O.Lielausis,
M.Pukis, L.Ulmanis, R.Valdmane
Institute of Physics, Latvian SSR Academy of Sciences
229021 Riga-Salaspils USSR

ABSTRACT. A topical problem in the MHD machine theory is connected with computation of machine parameters in the case of complex velocity profiles. Comparison with experiments shows that the simplest ID turbulent model (OTM) of duct flow can serve as a base for such calculations, especially in the presence of longitudinal partitions inside the channel. But even in the case of a traditional undivided duct, when the velocity profile is developing along the channel length, the results remain acceptable. A modification of the OTM by introduction of some empirical correcting factors can be recommended as a logical next step.

1. INTRODUCTION

It is undoubtedly clear today that the flow in the channels of MHD machines, caused by generally nonuniform force fields, can be strongly non-homogeneous. In the case of induction machines, the relevant considerations are somewhat simplified due to the fact that a definite velocity component, the longitudinal one, can be a priori indicated as predominating (according to the typical geometry of the channel). It is the mean presumption in the ID flow models /1,2/. The flow is assumed to follow in thin layers parallel to the mean velocity. The local velocities can be different and even opposite directed. The pressure gradient, the e.m. forces, the turbulent pressure losses and the forces accelerating / decelerating the fluid from the mean velocity up / down to the local one are taken into account. In the presence of longitudinal partitions inside the channel, the OTM can be treated as a direct model of the real flow. The application of OTM to undivided channels is based on the idea that the averaged longitudinal velocity components are responsible for the mean energy transfer relations inside the channel.

Among the results obtained by means of OTM the prediction and evaluation of instability conditions for the uniform velocity profile in large annular pumps /3/ can be mentioned. This instability results in a peculiar pump "stalling". It should be remembered that similar phenomena are typical for other hydraulic machines too, if they have a region with a positive slope (pressure increasing with flow rate) in their p(Q)-characteristics /4/.

J. Lielpeteris and R. Moreau (eds.), Liquid Metal Magnetohydrodynamics, 317–323.
© *1989 by Kluwer Academic Publishers.*

2. MATHEMATICAL MODEL OF THE INDUCTION PUMP

Dimensional parameters: g, h_h, h_w, h_c, h_s – height of gap, duct, duct wall, side bars and primary windings; 2d – duct width; D – annular duct diameter; σ_h, σ_w, σ_c, σ_s – conductivity of LM, walls, side bars and winding zone, ϑ – LM density, A – line current density in winding zone, τ and t_s – pole and slot pitch; Ω – angular frequency.

Similarity factors: $t_0 = \Omega^{-1}$ (for time t); $l_0 = \tau/\pi = a^{-1}$ (for distance l); $v_0 = q_0 = \Omega/a$ (for local velocity v and mean velocity q); $J_0 = A_0 = A$ (for induced j and primary a line current density); $B_0 = \mu_0 A/ag$ (for magnetic flux density d); $p_0 = n\tau AB_0/2h$ (for pressure p); $Q_0 = 2dhq_0$ (for flow rate Q); $P_0 = p_0 Q_0$ (for power P); $U_0 = \Omega B_0 v_0 2d/\sqrt{2}$ (for voltage u).

Dimensionaless groups and coefficients: $R = 4ah/\lambda_h$ and $M^2 = 2A_0 B_0/\lambda_h \varrho v_0^2$ -inertia to friction and inertia to e.m.forces ratios; $\epsilon_k = \mu_0 v_0 \sigma_k h_k/ag$- - goodness factors for LM, duct walls, side bars and primary windings (k=h,w,c,s – respectively); $\epsilon = \epsilon_h(1 - v) + \epsilon_w$ – local magnetic Reynolds number; $L = a^2 g t_s \lambda$ – primary circuit inductance; w – number of turnes on a parallel group and phase; λ_h – hydraulic friction coefficient; λ – magnetic dispersion coefficient; n – number of poles; z and I – parameters for loop resistence and inertia; k_1 – parameter for inertia losses in the channel [2].

A following system of dimensionless non-stationary equations was solved:

$$R\delta v/\delta t + M^2 \Delta p = -(\mathrm{Re}\overset{*}{a}\dot{b} + v\epsilon_w|\dot{b}|^2)M^2 \epsilon_h/(\epsilon_h+\epsilon_w)^{-1} - v|v| -$$

$$- k_1(v|v| - q|q|)M^2/\pi n \tag{1}$$

$$p = zq|q| + I \delta q/\delta t \tag{2}$$

$$(\epsilon_h + \epsilon_w) \dot{\delta b}/\delta t = \delta^2 b/\delta y^2 - (1 + i\epsilon)b - i\dot{a} \tag{3}$$

$$L\dot{\delta a}/\delta t + ((\epsilon_h+\epsilon_w)^{-1} + \epsilon_s^{-1} + iL)\dot{a} = \dot{u} - i(\epsilon_h+\epsilon_w)^{-1}\delta b/\delta y|_{y=ad} -$$
$$-(2ad)^{-1}\int_{-ad}^{ad} \dot{b}(i+\epsilon_h v)(\epsilon_h+\epsilon_w)^{-1}dy \tag{4}$$

According to OTM,(1) represents the force balance in the case of an ideal contact between the LM and the wall. (2) expresses the pressure losses in the external hydraulic system. (3), describing the magnetic flux, is independent of the way the wall shunting effect is introduced. (4) represents the voltage balance (on a parallel group and phase).

Boundary conditions depend on the machine design. In the case of an annular pump, periodical conditions along its circumference are assumed. In the case of a flat pump without side bars, the transverse component of the induced current is equal to zero on the side walls. In the presence of side bars with a width of a(c-d) following conditions at y = ad are used: $\dot{b} = i(j_1 + \dot{a})$; $\delta b/\delta y = \pm i(\epsilon_h + \epsilon_w)j_1/\epsilon_c$ tanh a(c - d).

The solutions of (1-4), corresponding to various points of the stationary p(Q)-curve at $\dot{u}$=const., were evaluated by changing the parameter z. This corresponds to a gradual opening of some ideal valve.

In the case of an annular pump, the system (1-4) has a uniform stationary solution, corresponding to a uniform velocity profile. An analytic study enables to draw a curve (5) of its <u>neutral stability</u>, which represents a generalization (conducting walls are taken into account) of the original curve from /3/:

$$p_m = \frac{(\epsilon_h + \epsilon_w)(1 + k_1 R/\pi n)}{\epsilon_h(1 - 2q_m \epsilon_w/(1 + \epsilon^2)^{1/2}} \cdot \frac{2|q_m|(1 - q_m)(T_m^2 + \epsilon^2)}{M^2(\epsilon^2 - T_m)} - \frac{q_m|q_m|}{M^2} \tag{5}$$

where $T_m = 1 - (2\tau/D)^2$.

In the case of a flat pump (1-4) has no uniform solutions. The initial profile is specified by recirculating zones adjacent to the side walls, therefore, it is more convenient to compute the boundary of stability by direct solution of the system. It can be determined as the transition point from the initial profile to that with a back flow in the midle of the channel.

3. COMPARISON WITH EXPERIMENTS

3.1. Flat pump with longitudinal partitions in the channel.

At the very begining of R&D on electromagnetic pumps a large scale experiment /5/ on a Na loop was performed covering a wide range of parameters (Table 1). On the p(Q)-curves crisis like phenomena were clearly spotted (Fig.1). Today it is demonstrated that this peculiar pump "stalling" reflects the loss of stability of the uniform velocity profile. The channel under consideration with a size of $(1.7 \times 42.5 \times 101)cm^3$ was devided by 7 partitioning walls in equal subchannels and provided with side bars. There are all grounds to assume that the pole number n=8, number of slots per pole and phase q=1. The dimensions of the stator are not clearly given and they are chosen somewhat arbitrary according to the general pump design experience. Dimensionless parameters of the computed versions are presented in Table 1. Values R=const.=27.5 and M^2=const.=34.1 are used (for the first version). Figure 1 shows the comparison of evaluated curves with the experimental ones. As the pump parameters are not fully defined, the calculated and experimental curves, corresponding to the first version, were matched in points indicated by arrows. The obtained scaling factors (0.8 for pressure and 0.9 for flow rate) were used to correct the evaluated data for all other versions.

In Fig.2., an additional example of comparison is presented - the measured /6/ and computed by means of OTM /10/ profiles of velocity along the channel width.The two theoretical curves differ in coeficients k_1, characterizing the inertia losses. The channel was similar, with a size of $(2.0 \times 42.6 \times 108) cm^3$, having 7 partitions and no side bars.

320

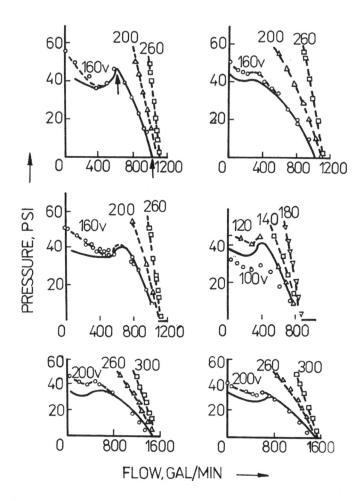

Figure 1. Experimental /5/ and theoretical (——) p(Q)-characteristics of 1200-gpm flat induction pump.

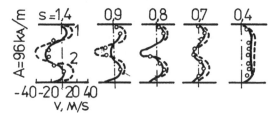

Figure 2. Experimental and theoretical (— k_1=1;--k_1=0) velocity profiles along the channel width.

TABLE 1

Freq. (hz)	Temp. (^{0}F)	ϵ_h;	ϵ_w;	ϵ_s;	ϵ_c;	L;	Volt. (V)	Insulat
1 40	300	3.0	.17	4.48	9.0	.81	160	+
2 40	700	1.8	.17	2.70	5.4	.81	160	+
3 40	300	3.5	.20	5.20	10.4	.70	160	-
4 30	300	2.0	.14	3.00	6.0	.81	120	+
5 60	700	2.7	.17	4.10	8.1	.81	200	+
6 60	700	3.1	.20	4.60	9.4	.70	200	-

3.2 Annular pump with an undivided channel.
The largest to date tested e.m. pump /7/ has a traditional undivided
annular channel. In such channels development of the nonuniformity of
the profile along the channel length, neglected by OTM, is clearly demonstrated /8/. In this case, the aim of the comparison is to make out
whether there is a sense at all to try to develop some engineering methods using OTM as a basis. The idea is not groundless. In channels under
discussion the OTM is proved well in such an important question as determination of the boundary of instability /8/.
    The pump /7/, delivering 3500 m³/h at 3 atm., which can circulate
Na in the secondary loop of the fast reactor BN-350, is characterized by
following dimensionless parameters: R=184; M=9.86; n=12;L=1.26; ϵ_h=3.75;
ϵ_w=0.195; ϵ_s=7.87. The high relative thickness of the walls
(h_w=6 mm; h_h=26 mm) is to be mentioned. It means a correct determination
of the losses in the walls and of the wall shunting effect is essential,
especially in the presence of a nonuniform profile. The comparison for
the efficiency, power factor and line current at U=const.=650 V is presented in Fig.3. In Fig.4, the p(Q)-curves are shown. The agreement for
the main part of parameters can be characterized as very good (even a
bit "too good").
    As noted in /9/,OTM allows to indicate such a delicate phenomenon
as hysteresis on the p(Q)-curve. The computed curve in Fig.4 reveals the
same feature. In addition, the computation process in the direction of
increasing flow rates was less stable. Qualitatively similar processes
were observed during the experimental study of instabilities in an annular pump /8/. Another result from /8/, worth to emphasise is that after
the introduction of longitudinal partitions in the channel the fluctuations of parameters in the region of the stability boundary were practically damped out. The question arises if it is not high time to accept some
guide-apparatus as a standard element of the main part of MHD channels,
at least in the form of simple longitudinal partitions.

4. DISCUSSION

In /2,11/ it was shown that OTM enables to improve the determination of
parameters in the case of traditional small pumps with flat channels.
The above examples allow to suggest that OTM can be used as a basis for
calculations in the case of large pumps characterized by high electro-

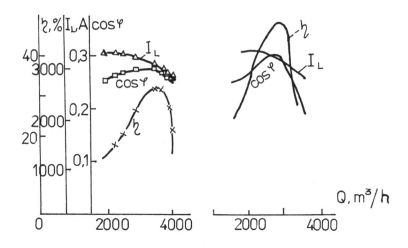

Figure 3. Experimental /7/ (left side) and theoretical characteristics
of a large annular pump at U=const.=650 V: η -efficiency; I - current;
cos φ - power factor.

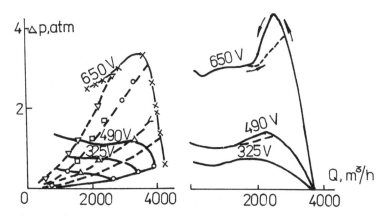

Figure 4. Experimental /7/ (left side) and theoretical
p(Q)-characteristics of a large annular pump.

magnetic interaction parameters. Obviously, it is necessary to work on the selection of systems of empirical correcting factors, depending on the design of the pump.But it seems that the number of such coefficients will be rather limited.

Nowadays much interest is devoted to more complicated 2D duct flow models. As a matter of fact, they will allow to describe the phenomena under discussion more adequately. But it will be a long way to develop some engineering methods on the basis of such models. The selection of rational flow models will be strongly influenced by the presence of different guide-apparatus.

5. REFERENCES

/1/ GAILITIS A., LIELAUSIS O.,1971, 'On the initial hydraulics of MHD-machines at nonuniform force distribution', Magn. Gidrodin., No.2, 123.
/2/ VALDMANE R., LIELAUSIS O., ULMANIS L., 1983, 'Model of nonuniform flow in the channel of induction pump', Magn. Gidrodin., No.2, 98.
/3/ GAILITIS A., LIELAUSIS O., 1975, 'Instability of uniform velocity distribution in induction MHD-machines', Magn. Gidrodin., No.1, 87.
/4/GREITZER E.M., 1981, 'The stability of pumping systems. - The 1980 Freeman scholar lecture', J. of Fluid Engeneering, V. 103, No. 2.
/5/ BARNARD J., COLLINS G., 1951, 'Test of 1200-gpm linear A-C electromagnetic pump', Knolls Atomic Power Lab., Report AECD-30600.
/6/ VALDMANE R., KRISBERGS R., LIELPETER J. et al., 1987, 'Local characteristics of flow in channel of induction machine at large electromagnetic interaction parameters', Magn. Gidrodin., No.3, 99.
/7/ ANDREEV A.M., BEZGACHEV E.A., KARASEV B.G. et al., 1988, 'Electromagnetic pump CLIP-3/3500', Magn. Gidrodin.., No.1, 61.
/8/ KIRILOV I.R., OSTAPENKO V.P., 1987, 'Local characteristics of induction pump at $R_m s > 1$', Magn. Gidrodin., No.2, 95.
/9/ VALDMANE R., KRISBERGS R., LIELPETER J. et al., 1977, 'Integral characteristics of induction machine at large electromagnetic interaction parameters, Magn. Gidrodin., No.4, 107.
/10/ VALDMANE R., LIELAUSIS O., ULMANIS L., 1985, 'Ccomputation of a nonuniform flow in the channel of induction pump with longitudinal partitions'.Magn. Gidrodin., No.4, 85.
/11/ KALNIN A., MIKRJUKOV Th., PETROVICHA P. et al., 1971, 'Characteristics of a flat induction pump in the case of nonuniform force distribution along the channel width'. Magn. Gidrodin., No.4, 94.

EXPERIMENTAL AND THEORETICAL STUDIES ON THE STABILITY OF INDUCTION PUMPS AT LARGE Rm NUMBERS

J. RAPIN, Ph. VAILLANT
NOVATOME
B.P. 3087
69398 LYON CEDEX
France

F. WERKOFF
CEA/IRDI/DEDR/DTE
CENG/STT/LPML
85 X 38041 GRENOBLE CEDEX
France

ABSTRACT. In large size electromagnetic pumps flow inhomogenities can arise and perturb the operating conditions. We present some experimental results from two annular pumps with a maximum sodium delivery of about 700 m³/hour. Comparison is also made with a numerical model in which the flow is assumed to be homogeneous. The discrepancies are less than that is expected when examined at the light of an analytical expression for the onset of azimuthal inhomogenities. Finally, we present a numerical model which gives the velocity profile in the radial direction. A linear stability analysis indicates that such profiles are always stable in practical conditions.

1. INTRODUCTION

In fast reactors, as well as in the laboratory loops, Electro Magnetic Pumps (EMP) with alkali metals have almost completely replaced mechanical pumps where the deliveries are lower than 200 m³/h. The situation is very different for higher delivery (in the main loops of the fast reactors) where a predicted instability [1] and longitudinal edge effects which limit the energy efficiency, have dissuaded from using EMP up to now. However the trend of the new concepts of fast reactors towards simplification, leads to a renewed interest for the large size induction EMP [2]. In France a R.D. program was initialized and financially supported by the Commissariat à l'Energie Atomique. This leads to the realisation by the Novatome society of a high delivery EMP prototype, coupled with a cyclo-converter which allows to adjust the frequency of the supply voltage [2]. The preliminary operations were performed in november 1987 and are presented in the following paragraph.

2. THE FIRST EXPERIMENTAL RESULTS OF THE EMP PROTOTYPE

We give in figures 1 and 2 the characteristic curves for intensity of current by phase of 295 Amps, two sodium temperatures (T = 200°C) and T = 400°C) and a few values of the frequency, between 4 and 12 Hz. The pump efficiency η_e = the product of the pressure increase by the delivery over the supply electric power, has a maximum for T = 400°C,

325

J. Lielpeteris and R. Moreau (eds.), Liquid Metal Magnetohydrodynamics, 325–331.
© 1989 by Kluwer Academic Publishers.

326

ν = 10 Hz and 625 m³/h of delivery. Hence we include the power, needed by the cyclo-converter and find η_e = 0.46 or we do not include it and then find η_e = 0.485.

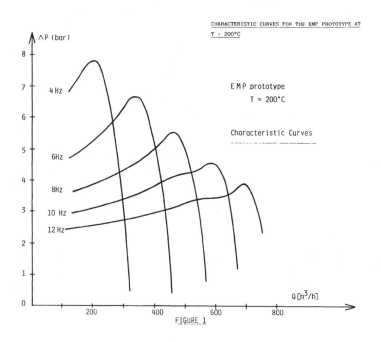

FIGURE 1

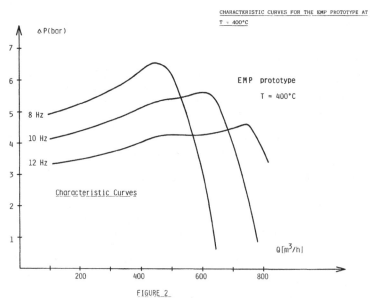

FIGURE 2

At the present stage of experimentation, it seems that instabilities [3] do not hinder normal running conditions. If in some cases, we observed mechanical vibrations of the whole hydraulic loop, they can also be due to the fact that the electromagnetic forces are no longer averaged over a period. For instance, with ν = 4 Hz, T = 200°C and a typical 0.15 Tesla value of the effective magnetic field B, the interaction parameter built with the pulsation ω = 2 $\pi\nu$ is :

$$N_\omega = \frac{\sigma B^2}{\rho \omega} \# 7 \tag{1}$$

where σ and ρ are the sodium electrical conductivity and specific mass, respectively.

3. AZIMUTHAL FLOW HOMOGENITY

A characteristic number of a MHD flow in an induction machine is the magnetic Reynolds number :

$$R_m = \frac{\mu\sigma}{k} \left(\frac{\omega}{k} - V\right) \tag{2}$$

where μ is the magnetic permeability, k the wave number and V the average longitudinal velocity. It was pointed out in [1] that if R_m is greater than a critical value, homogeneous flows become unstable and thereby, the performances of the pumps diminish. The critical value is equal to the unity, increased by correction terms due to internal friction losses and finite radius effect [1].

In order to identify the effect of some possible azimuthal flow inhomogenities, we have considered results of the high delivery EMP prototype, results of a standard IA1401 Novatome EMP and compared them with a 2 D finite elements numerical model [2]. In this model the fluid is assumed to move with a constant velocity along the radial and azimuthal directions. The results are reported in figures 3 and 4. It can be observed, as typical of high R_m flows [1], that for a given increase pressure, there exist two different delivery values, the upper one corresponding to the best efficiency.

Moreover, it appears that experimental results are generally not far away from the theoretical ones, except in limited regions where the experimental performances are reduced. In order to connect these results with theories on azimuthal inhomogeneities, we have considered the linear stability analysis, removing the assumption that the induced current could be lower than the inductor current. For the most susceptible (lower) mode, we obtained an onset of instability for a critical Reynolds number, somewhat less severe than in [1] and which depends on the pump pressure increase ΔP and delivery Q :

$$R_{mc} = \left(1 + \frac{1}{k^2 R^2}\right) \left(\frac{1 + x}{1 - x}\right)^{1/2} \tag{3}$$

with

$$x = \frac{2 (Q_s - Q)}{\Delta P + C_{fr} Q^2} C_{fr}. \tag{4}$$

ΔP(bar)

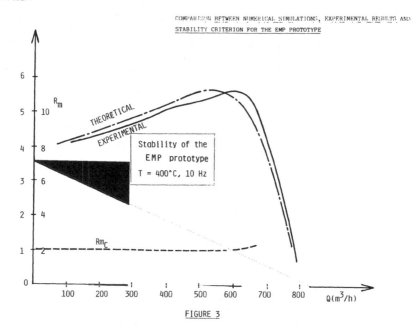

COMPARISON BETWEEN NUMERICAL SIMULATIONS, EXPERIMENTAL RESULTS AND
STABILITY CRITERION FOR THE EMP PROTOTYPE

Stability of the
EMP prototype
T = 400°C, 10 Hz

FIGURE 3

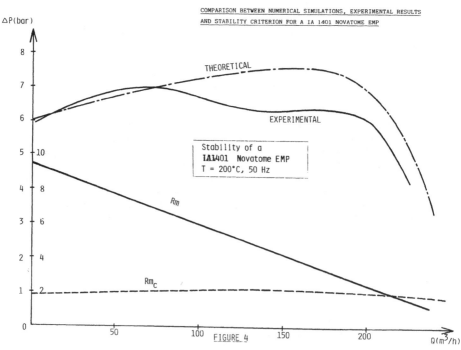

COMPARISON BETWEEN NUMERICAL SIMULATIONS, EXPERIMENTAL RESULTS
AND STABILITY CRITERION FOR A IA 1401 NOVATOME EMP

Stability of a
IA1401 Novatome EMP
T = 200°C, 50 Hz

FIGURE 4

In (3) and (4) R stands for the radius of the annular channel and Q_s the delivery which corresponds to the synchronous velocity ω/k, while friction losses in the channel pump are expressed in the form :

$$\Delta P_{fr} = C_{fr} \, Q^2, \tag{5}$$

with C_{fr} experimentaly measured for the IA1401 pump and theoretically estimated for the EMP prototype.

In figure 3 and 4 we also report the values of R_m and R_{mc}. It appears that for the IA1401 EMP, R_m is further from R_{mc} than for the high delivery EMP prototype. Correspondingly, experimental results are further from the ideal computations on homogeneous flow.

4. RADIAL FLOW HOMOGENITY

In order to work with EMP of 8 or 10 000 m^3/h, the height of the liquid channel must be considerably greater than in present pumps. Such a situation is favourable to the apparition of flow inhomogeneities, which in turn can modify the magnetic field penetration and even, in some cases, the creation of an instability, similar to the runaway effect of flux expulsion reported in [4].

In order to estimate that effect, we have considered the one dimensional induction and momentum equations :

$$\rho \frac{\partial V}{\partial t} + \frac{\partial P}{\partial z} = \eta \frac{1}{r} \frac{\partial}{\partial r} \left(r \frac{\partial V}{\partial r} \right) + F \tag{6}$$

$$\frac{1}{r} \frac{\partial}{\partial r} \left(r \frac{\partial A_\theta}{\partial r} \right) - \frac{A_\theta}{r^2} + \frac{\partial^2 A_\theta}{\partial z^2} - \mu\sigma \left(\frac{\partial A_\theta}{\partial t} + V \frac{\partial A_\theta}{\partial z} \right) = 0 \tag{7}$$

where V is the longitudinal component of the fluid velocity, A_θ the azimuthal component of the vector potential ($\vec{B} = \overrightarrow{rot} \, \vec{A}$) and is such that $A_\theta (t, r, z) = A_\theta^* (r) e^{j(\omega t - kz)}$, η is the dynamic viscosity and F is the Laplace force :

$$F = \sigma \left(\frac{\partial A_\theta}{\partial t} + V \frac{\partial A_\theta}{\partial z} \right) \frac{\partial A_\theta}{\partial z}$$

The boundary conditions for eq. 6 are $V = 0$ in $r = r_2$ and $r = r_3$, the radii of the internal and external channel walls. For equation (7) $\frac{\partial A_\theta}{\partial r} = 0$ in $r = r_1$ at the contact of the internal magnetic core and $\frac{\partial A_\theta}{\partial r} = - \mu_0 J$ in $r = r_4$, at the contact of the stator, schematized by an equivalent surface current J.

Equations (6) and (7) are first time averaged and then solved with an imposed pressure gradient $\frac{\partial P}{\partial z}$, while J was adjusted in an external loop, to obtain a desired delivery value Q. We report in figure 5 results for $Q = 600 \, m^3/h$ and $\frac{\partial P}{\partial z} = 2$ bars, which corresponds to experimental characteristics of the EMP prototype whose active length is 2.25 m. The velocity profile shows very thin layers : less than 1 mm.

330

At the plateau of the curve the velocity is almost constant from 6.45 m/s at the internal edge to 5.81 m/s at the external edge. This corresponds to amplitudes of the radial magnetic field of 0.146 and 0.157 T respectively.

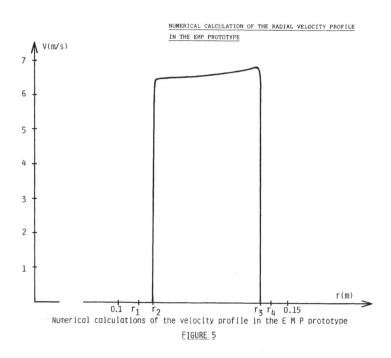

Numerical calculations of the velocity profile in the E M P prototype

FIGURE 5

In order to estimate the probability for a runaway effect |4| to occur with such a flat velocity profile, we performed a linear stability analysis on the set equations (6) and (7). The perturbations are assumed of the form $e^{j(\Omega t - Kr)}$, where K is real and Ω complex.

We found the following onset for instability :

$$\eta K^2 \left[(k^2+K^2)^2 + (\mu\sigma)^2 (\omega-Vk)^2\right] + \frac{\sigma}{2} B^2 (k^2+K^2)^2 - \frac{(\mu\sigma)^2}{2}(\omega-Vk)^2 \sigma B^2 < 0$$

(8)

We find that the stability is always guaranteed for infinitively short perturbations (K → ∞). For greater wave lengths, limited by the walls, we find that if all the other characteristics would be kept constant, the high delivery EMP prototype could be unstable for a channel height greater than 0.4 m. This condition is obviously very far away from practical conditions in EMP.

5. CONCLUSION

We have reported experimental results, up to 700 m³/h and theoretical analysis. Both of them lead us to conclude that correctly designed high delivery EMP, with stable running conditions are quite possible.

6. REFERENCES

[1] GAILITIS A. and LIELAUSIS O., 1975, 'Instability of homogenous velocity distribution in an induction-type MHD machine', Magnitnaya Gidrodinamika, N° 1, P.87.

[2] DEVERGE C., LEFRERE J.P., PETURAUD P. and SAUVAGE M., 1984, 'The use of electromagnetic pumps in large LMFBR plants', Liquid Metal Engineering and Technology. Proceedings of the 3rd International Conference - Oxford, Vol. 3, P.209.

[3] KEBADZE B.V. et al., 1979, 'Investigation of unstable working conditions of a cylindrical linear induction pump', Magnitnaya Gidrodinamika, N° 4, P.89.

[4] KAMKAR H. and MOFFATT H.K., 1982, 'A dynamic runaway effect associated with flux expulsion in magnetohydrodynamic channel flow', Journal of Fluid Mechanics, Vol. 121, P.107.

3500 M^3/H MHD PUMP FOR FAST BREEDER REACTOR

B.G. KARASEV, I.R. KIRILLOV, A.P. OGORODNIKOV
Efremov Institute of Electrophysical Apparatus
189631 LENINGRAD
USSR

ABSTRACT. This paper deals with the results of the design and liquid metal tests of the electromagnetic pump CLIP-3/3500 on 3500 m^3/h. The pump is intended for installation in one of the second loops of BN-350 commercial reactor in the Mangyshlak power plant. The characteristics of the pump which illustrate its suitability for exploitation are stated. Finally, the paper describes the powerful electromagnetic pumps working under large values of electromagnetic and MHD-interaction parameters.

Introduction

In the USSR there has been a steady interest in the development of high power electromagnetic pumps (EMP) for the main circuits of fast breeder reactofs [1]. This interest periodically also appears in other countries, e.g. USA [2], France [3]. It is caused by the potential advantages of EM pumps over mechanical pump, owing to the absence of rotating parts, which results in higher reliability, simplicity and convenience of maintenance and control.

The USSR has a successful history of EMP operation in the main circuits of experimental fast reactors. The next step for EMP installation in the main circuits of fast reactors should be a semicommercial operation of a rather powerful EM pump to verify its potential advantages. For this purpose, in 1985, the Efremov Institute designed and manufactured the pump CLIP-3/3500. This pump is intended for one of the secondary loops of BN-350 commerical reactor in the Mangyshlak power plant.

1. CLIP-3/3500 DESIGN PARAMETERS AND TEST RESULTS

The general view of the pump and its construction plan are given in Fig. 1a, b. Its flow passage has a 180° turn which makes it possible to understand one of EMP's important advantages : the possibility to disassemble its parts (inductor) requiring repair or replacement without loss of liquid metal circuit tightness. The existing arrangement of pipes in the BN-350 second loop caused us to accept the unusual direction of flow in the pump's duct as shown by arrows in Fig. 1b. The winding is water cooled through the backs of inductor packets which is allowed for pumps not immersed in sodium.

To remove gas while filling the loop with metal top part of the duct can be connected with the expansion tank through a valve. Main design dimensions and parameters of the pump are as follows : duct mean diameter d = 0.95 m, duct height 2b = 26 mm, nonmagnetic gap 2δ = 39.5 mm, duct outer wall thickness b_0 = 6 mm, inner wall thickness b_i = 1.5 mm, pole pitch τ = 156 mm, number of

333

J. Lielpeteris and R. Moreau (eds.), Liquid Metal Magnetohydrodynamics, 333–338.
© *1989 by Kluwer Academic Publishers.*

 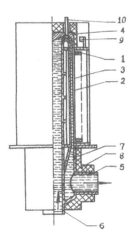

Figure 1. General view (a) and construction scheme (b) of pump CLIP-3/3500 : 1 - inductor ; 2 - inner magnetic structure ; 3 - duct ; 4 - adapter ; 5 - outlet tube ; 6 - inlet tube ; 7 - receiving tank ; 8 - electric heating with thermal insulation ; 9 - current supply leads ; 10 - drain tube.

poles $2 p_n = 12$, supply voltage frequency $f = 50$ Hz. The pump overall dimensions : height - 5 m, diameter - 1.8 m, total mass - 17 t, removable part (inductor) mass - 10 t.

Tests of the pump in liquid sodium metal at a temperature of 300° were carried out in 1986 [4]. The experimentally obtained nominal regime is characterized by the following parameters : flowrate Q = 4600 m^3/h, developed pressure U = 650 V and I = 3000 A, power factor 0.3 efficiency η = 30 %.

Listed parameter values are rather close to design values (calculation errors do not exceed 9 %, for efficiency at the same flowrate - 3.5 % absolute).

Experimental hydromechanical (outer) characteristics of the pump at variable hydraulic resistance of the loop and regulating characteristics at constant hydraulic resistance are shown in Fig. 2. In the range of presented flowrates and pressures pump operation is stable and the amplitude of parameter obtained during the experiment are : pump developed pressure - 0.32 MPa ; flowrate - 4200 m^3/h. Maximum efficiency is reached at Q = 3600 m^3/h. The value of power factor in this point is close to maximum.

Heat tests showed that in rated operating conditions the winding temperature did not exceed 280°, which provides highly reliable service life of no less than 30000 hours. The studies of mechanical stresses, cavitation, vibration and noise characteristics showed that the pump met the operation requirements of fast reactors main circuits (measured vibration level was smaller by a factor of 3 than that of a mechanical pump). As a result of tests the pump CLIP-3/3500 was accepted for installation in the BN-350 reactor.

3. Powerful EMPs

The development and production of powerful EMP have not only practical but also great theoretical interest. Practice has shown the unique possibility to investigate MHD-flows in a travelling magnetic field at high values of electromagnetic and MHD interaction parameters. Operation modes of the pump CLIP-3/3500 are characterized by the following values of these parameters :
- magnetic Reynolds number or electromagnetic interaction parameter, defining the ratio of induced and resultant magnetic field induction modules,

$$0.15 \leqslant R_m s = \mu_0 \sigma \omega b s / (\alpha^2 \delta) \leqslant 2,$$

- modified parameter of MHD-interaction which is equal to the ratio of electromagnetic pressure at s = 1 and pressure friction drop at turbulent flow with synchronous velocity

$$Ha^2 = 4B^2_{mo} \sigma b / (\rho \lambda v_s) \leqslant 180 ;$$

- parameter of MHD-interaction which is equal to the ratio of electromagnetic and inertial forces, expressed through the flow local (N_1) and integral (N_2) characteristics

$$0.1 \leqslant N_1 = \sigma B^2_m s \tau / [\pi \rho v_s (1 - s)^2] \leqslant 5 ;$$

$$2 \leqslant N_2 = \sigma B^2_m s L k / [\rho v_s (1 - s)^2] \leqslant 20 ;$$

Here $\omega = 2\pi f$, s - slip, $\alpha = \pi/\tau$, B_m, B_{mo} - induction amplitude in operating and idle (without liquid metal) conditions, respectively : ρ - density, λ - friction coefficient, v_s - synchronous velocity, L - active length of the inductor, k - factor. taking into account various end effects.

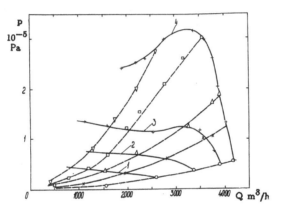

Figure 2. Outer (1-4) and regulating (dots) characteristics of pump CLIP-3/3500 : 1-U = 200 V ; 2-325 V ; 3-490 V ; 4-650 V.

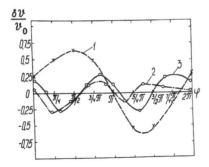

Figure 3. Velocity distribution over the duct azimuth in pumps CLIP-3/150 (curve 1), CLIP-5/850 (2), CLIP-6/900 (3) at $N_2 \approx 10$.

Under $R_m > 1$ and $Ha^2 > 1$ outer (hydromechanical) characteristics of the EMP are non-monotonic (Fig. 2, curve 4). The same developed pressure can be obtained at different flowrates (metal velocities in the pump duct). Theoretical studies of these operation modes made by soviet investigators [5, 6], show the instability of uniform and the stability of nonuniform velocity distribution in MHD-machine duct in pump and generator operating conditions. A number of studies have define the boundaries of uniform flow instability and analysed the stability of the non-uniform flow of different types. Nonuniform distribution of the pump duct velocities leads to a reduction of its pressure and power transfered into secondary medium compared to uniform flow case. Under certain conditions this may cause pump parameter pulsations.

Experimental studies of the above mentioned operating conditions have been made on a number of EMPs with cylindrical ducts and rated flowrate of sodium from 150 to 1200 m^3/h [1, 7, 8]. Direct measurements of velocity were made with special sensors (Pitot-Prandtl tubes) distributed around the circumference in several cross-sections along the duct length.

Nonuniform velocity distributions over duct azimuthal close to sinusoidal were measured in all these pumps. Deviation of local velocity from mean velocity $\delta v = v - v_0$ given in Fig. 3 refers to the duct cross-section at the inductor outlet (for CLIP-5/850 to the cross section at a distance of $1.5\,\tau$ from the outlet inside an active zone). The number of maximums and minimums in $\delta v/v = f(\varphi)$ curve and their positions at the circumference are obviously defined by external disturbances : conditions at the flow inlet (outlet) to the annular duct and nonuniform distribution of the external magnetic field over azimuth. The movement in space of the $\delta v/v = f(\varphi)$ distribution, relative to fixed sensors while turning the inductor, whose partly absent teeth caused a nonuniform distribution of external magnetic field over φ, was experimentally demonstrated for the pump CLIP-5/850 [1].

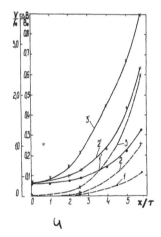

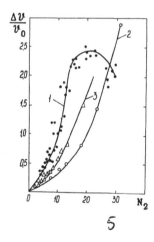

Figure 4. Distribution of velocity (curves 1-3) and magnetic field induction (1'-3') disturbances along the pump CLIP-3/150 length : 1,1'-s = 0.32, N_2 = 4.9 ; 2,2'-s = 0.48, N_2 = 9.6 ; 3,3'-s = 0.62, N_2 = 19.7.

Figure 5. Dependence of velocity disturbance at pump outlet upon N_2. Curves : 1 - CLIP-3/150, duct without baffles ; 2 - the same, duct with baffles ; 3 - CLIP-6/900, duct without baffles.

In the pumps CLIP-3/150 and CLIP-6/900 external disturbances may result from asymmetry of two-layered disk coils producing the local disturbances of the external magnetic field over one azimuthal line in CLIP-3/150 and two in CLIP-6/900.

It is found that the azimuthal distribution of resultant magnetic field follows exactly the velocity distribution. This makes it possible to use field measurements for qualitative description of velocity profile change along the machine length. The azimuthal velocity profile develops from the pump inlet to outlet and achieves maximum nonuniformity in the pump outlet part. In the duct divided by a sufficient number of longitudinal electroconducting baffles, the velocity profile is developed at a short inlet part and remains unchanged over the major part of the duct length. Fig. 4 shows the distribution of velocity (curves 1-3) and resultant magnetic field induction (curves 1'-3') disturbances in three cross-sections along the pump CLIP-3/150 length ($\Delta v = (\delta v)_{max} - (\delta v)_{min}$; $x/\tau = 0$ - inlet, $x/\tau = 6$ - outlet).

Nonuniformity of azimuthal velocity distribution increases with increasing s and Ha^2 up to the specified values of these parameters, and for further increases the distribution character is changed and parameter pulsations appear. For a pump all measured values of pole pairs can be expressed as a function of one parameter N_1 or N_2 (Fig. 5).

The fact that Δv smoothly approaches zero at N-->0 shows that in practice there is no sharp boundary in transition from uniform to nonuniform flow, i.e. no boundary theoretically investigated in a number of works.

In the duct with longitudinal baffles flow non uniformity is less than that in the duct without baffles at the same values of N. Pump parameter pulsations (flowrate, pressure, supply current) observed in our experiments at $N_2 \geq 15$-20 took the form of low frequency oscillations (f = 0.1-1.5 Hz) with an amplitude of up to 3.5 % in CLIP-3/150 and 18 % in CLIP-5/850.

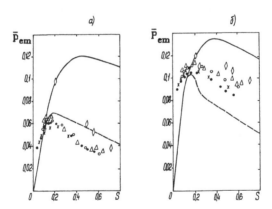

Figure 6. Dependences of relative electromagnetic pressure (a) and power (b) upon slip in pump CLIP-3/3500.

Another negative consequence of nonuniform flow in the duct is the decrease in densities of mechanical and electromagnetic power and pump efficiency. Fig. 6a, b show experimental values of electromagnetic pressure p_{em} developed by the pump and electromagnetic power P_{em} transferred into the secondary medium. Base values are pressure and power, respectively, which are calculated on the external magnetic field induction, corresponding to experimentally measured phase current, and slip s = 1 [1]. As seen from Fig. 6, the experimental data, beginning from some values of s, are arranged much lower than those calculated in electrodynamic approximation (solid curves) [1]. As seen from Fig. 6

the experimental data, beginning from some values of s, are arranged much lower than those calculated in electrodynamic approximation (solid curves) [1]. Calculation using semiempirical dependences taking into account nonuniform flow (dotted curves) agrees more with the experiment, but also requires further improvement.

4. Conclusion

In designing powerful EMPs the features of their operation discussed above are to be taken into account. Rated operating conditions should be chosen with a number of restrictions : $R_m s < R_{m\ crit}$, $Ha^2 < (Ha^2)_{crit}$, $N < N_{crit}$, Ha^2 and $R_m s$ are to be such, that an operating point is at the increasing part of p(s) - pump characteristic. The results of this paper can be used for estimating allowable values of N with reasonable flow nonuniformity. For CLIP-3/3500 the corresponding parameters at rated point are : $s = 0.176$, $R_m s = 0.66$, $Ha^2 = 154$, $N_1 = 0.25$, $N_2 = 5.8$. In this point there are no large deviations from data calculated in electrodynamic approximation and practically no parameter pulsations.

The use of feeding current frequency less than 50 Hz simplifies the fulfilment of the above mentioned conditions and allows the pump efficiency to increase to some extent in comparison with 50 Hz case.

If the pump should operate in slip zone, i.e. $R_m s > 1$, it is necessary to take special measures, reducing flow nonuniformity and supressing parameter pulsations. Among these are the installation of longitudinal electroconducting baffles and the use of special windings [9].

References

[1] GLUKHIKH V.A., TANANAEV A.V., KIRILLOV I.R., 1987, Magnetohydrodynamics in Nuclear Energetics, M., Energoaromizdat.

[2] KLIMAN G.B., 1979, "Large Electromagnetic Pumps", Electrical Machines and Electromechanics, vol. 3, 129.

[3] DEVERGE C., LEFRERE J.P., PETURAID P., SAUVAGE M., 1984, "The use of Electromagnetic Pumps in Large LMFBR Plants", In : Liquid Metal Engineering and Technology, vol. 3, 245, Proc. 3-d Intern. Conf. Oxford, April 1984.

[4] ANDREEV A.M., BEZGACHEV E.A., KARASEV B.G. et al, 1988, "Electromagnetic Pump CLIP-3/3500', Magnitnaya Gidrodynamica, N 1, 61.

[5] GAYLITIS A., LIELAUSIS O., 1975, "Uniform Velocity Distribution Instability in Induction MHD Machine", Magnitnaya Gidrodynamica", N 1, 87.

[6] VOLCHEK B.V., GEKHT G.M., TOLMACH I.M., ALKIN A.I., 1976, "Hydrodynamic Instability and Stationary Flows in Induction MHD Pumps Coaxial Channel", Magnitnaya Gidrodynamica, N 2, 62.

[7] KIRILLOV I.R., OSTAPENKO V.P., 1987, "Cylindrical Induction Pump Local Characteristics under $R_m s > 1$", Magnitnaya Gidrodynamica, N 2, 95.

[8] KIRILLOV I.R., OSTAPENKO V.P., 1987, "Cylindrical Induction Pump Integral Characteristics under $R_m s > 1$", Magnitnaya Gidrodynamica, N 3, 115.

[9] KIRILLOV I.R., OGORODNIKOV A.P., OSTAPENKO V.P. PRESLITSKIY G.V., 1981, "Suppression of Instability in Linear Induction Pumps under $R_m s > 1$ with Outer Magnetic Field Modification", Magnitnaya Gidrodynamica, N 4, 105.

SELF-EXCITATION OF LIQUID METAL MHD GENERATORS

Ph. MARTY, A. ALEMANY
Institut de Mécanique de Grenoble
B.P. 53 X
38041 - GRENOBLE Cedex
France

ABSTRACT. This paper gives recent results obtained on the study of liquid metal MHD generators working under self-excitation. Both theoretical and experimental investigations have been made on induction and conduction generators. For each type of machines, an analytical expression of the critical velocity allowing self-excitation is given and the experimental results obtained on a mercury loop are presented. These results are in fair agreement with the theoretical predictions.

1. INTRODUCTION

In Liquid Metal MHD Conversion, the problem of self-excitation of the generator arises when the electrical power has to be furnished in remote places (space applications for example). This means that the generator has to start up in a complete autonomous way without the help of any external power supply. This question has been experimentally investigated by some researchers/1, 2, 3, 4/ but no exact analytical expression of the critical velocity that the liquid metal has to reach has been proposed until now.

This paper consists of a description of theoretical, numerical and experimental works carried out on both induction and conduction generators.

2. THE INDUCTION GENERATOR

The induction process makes use of a sliding magnetic field and the electric current is created by direct magnetic coupling between the liquid metal vein and the windings of the inductor.

Let us consider the induction generator presented in Ref. /5/. It is composed of a liquid metal stream flowing in a rectangular channel located in the gap of the magnetic field inductor. The latter is composed of two identical rectangular iron blocks with open slots to set the polyphase electrical windings necessary to create the primary sliding magnetic field. The reactive power necessary to magnetize the

339

J. Lielpeteris and R. Moreau (eds.), Liquid Metal Magnetohydrodynamics, 339–346.
© *1989 by Kluwer Academic Publishers.*

gap is provided by capacitors, C, connected in parallel with each phase of the inducting coils (figure 1). The problem consists of finding the expression of the "critical" velocity of the fluid, V_c, which will allow self-excitation. In this section we will present an analytical model of the infinitely long induction generator leading to expression for V_c. Then, the consequences of the finite length of the generator will be studied by the use of a numerical model. Finally the results obtained with an experimental mercury loop will be described.

2.1. Analytical expression of the critical velocity

We consider a liquid metal of electrical conductivity σ_f and density ρ, flowing at constant velocity V (fig. 1) in the x direction of an infinitely long channel of thickness a and width l. The thickness ε of the side walls, which have an electrical conductivity σ_w leads to a value e of the gap such as : $e = a + 2\varepsilon$.

Because of the two copper bars located on both sides of the flow, the electrical current density, j, can be assumed to be directed along the z axis only, whereas the magnetic field, B, lays in the x-y plane.

The inducting coils are frequently built in the form of three equally spaced windings and fed by a three-phases alternative set of currents. Previous works of the authors |5| have shown that the discrete distribution of the primary currents can be looked at as a continuous current sheet for low values of the magnetic Reynolds number, Rm :

$$Rm = \frac{\mu_o \cdot \sigma_f \ (V - Vs)}{k} \ll 1 \tag{1}$$

where μ_o is the magnetic permeability of the fluid, and k is the wave number such as : $k = 2\pi /\lambda$ (λ is the wavelength of the windings).

Under this condition, the equivalent primary current sheet can be written :

$$i = i_o \ e^{j(\omega t - kx)} = \frac{3 \ k.I}{\pi} \ e^{j(\omega t - kx)} \tag{2}$$

where ω is the pulsation of the electrical parameters.

Once integrated along the gap, e, the induction equation :

$$\mu_o \cdot \sigma_f \cdot \frac{\partial By}{\partial t} = \mu_o \cdot \sigma_f \cdot V \cdot \frac{\partial By}{\partial x} + \frac{\partial^2 By}{\partial x^2} + \frac{\partial^2 By}{\partial y^2} \tag{3}$$

becomes

$$\mu_o \ (a \ \sigma_f + 2\varepsilon \sigma_w) \ \frac{\partial B_y}{\partial t} + \mu_o a V \sigma_f \ \frac{\partial B_y}{\partial x} - e \ \frac{\partial^2 B_y}{\partial x^2} = - 2 \left(\frac{\partial B_x}{\partial x} \right)_{y = \pm e/2} \tag{4}$$

Using Ampere's theorem over a contour surrounding the current sheet leads to the expression of B_x :

$$(B_x)_{y = \pm e/2} = \mu_o \cdot \frac{i}{2} \tag{5}$$

In the frame of low magnetic Reynolds number, the solution of equation (4) satisfying spatial and temporal periodicity conditions is :

$$B_y = - j(\mu_o i_o /ke) \cdot (1 + j \ Rm^*) \cdot e^{j(\omega t - kx)} \tag{6}$$

where Rm^* = $\dfrac{\mu_o.\sigma_f.a(V-V_s)}{k.e}$ − $\dfrac{\mu_o.\sigma_w.2\epsilon.V_s}{k.e}$

In order to find the solution of the self-excitation regime equation (6) has to be linked to the equation of the external electrical circuit. This equation writes :

$$RC \frac{d^2\varphi}{dt^2} + r\ RC\ \frac{dI}{dt} + . \frac{d\varphi}{dt} + (r + R)I = 0 \qquad (7)$$

where φ is the magnetic flux per phase of the inductor. The latter can be obtained by integrating the y component of the magnetic field over a polar pitch :

$$\varphi = 2\ p\ N \int_{\lambda/2}^{\lambda} B_y.\ell.dx \qquad (8)$$

where $2p$ is the number of poles of the generator and N is the number of turns per pole.

If we remark that $\dfrac{d\varphi}{dt} = j\omega\,\varphi$ and $\dfrac{dI}{dt} = j\,\omega\,I$,

equations (6, 7, 8) give the dispersion equation :

$$- \omega^2\ RC\ \mathcal{L}_e(1 + jR_m^*) + j\omega(rRC + \mathcal{L}_e(1 + jR_m^*)) + r + R = 0 \qquad (9)$$

where $\mathcal{L}_e$ is the effective self-inductance for each phase without the secondary liquid metal circuit :

$$\mathcal{L}_e = \frac{3\ p.\mu_o.\ell.\lambda.N^2}{\pi^2.e} \qquad (10)$$

Because of the magnetic leakage between the slots, a fraction of the magnetic flux will not contribute to the linkage between the magnetic field and the metal flow. To take account of the influence of this leakage, we must distinguish between the flux effectively crossing the gap, characterized by the effective self-inductance, $\mathcal{L}_e$,and the total self-inductance $\mathcal{L}_t$. Therefore, the dispersion equation (9) has to be substituted by :

$$- \omega^2 RC(\mathcal{L}_t + jR_m^*\mathcal{L}_e) + j\omega(rRC + \mathcal{L}_t + j\mathcal{L}_e R_m^*) + r + R = 0 \qquad (11)$$

If we use the complex notation for ω :

$$\omega = \omega_r + j\ \omega_i \qquad (12)$$

where ω_r is the pulsation of both electric and magnetic fields and ω_i their growth rate in the transitory regime, the stable regime of operation of the generator ($\omega i = 0$) will also satisfy the dispersion equation (11) whose solution is :

$$\omega_r^2 = 1/\mathcal{L}_t C - 1/R^2 C^2 \qquad (13)$$

$$R_m^* = r/\mathcal{L}_e.\omega_r + \mathcal{L}_t/\mathcal{L}_e.RC.\omega_r \qquad (14)$$

These solutions caraterize the critical velocity of the flow corresponding to the self-excitation threshold :

$$V_c = V_s \left(1 + \frac{\sigma_w}{\sigma_f} \cdot \frac{2\,\varepsilon}{a} \right) + \frac{\mathcal{L}_t}{\mathcal{L}_e} \frac{e}{V_s \cdot \mu_o \cdot \sigma_f \cdot a} \left(\frac{r}{\mathcal{L}_t} + \frac{1}{RC} \right) \qquad (15)$$

2.2. Description of the experimental installation

The installation (see ref. 5) is a stainless steel closed loop, about 2 meters in height, in which a flow of mercury is maintained by a centrigugal pump. The mercury flowrate can be measured by an electromagnetic flowmeter. The experimental generator comprises a channel of a rectangular cross-section situated between two magnetic yokes which support the induction coils of the stator.

2.2.1. The Channel The test-channel is built of non-magnetic stainless steel, 0,5 mm in thickness, in order to limit the possibility of shunting the induced currents through the generator walls. The interior cross-section of the channel is : a x 1 = 5x50 mm^2 and its length L is 600 mm. To permit the currents induced in the fluid to develop perpendicularly to the flow, a copper side-bar was placed on each side of the channel. The ratio of the electrical conductivities of copper and mercury (σcopper $\simeq$ 60. σ mercury) allows us to justify the use and validity of the two-dimensional theoretical model presented above.

2.2.2. The Stator. The characteristics of the magnetic circuit and of the winding were chosen in such a way as to permit self-excitation of the generator at a velocity well within the capabilities of the installation. These were :
* wavelength : 0.28 cm ;
* number of poles per phase : 4 ;
* gap : 6 mm (i.e. 5 mm of mercury and 2 x 0.5 mm of stainless steel walls) ;
* Number of slots on each half of the stator : 24
* Width of a tooth : 10 mm
* Width of a slot : 13 mm
* Depth of the slots :
  Nos 1 to 6 and 19 to 24 : 50 mm
  Nos 7 to 18 : 62 mm
* Height of the filling :
  Nos 1 to 6 and 19 to 24 : 27 mm
  Nos 7 to 18 : 54 mm
* Number of conductors in slots Nos 1 to 6 and 19 to 24 : 90
* Number of conductor in slots Nos 7 to 18 : 180
* Total numbers of turns per pole and per phase : 360
* Cross-sectional area of a copper conductor : about 2 mm^2
* Length of conductor necessary for the winding of 1 phase : about 700m
* D.C. resistance of one phase : 6
    In reference /5/ it was shown that the finite length of the generator is responsible for an important inbalance between the three phases of the windings and, then of the inductances of the coils. A solution to fight this problem consists of adding additional

compensative coils in the "extreme" phases (upstream and downstream). The exact calculation of the optimum value of the number of turns N_{comp} of the compensative coils has been determined by the use of a finite elements software (FLUX 2D).

2.3. Description of the experimental installation

For a value of N_{comp} equal to 30, self-excitation has occured for a critical velocity, Vc, of the fluid equal to 10,5 m/s. Figure 2 shows the evolution of an initial impulse of 7 volt dc applied to one of the phases when $R = \infty$. The steady state voltage is 13,9 Hertz, 220 V between phases.

Substituting the experimental values of the parameters of the generator and external circuit into equation (15) (C = 200 μF ; R = $= \infty$; $\mathcal{L}_t/\mathcal{L}_e \approx 2,7$; $\mathcal{L}_t$ = 650 mH) gives the theoretical value : Vc = = 10,4 m/s very close to the experimental measured velocity.

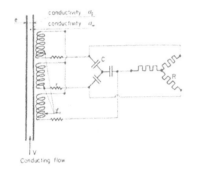

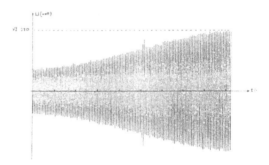

Fig. 2 – Transient regime of the self-
excitation : frequency $\approx$ 13 Hz

Fig. 1 – Diagram of the electrical connexion of a three-
phased self-excited MHD
induction generator

3. THE CONDUCTION GENERATOR

It is composed of a stainless steel channel with insulated walls and copper electrodes to collect the electrical current which will feed the magnet coil and the load. The figure 3 shows the equivalent electrical circuit of our experimental conduction generator where w is the distance between the electrodes, k is an "attenuation" coefficient taking into account the non-uniform profile of the velocity. R_i is the internal resistance of the flow including contact resistances between mercury and copper and R_E is the end effect shunting resistance (see /6/ for example). R_L is the total load resistance including the magnet coil resistance as well as the connexion bars resistance.

344

With this model the self-excitation critical velocity can be written as :

$$V_c = \frac{(R_L + R_i /\!\!/ R_E)(1 + R_i/R_E).E}{k \, w \, \mu_o \, N} \tag{16}$$

where $R_i /\!\!/ R_E = R_i R_E / R_i + R_E$ and where E is the magnet air gap and N the number of turns of the coil. Introducing the load factor ϕ defined as the load voltage divided by $V B w$:

$$\varphi = U/V \, B \, w \tag{17}$$

leads to a simple expression of the electrical efficiency, η_{el}, defined as the ratio between the power given to the load and the power extracted to the flow by the $j \times B$ forces :

$$\eta_{el} = \varphi \left(1 - \frac{\varphi}{1 - \varphi} \cdot \frac{R_i}{R_E} \right) \tag{18}$$

Then the efficiency is zero for $R_L = 0$ and $R_L = \frac{1}{1 + R_i/R_E}$ (see figure 4) In our experimental generator the measured values of k and R_i/R_E are :

$$\begin{cases} k = 0,69 \\ R_i / R_E = 0,052 \end{cases} \tag{19}$$

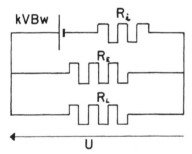

Fig.3 – Equivalent electrical circuit of a self-excited conduction generator

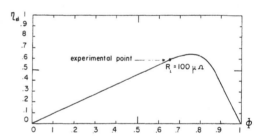

Fig. 4 – Electrical efficiency VS. the load factor

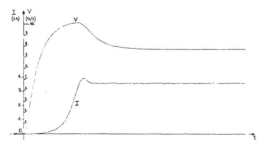

Fig. 5 – Evolution in time of the flow velocity and of the electric current in the load after the starting of the pump ($t = 0$)

The value of R_i/R_E is in global agreement with the value predicted by GHERSON an LYKOUDIS /6/ where :

$$R_1/R_E = \frac{w}{L} \cdot \frac{2.\text{Ln } 2}{\pi} = 0.08 \qquad (20)$$

Figure 5 shows the"natural" evolution in time of the electrical current in the channel and of the flowrate once the pump has been started.

Once the stationary regime is obtained, the final voltage drop to the magnet coils ends is $\simeq$ 0,38 Volt and the electric current delivered by the fluid is approximatively 3700 A. The magnetic field is about 1,4 Tesla.

The measured critical velocity is V_c = 7,98 m/s corresponding to a load factor φ = 0,65 and an electrical efficiency η_{el} = 57 %

4. CONCLUSION

An expression for the critical velocity to obtain self-excitation in induction and conduction LMMHD generators has been proposed and experimentally verified. However, more extensive works could be done :
* Concerning induction generators a theoretical and numerical study should allow to take into account the ends effects which previously have been described (see /7/ and /8/). On the experimental point of view a Nak flow experiment is under construction in the frame of a joint project with the Beer Sheva University, Israel.
* Concerning conduction generators, a study of the transient regime between the starting up of the flow and the stationary state could be useful in view of space applications.

REFERENCES

1. GLUKHIKH V.A., KIRILOV I.P., 1966, 'Experimental investigation of an asynchronous self-excited liquid metal MHD generator'. Magnit. Gidrodin, Vol. 2, n° 4,   107
2. ULBER M., SCHULZ T., 1967, 'Experimental study of a liquid-metal, self-excited MHD induction generator' Eighth Symposium on Engineering Aspects of Magnetohydrodynamics, American Institute of Aeronautics and Astronautics, New York, 33
3. CERINI D.J., 1974, Nak-Nitrogen Liquid metal MHD Converter tests at 30 kW". AIAA.J., vol 12, n° 1, 78
4. ELLIOTT D.G. et al., 1966, 'Theoretical and experimental investigation of liquid metal MHD power generation' Symposium on Electricity from MHD, Salzbourg, vol. II, SM-74/177, 995
5. JOUSSELLIN F. ALEMANY A., WERKOFF F. MARTY ph., 1988, 'MHD induction generator at weak magnetic Reynolds number : Part 1.- Self-excitation criterion and efficiency' Journal de Mécanique Théorique et Appliquée

346

6. GHERSON P., LYKOUDIS P., 1979, 'Analyticalstudy of end effects inliquid metal MHD generators', School of nuclear engineering, Purdue Univ., West Lafayette, Ind. US.

7 .POLOUJADOFF M., 1980, The theory of linear induction machinery,, Clarendon Press-Oxford UK

8. YAMAMURA S., 1972, Theory of linear induction motors, University of Tokyo Press

COMPREHENSIVE STUDY ON THE MHD PHENOMENA IN THE METAL POOL WITH THE
SINGLE-PHASE INDUCTION COIL

E. Takeuchi, J. Sakane, H. Yano, I. Miyoshino, T. Saeki, and
H. Kajioka
Nippon Steel Corporation
1-1-1 Edamitsu, Yawata-higashi-ku,
Kitakyushu city 805
Japan

ABSTRACT. The behavior of liquid metal with the single-phase induction
coil has been studied by measuring the magnetohydrodynamic phenomena in
the mercury pool and by performing the numerical analyses, in which the
electromagnetic fields are coupled with the velocity fields. The theoretical analysis has generally explained the experimental results : the
distribution of electromagnetic pressure and the resulting both meniscus
shape and flow pattern in the mercury pool.

1. INTRODUCTION

Sufficient understanding of MHD phenomena in the metal pool surrounded
by the single-phase induction coil is a key for the fine control of the
refining process using a coreless induction furnace and, more recently,
of the electromagnetic casting process. A number of experimental works
[1-5] and numerical analyses [2],[3],[5-12] have been carried out in
order to quantify both the stirring characteristics and the shaping
characteristics in these operation systems.

In this comprehensive study, the important parameters describing
MHD phenomena ; the electromagnetic fields, the electromagnetic pressure, and the resulting both meniscus shape and velocity fields in the
experimental system employing mercury with the low-frequency (60Hz) induction coil are measured by means of various kind of local-measurement
devices. Measured data have been compared with the predictions performed by the mathematical analyses.

2. SIMULATION EXPERIMENTS

2.1. Experimental System and Conditions

Schematic illustration of experimental apparatus and experimental conditions are shown in Fig. 1 and Table 1 respectively. System consisted
of a glass vessel with a water-cooled induction coil. Experiments were
carried out for the various level of meniscus in the range of the coil
thickness ($-40\text{mm} \leq z \leq 40\text{mm}$).

J. Lielpeteris and R. Moreau (eds.), Liquid Metal Magnetohydrodynamics, 347-354.

348

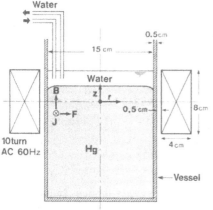

Fig. 1 Experimental apparatus.

Table 1 Experimental conditions.

Induction coil	Outer diameter;246(mm) Inner diameter;170(mm) Coil height ; 80(mm) 10turn(5turn-parallel) Supplied current; 2200A(max) 60Hz, single phase
Mercury sump	Inner diameter;150(mm) Bottom of the sump; 73mm below the end of the coil Meniscus level; $-40 \leq z \leq 40$mm

2.2. Measurement Technique

2.2.1. <u>Electromagnetic Fields</u> Fig. 2(a) shows the structure of the probe for the simultaneous determination of the magnetic flux, the induced current in the mercury pool, and the phase difference between them. The r- and the z- component of the magnetic flux were determined by the measurements of electric potential induced in the search coils which had symmetric axes in the r- and the z-direction respectively. Potential sensor located at the bottom of these search coils permitted the simultaneous determination of the density of eddy current induced in the mercury pool.

2.2.2. <u>Electromagnetic Pressure</u> The pressure distribution in the mercury pool was measured by a manometer made of glass tube. Because the mercury in the glass tube is insulated electrically from the eddy current induced in the bulk mercury, the height of mercury in the glass tube practically indicates the local pressure at the tip of the probe. Electromagnetic pressure was derived from the pressure distribution measured in the mercury pool.

2.2.3. <u>Meniscus Shape</u> One of the distinctive phenomena resulting from the electromagnetic pressure is the shaping of mercury surface. The meniscus-shape measuring system consisted of the needle electrode traveling both vertically and horizontally by driving unit, and the displacement sensor. On account of the turbulent meniscus shape, the measurements were iterated to yield the average shape.

2.2.4. <u>Velocity Fields</u> Another major influence of the electromagnetic pressure is the fluid flow in the mercury pool. Determination of the magnitude and the direction of the local velocity was performed with the probe developed by Vivés [13], which is schematically illustrated in Fig. 2(b). The output signal from the sensor was processed numerically to erase the electric noise. One example of the processing condition

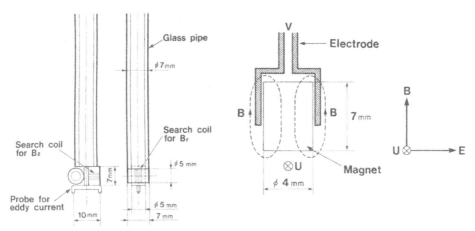

(a) magnetic flux and induced current (b) velocity

Fig. 2 Schematic illustration of measurement probes.

was; the sampling interval = 0.1s, the integration time = 0.08s, and the number of sampling = 301 times.

3. NUMERICAL CALCULATIONS

3.1. Electromagnetic Fields and Electromagnetic Pressure

Electromagnetic fields in the experimental system are governed by the set of Maxwell's equations. Additionally Ohm's law and Kirchoff's first law are introduced.

According to Tarapore and Evans [6], the eddy current induced in the controlled system shown in Fig. 3 becomes;

$$J_\theta(r,z) = -j\,\omega\sigma\mu \int_0^R \int_0^H J_\theta'(r',z')f(r,r',z,z')dr'dz'$$

$$-j\,\omega\sigma\mu \sum_{i=1}^{Nc} Ic(rc,zc)f(r,rc,z,zc) \qquad (1)$$

The factors $f(r, r', z, z')$ are the function of mutual inductance [14] between circular current at (r, z) and one at (r', z').

The r- and z-component of the magnetic flux are given by eqs.(2) and (3) respectively.

$$Br = -\frac{1}{\sigma\omega}\frac{\partial J_\theta}{\partial z} \qquad (2)\,, \qquad Bz = \frac{j}{\sigma\omega}\frac{\partial}{r\,\partial r}(rJ_\theta) \qquad (3)$$

The r- and z- component of time-averaged electromagnetic force are,

$$\overline{Fr} = \frac{1}{2}\,Re\,(J_\theta \cdot Bz^*) \qquad (4)\,, \qquad \overline{Fz} = -\frac{1}{2}\,Re\,(J_\theta \cdot Br^*) \qquad (5)$$

where the superscript asterisk denotes a complex conjugate quantity.

3.2. Meniscus Shape

Assuming that the profile of meniscus is affected by the static pressure, the electromagnetic pressure, and the surface tension, neglecting the term of dynamic pressure for the simplication as a first generation model, the equation governing the shape is derived as follows;

$$\rho g \Delta z = - \int \bar{F}r \; dr + \int \bar{F}z \; dz + \gamma \left(\frac{1}{R_1} + \frac{1}{R_2} \right) \tag{6}$$

3.3. Velocity Fields

Velocity distribution in steady state has been obtained by solving Navier-Stokes equation, eq.(7), and equation of continuity, eq.(8).

$$\rho(u \cdot \nabla)u = - \nabla \cdot P + \mu_{eff} \nabla^2 u + \rho g + \bar{F} \tag{7}$$

$$\nabla \cdot u = 0 \tag{8}$$

The effective viscosity appearing in eq.(7) is

$$\mu_{eff} = \mu_\ell + \mu_t \tag{9}$$

The electromagnetic force calculated by eqs.(4) and (5) is substituted into the last term on the right of eq.(7).

Eqs.(1) through (6) were solved, subject to the boundary conditions, using implicit finite-difference method assuming that the liquid metal is stagnant, first to obtain the meniscus shape. Then the velocity fields were computed with the k-ε model, in which the wall function

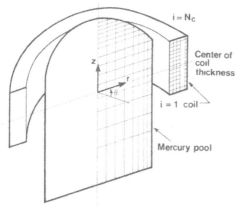

Fig. 3 Sectioned view of the mercury pool and the coil showing coordinate system in the numerical model.

Table 2 Properties and parameters used in the computation.

σ	(S/m)	1.04×10^6
ρ	(kg/m^3)	13.6×10^3
f	(Hz)	60
μ_0	(H/m)	$4\pi \times 10^{-7}$
μ_ℓ	(kg/m·s)	0.15×10^{-2}
γ	(N/m)	0.482

and the free slip condition were employed at the vessel wall and at the meniscus respectively.

Properties and parameters used in the computation are listed in Table 2.

4. EXPERIMENTAL RESULTS IN COMPARISON WITH THE NUMERICALLY CALCULATED RESULTS

4.1. Electromagnetic Fields

Fig. 4(a) $\sim$ (c) show the radial distribution of the magnetic flux, the induced current, and the phase difference at the center of coil thickness respectively. Note that the meniscus level located at the top of the coil. These parameters were normalized by the specific magnitude at the point 5mm inside the vessel wall. The calculated results of these parameters are in coincident with the measured results both in the magnitude and in the distribution.

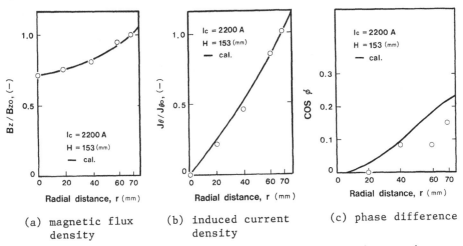

(a) magnetic flux density (b) induced current density (c) phase difference

Fig. 4 Comparison between the measured and the predicted magnetic fields in the mercury pool

4.2. Electromagnetic Pressure

Fig. 5(a) shows the pressure distribution both along the symmetric axis of the metal pool and along the vessel wall measured by the probe, shown in Fig. 3, when the meniscus level located at the top of the coil. The pressure at the central axis of the mercury pool changed by the gradient of ρg, as well as the static pressure distribution. On the other hand the pressure distribution along the vessel wall represented the non-linear change. Pressure gradient decreased above the coil center ($z=0$) and then increased to approach to the same pressure distribution at the central axis. The difference of these two distributions reveals the

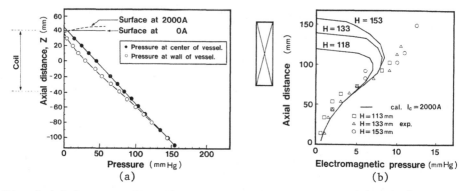

Fig. 5 (a) Pressure distribution in the mercury pool, (b) Comparison between the measured and the predicted axial distribution of the electromagnetic pressure.

electromagnetic pressure acting on the free surface and the wall side of the mercury pool. The distributions of the electromagnetic pressure evaluated in this manner and the calculated by the numerical model are shown for the different meniscus levels in Fig. 5(b). The measured distribution has a peak in the vicinity of free surface. The calculated has shown the same tendency, however, the unambiguous discrepancy can be seen in the magnitude of the peak. This is possibly due to the effect of dynamic force which is not taken account of in the computation. The boundary conditions for numerical calculation, which do not involve the free surface condition, are also responsible for this discrepancy.

4.3. Meniscus Shape

Fig. 6(a) shows the measured meniscus profiles at different meniscus levels relative to the coil. The shape of meniscus was affected by the location of meniscus and showed its maximum change when the meniscus level located at the center of the coil height. The effect of surface tension was found to be almost negligible in the computation. Fig. 6(b) shows the dependence of the meniscus height on the relative meniscus location to the coil, which is defined by axial distance from the top of meniscus to the contact point of mercury with the vessel wall. The measured meniscus height became maximum at z=0, and was relatively higher at z > 0 than at z < 0. Similar tendency was obtained in the calculated result.

4.4. Velocity Fields

The flow distributions measured by the probe illustrated in Fig. 2(b) are shown in Fig. 7 with the calculated results. Distribution of flow clearly represented the two-counter rotational vortices in the calculated results as well as in the measured results. The centripetal flow is seen below the meniscus which corresponds to the location of the peak of

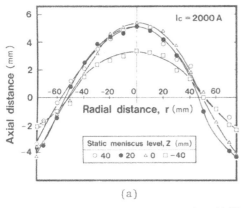 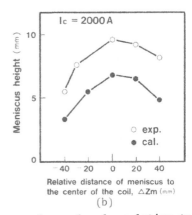

(a) (b)

Fig.6 (a) Meniscus shapes at the different meniscus levels relative to the coil, (b) Comparison between the measured and the predicted meniscus height.

electromagnetic pressure shown in Fig.5(b). The rising flow is seen at the axial center of the pool, which may affect the meniscus profile. Magnitude of velocity calculated was consistent with that of measured. These results reveal that the application of k-ε model to the electromagnetic field at the low-frequency is relatively effective.

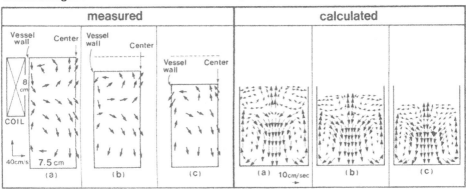

Fig. 7   Comparison between the measured and the predicted velocity profile in the mercury pool, (Ic=2200A).
(a) H=153mm, (b) H=133mm, (c) H=133mm.

5. CONCLUSION

A study involving the simulation experiments employing mercury sorrounded by induction coil and the theoretical analyses by numerical method was conducted in order to elucidate the MHD phenomena : the electromagnetic field, the electromagnetic pressure, the meniscus shape, and the velocity field. The calculated results were generally in consistent with the measured results, which permitted the quantitative explanation of MHD phenomena in the liquid pool.

354

LIST OF SYMBOLS

Br, Bz : magnetic flux density in r- and z- direction, Wb/m^2.

Bzm : peak value of magnetic flux density in z-direction, Wb/m^2.

Bzo : specific value of magnetic flux density in z-direction, Wb/m^2.

Fr, Fz : electromagnetic force in r- and z- direction, N/m^3.

g : acceleration due to gravity, m/s^2.

H : depth of mercury pool $(=Z+0.113)$, m.

Ic : current density in coil, A/m^2.

J_θ : current density in θ-direction, A/m^2.

$J_\theta m$: peak value of current density in θ-direction, A/m^2.

$J_\theta o$: specific vlue of current in θ-direction, A/m^2.

P : pressure, Pa.

R : radius of mercury pool, m.

v : voltage, volt.

u : velocity, m/s.

ΔZ : meniscus height, m.

γ : surface tension of mercury, N/m.

μ : magnetic permeability, H/m.

μ_ℓ : laminar viscosity, kg/ms

μ_t : turbulent viscosity, kg/ms.

ρ : density, kg/m^3.

σ : electric conductivity, S/m.

ω : angular frequency, rad/s.

ϕ : phase difference, rad.

REFFERENCES

[1] MOOR D. J. and HUNT J. C. R., 1982, Proc. 3rd Beer-Sheva Seminar on Liquid Metal Flows and Magnetohydroynamics, Prog. Astro. Aero., 84.

[2] EL-KADDAH N. and SZEKELY J., 1983, J. Fluid Mechanics, 133, 37.

[3] EL-KADDAH N., SZEKELY J., and CARLSSON G., 1984, Metall. Trans. B, 15B , 633.

[4] VIVES C. and RICOU R., 1986, Proc. Light Metals Committee Technical Session, 115th TMS-AIME Annual Meeting, New Orleans, 873.

[5] MEYER J. L., SZEKELY J., EL-KADDAH N., VIVES C., and RICOU R., 1987, Metall. Trans. B, 18B, 539.

[6] TARAPORE E. D. and EVANS J. W., 1976, Metall. Trans. B, 7B , 343.

[7] TARAPORE E. D., EVANS J. W., and LANGPELT J., 1977, Metall. Trans. B, 8B, 179.

[8] MEYER J. L., EL-KADDAH N., SZEKELY J., VIVES C., and RICOUS R., 1987, Metall. Trans. B, 18B, 529.

[9] SAKANE J., LI B. Q., and EVANS J. W., 1988, Metall. Trans. B, 19B, 397.

[10] TABERLET E. and FAUTRELLE Y. R., 1985, Prog. Astro. Aero., 100, 680.

[11] NAGANUMA A. and SANO K., 1987, Tetsu-to-Hagné, 73 , S681.

[12] KOBAYASHI S. and ISHIMURA S., 1987, Tetsu-to-Hagané, 73, S683.

[13] RICOU R. and VIVES C., 1982, Int. J. Heat Mass Transfer, 25, 1579.

[14] DUDLEY R. F., 1972, IEEE TRANS. on Industry Appl., 565.

GRAIN REFINEMENT IN CONTINUOUSLY CAST INGOTS OF LIGHT METALS BY ELECTRO-MAGNETIC CASTING

Charles Vivès
Université d'Avignon
33, rue Pasteur
84000 Avignon
France

ABSTRACT. The working principle of a new electromagnetic continuous cas-ting process is described. The main pecularities of this process lie, first, in the presence in the sump of a strong electromagnetically driven forced convection, which promotes the production of a fine equiaxed structure and, secondly, in the fact that the thickness of the segrega-tion zone tends to zero. The evolution of both the grain size and the thickness of the segregated cortical layer with the electric power input is presented.

I. INTRODUCTION

The addition of nucleating agent is a commonly adopted method in D.C. casting of aluminum alloys |1| and it is well known that inoculation of small amounts of grain refiners master alloys, as Al - 5 % Ti - 1 % B (At5 B), results in a significant reduction of the crystal size. Howe-ver, the addition of grain refiners may cause negative secondary effects reducing the quality of the metal. For instance, this technique can be detrimental to the quality of the finished product, due to the presence of conglomerates of Ti B2 in the metal. Moreover, the refinement may become practically ineffectual, when the efficiency of the filtering system is sufficiently high to retain a large part of the nuclei of Ti B2 before casting. Furthermore, high strength aluminum alloys, espe-cially those containing zirconium and lithium are historically difficult to grain refine with conventional titanium-boron-aluminum master alloy.

An alternative way, consisting of several dynamic methods producing a vigorous forced convection in the melt during freezing, leads to subs-tantial grain refinement. In these processes, heat and fluid flow are controlled by various externally applied forces. These methods primarily include the use of sonic, or ultrasonic, vibration and mechanical, or electromagnetic, stirring |1-3|. Among this techniques, the application of a time-varying magnetic field is of particular interest, because for-ced convection is generated in an easily controlable manner, while avoiding direct contact between the stirrer and the melt. The effects of electromagnetic stirring have been extensively studied on the solidi-

J. Lielpeteris and R. Moreau (eds.), Liquid Metal Magnetohydrodynamics, 355–361.
© 1989 by Kluwer Academic Publishers.

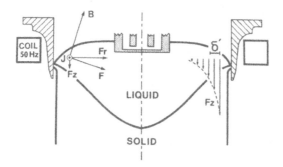

Figure 1.   Schematic of the CREM Process.

fication of steel, particularly in continuous casting |2|. However, the casting techniques of steel and aluminum have their own pecularities. For steel, the depth of the sump is of several meters, so it is relatively easy to arrange, around the solidifying ingot, several electromagnetic stirrers consisting, for instance, of a combination of linear and rotary induction motors. On the contrary, in the case of aluminum alloys, the sump depth is of the order of fifteen centimeters, i.e. of the order of the ingot mold height ; as a result, it is practically impossible to set such stirring systems, on account both of their bulkiness and of the small gap which exists, in the case of the conventional casting equipment, between the ingot mold and the cooling device. So, it seems that these electromagnetic refining processes have not been applied to the industrial production of light metals.
     This paper is devoted to the description of the working principle of a new casting technology of aluminum alloys : the CREM process (Casting, Refining, ElectroMagnetic), which is primarily characterized by the fact that the presence of a strong forced convection of electromagnetic origin leads to a significant reduction of the grain size.

2.   THE WORKING PRINCIPLE

In this process, the liquid metal could be inductively stirred, using an inductor supplied with a 50 Hz alternating current. The coil surrounds the ingot mold and assumes approximately its shape (Fig. 1). Under the effect of the periodic current, the inductor generates in the melt a variable magnetic field which, in turn, gives rise to an induced current. Thus, the melt is subject to electromagnetic body forces caused by the interaction of the eddy currents $\vec{J}$ and of the magnetic field $\vec{B}$.
     This process is characterized by the presence of an electromagnetic fringe effect consisting of a pronounced inclination of the magnetic field lines towards the axis of symmetry of the ingot (Fig. 1). This phenomenon is mainly explained by the occurence of a dissymetry of the electric current distribution, caused by the peripheral layer of solidifying metal |4|. Indeed, the magnetic field is here generated by the electric currents which are flowing both through the inductor and the

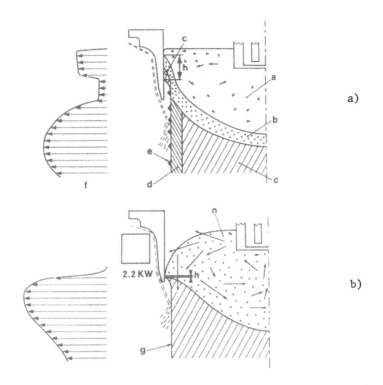

Figure 2. Evolution of the CREM process with the total power input (2214 aluminum alloy billet of 320 mm diameter) : (a) conventional casting (D.C.), P = 0 ; (b) CREM, P = 2.2. Kw. (a) liquid metal, (b) pasty zone, (c) solidified metal, (d) segregation zone, (e) exudations, (f) estimated heat flux profile, (g) smooth surface, (h) height of solidifying metal in contact with the ingot mold, (i) inductor, (n) nuclei.

solid and liquid part of the slab. The $\vec{J} \times \vec{B}$ force consists of a time-independent component and an oscillatory component of frequency 2N, where N is the frequency of the applied coil current. The time mean electromagnetic body force $(\vec{J} \times \vec{B})$ may be resolved into a radial component (principally irrotational) and a vertical component (primarily rotational). The potential forces, balanced by a pressure gradient, result in the formation of a convex surface meniscus, while the rotational forces are responsible for an electromagnetic stirring, similar to that encountered in a coreless induction furnace. The action of such a vigorous forced convection results in refining of the grain and a higher degree of homogeneity in crystallization.

Figure 2 depicts the behavior of a 2024 aluminum alloy billet of 320 mm diameter when the total power input (or the magnetizing force) increases. It appears that both the height of the dome-shaped free surface and the stirring intensity increase with the excitation while, at the same time, the height of contact h between the mold and the metal is gradually reduced.

During conventional continuous casting, the phenomenon of detachment from the mold arises. The result is a decrease in the escape of heat and the formation of exudations (Fig. 2(a)), which are markedly richer in alloying elements. Moreover, a costly scalping operation is necessary to remove the disturbed and segregated region usually seen in the peripheral crust.

It is well known that the early stages of a casting are considerably easier as the level of the liquid metal in the mold is higher ; however, in normal working, it is preferable to cast with a low level of metal head. The decreasing of the height of metal h in contact with the ingot mold is indeed followed by a modification of the heat flux profile which, in turn, results both in a progressive decay of the size of the exudations and in a diminution of the thickness of the segregation zone.

In this new process, the gradual enhancement of the magnetizing force allows a decrease in the level of the contact line between the liquid metal and the mold, until the metallic region in contact with the wall is practically reduced to a circular line (Fig. 2(b)). Figure 2 shows that the action of progressively lowering the contact line has a marked impact on the heat flow distribution. Under these circumstances, the ingot surface may become very smooth and the thickness of the segregated corticale layer tends to zero.

The mean velocity field has been experimentally explored in a half vertical cross-section of the sump and the flow patterns, obtained for different values of the electric power input P, are schematically sketched in Fig. 2. Figure 2(a) corresponds to a conventional casting and shows the presence of two main vortices. The existence of these loops is mainly explained by a viscous friction phenomenon caused by the radial and horizontal pouring jet running out of the float. It should be mentioned that the velocities are low in the proximity of the mushy zone (of the order of 1 cm.s^{-1}) and that the maximum velocities (10 cm.s^{-1}) have been detected in the immediate vicinity of the dispenser outlet. When the power P increases (Fig. 2(b)), the relative importance of the viscous effect induced by the jet is progressively attenuated. The electromagnetically driven flow is now characterized by the occurence of a main cell which fills the entire half cross-section. Experiments shows that the velocity, measured at a given point, is nearly proportional to the magnetizing force applied to the coil and to the square root of the total power P. At full power, the peak velocity can exceed $50.\text{cm s}^{-1}$; under this circumstance, the flow is obviously strongly turbulent.

Finally, thermal measurements prove that the temperature field in the sump is strongly affected by the electromagnetic stirring. This intense forced convection promotes the evacuation of superheat [3] ; the bulk liquid is now isothermal to within 1°C and its temperature is lower than that of the liquidus. Moreover, for the case of aluminum alloys vigorously stirred and characterized by a wide freezing range (alloy 2214, for instance), the pasty zone may occupy the whole volume of the pool (Fig. 2(b)).

3. METALLURGICAL RESULTS

Figure 3 shows the qualitative difference in grain size of 2214-51 alloy
slabs (700 x 200 mm cross-section). These ingots were produced using a
graphite mold and without inoculation of grain refiners master alloys,
in conventional casting (Fig. 3(a)) and by the CREM process (Fig. 3(b)).
Inspection of these macrographies established that the average grain size was, respectively, of the order of 3 and 0.18 mm. The variation of
the average grain size d, as a function of the overall power consumption P, is depicted in Fig. 4. It can be seen that d decreases,
first radidly and, then, asymptotically with P, while the decided grain
refinement obtained by this method is of the order of a tenfold reduction
of the average grain diameter (the number of equiaxed crystals is multi-
plied by about 1000). Therefore, the efficiency of the CREM process is
comparable with that obtained by nucleating addition.

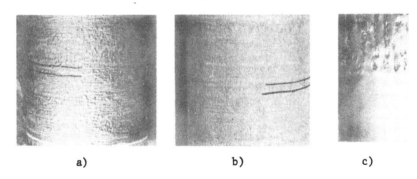

a)   b)   c)

Figure 5. Surface aspect of 2214 aluminum alloy billets of 320 mm diameter : (a) conventional casting. P = 0 ; (b) CREM process, P = 2.2 Kw ;
(c) upper part, P = 2.2 Kw, lower part, P = 3.15 Kw (magnification).

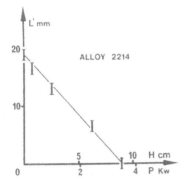

Figure 6. Thickness of the peripheral segregation zone as a function of
the total electric power input P, or of the dome height H .

Figure 3. **Macrostructures of 2214-51 aluminum** alloy ingots containing zirconium, (slabs of 700 x 200 mm cross-section) ; left hand side : conventional casting, P = 0, right hand side : CREM process, P = 1.9 Kw.

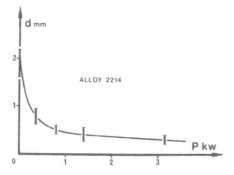

Figure 4. Evolution of the mean grain size with the total power input (2214 aluminum alloy billet of 320 mm diameter).

The refinement is mainly attributed to grain multiplication : the suspended nuclei localized in the near vicinity of the interface are carried away and dispersed in a slightly undercooled melt. Under these conditions, the crystallyzation takes place simultaneously in most of the sump, around a number of floating nuclei, and this increasing nucleation results in the appearance of a fine grained equiaxed structure.

Finally, it should be noted that, for the case of slabs, the electromagnetic and velocity fields are 3-D and non-uniform, particularly in the corner regions. This situation may give rise to some defects concerning the grain homogeneity, as well as the shape and surface aspect of the slabs. The use of corner electromagnetic screens has proven to be an efficient solution allowing to smooth out the non-uniformity of the magnetic field. On the other hand, the use of molds with rounded corners was recommended, in order to eliminate the pronounced weakening of the electric current density in the vertex regions.

The ingot surfaces presented in Fig. 5 have been yielded, without inoculation of nucleant agent, from 2214 aluminum alloy billet cast, either in conventional casting (Fig. 5(a)), or from the CREM process (Figs 5(b) and (c)). It appears that the circular ripples, initially observed on the ingot surface (P = 0), are gradually reduced as P increases and then replaced by less pronounced vertical striations (P = 2.2 Kw). These grooves totally vanish, when P reaches 3.15 Kw ; in this latter circumstance, the height of metal h in contact with the mold is practically reduced to zero, while the level of metal is very low in the mold (Fig. 2(b)). Inspection of Fig. 5(c) reveals that the CREM process improves substantially the ingot surface quality which is, for P = 3.15 Kw, similar to that produced by the conventional continuous electromagnetic casting technology (EMC). Figure 6 shows that the peripheral segregation zone varies linearly with the power consumption and is reduced to zero, when P reaches 3.15 Kw.

4. CONCLUSION

This basic study shows that the CREM process is an efficient dynamic method of grain refinement, eliminating the need for nucleant agents. Moreover, the easy adjustment of the level of the contact line of the metal with the ingot mold provides attractive additional effects, consisting in a substantial improvement of the surface aspect and in a marked reduction of the thickness of the cortical segregated layer. These properties result in substantial savings on scalping and edge triming.

REFERE?CES

|1| ABBASCHIAN G. & DAVID S., 1983, 'Grain Refinement in Casting and Welds', Met. Soc. of A.I.M.E., 3.
|2| TSAVARAS A. & BRODY H., 1984, 'Electromagnetic Stirring : Stepping Stone to improved Continuously Cast Products', Iron Steel Int., Vol. 2, 29.
|3| VIVES Ch. & PERRY Ch., 1986, 'Effects of Electromagnetic Stirring during the Controlled Solidification of Tin', Int. Journal of Heat and Mass Transfer, Vol. 10, 21.
|4| VIVES Ch. & RICOU R., 1986, 'Fluid Flow Phenomena in Continuous Electromagnetic Casting', Light Metals, Met. Soc. of A.I.M.E., 873.

LIQUID METAL FLOW CONTROL USING A.C. FIELDS

D. C. LILLICRAP
ELECTRICITY COUNCIL RESEARCH CENTRE
CAPENHURST
CHESTER, CH1 6ES
CHESHIRE.

ABSTRACT. The paper investigates the use of AC magnetic fields for controlling the flow of liquid metals through dispensing nozzles and valves. Flow rates are governed principally by the ratio of the magnetic pressure to the head pressure arising from the depth of metal above the nozzle. Measured flow rates are presented for aluminium and iron. The flow can be stopped completely when the magnetic pressure exceeds the head pressure.

1. INTRODUCTION

The process of casting metals invariably involves the step of pouring molten metal from a dispensing vessel into a mould. In continuous casting processes the metal runs through a nozzle in the base of a tundish into a water cooled mould from which the solidifying strand of metal is contstantly withdrawn. The majority of steel is now cast by this process, and on many plants the flow through the nozzle is regulated by raising and lowering a stopper above the nozzle inlet. More complex shapes are made by pouring metal into sand moulds. This operation can be carried out by manually pouring from a ladle, but the trend is towards automated lines producing several hundred castings an hour. Automatic pouring is essential on these lines and several schemes are in use. However, in nearly every case the metal is dispensed through a nozzle and the flow rate is controlled by a stopper.

Nozzles and stoppers are now widely used for flow control in casting processes but nevertheless suffer from a number of problems. Flow rates vary over a wide range for relatively small movements of the stopper and therefore a fast-acting and precise control of the stopper position is required. Thus the apparently simple stopper and nozzle requires sophisticated, and therefore expensive, drive and control systems. Furthermore, the stopper is constantly submerged in the metal and can suffer from erosion and slag build-up problems.

Electromagnetic flow control [1] offers a way around these difficulties. By placing a medium frequency induction coil around

363

364

the dispensing nozzle, electromagnetic forces can be produced in
the metal stream, capable of regulating the flow rate and in special
circumstances shutting off the stream altogether. Numerous arrangements of the coil and nozzle have been proposed. As far as the
author is aware, the earliest description of such a device is given
in a patent by Birlec Limited [1] but in more recent years Garnier
et al [2] [3] [4] [5] have described several devices based on similar
principles.

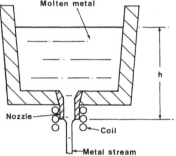

Figure 1 NOZZLE AND COIL FOR ELECTROMAGNETIC
FLOW CONTROL

As the electromagnetic 'valve' has no moving parts, the problems
encountered with stoppers do not arise. Furthermore, the flow
rate is controlled simply by varying the coil current, allowing
the 'valve' to be readily integrated into automated casting processes.
This paper describes work carried out at the Electricity Council
Research Centre, Capenhurst, Chester, England, into the performance
of some electromagnetic flow control devices and makes a comparison
with a simple theory.

2.   SIMPLE FLOW CONTROL THEORY

Figure 1 is a schematic diagram of a vessel containing molten metal
which is discharged through a nozzle in its base. An induction
coil carrying a medium frequency current is placed around the nozzle
as illustrated. The magnetic field produced by the coil induces
circumferential currents in the metal stream, which, because of
the 'skin' effect, are confined to the outer regions of the stream.
The interaction of these currents with the net magnetic field produces
an electromagnetic body force in the metal directed almost radially
towards the centre of the stream. These forces produce an overpressure in the stream which in turn causes the stream to slow
down. Hence, by varying the current to the coil the mass flow
through the nozzle can be controlled.
     In principle, the mass flow through the nozzle can be computed
as a function of the magnetic field strength. The first step would
be to solve the electromagnetic field equations to find the equilibrium shape of the stream and the distribution of electromagnetic
forces within the metal. These forces, plus gravitational forces,
are the body forces in the Navier Stokes equations governing the
flow through the nozzle. The Reynolds numbers for the stream are
sufficiently high for the stream to be turbulent and a reliable
turbulence model is therefore required.

The above procedure represents a considerable computational effort, particularly if accurate quantitative data are required. A much simpler model is therefore adopted to identify the controlling parameters, which are then used to correlate experimental data. The first simplification is to assume that the stream separates from the wall of the nozzle at the point of maximum field strength, usually on the mid-plane of the coil. Secondly, after separation the stream is given an assumed fixed shape. Standard finite element techniques are then used to calculate the electromagnetic force distribution. This simple model has been used to compute the distributions of magnetic induction $\underline{B}$ and current density $\underline{J}$ produced by a four turn coil placed around a stream of aluminium in a 13.5mm diameter nozzle. The current frequency was 22 kHz. The electro-magnetic body force $\underline{F}$ is given by:

$$\underline{F} = \underline{J} \times \underline{B} \text{ per unit volume} \tag{1}$$

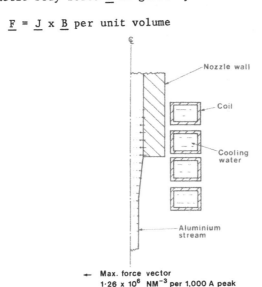

Max. force vector
$1 \cdot 26 \times 10^6$ NM^{-3} per 1,000 A peak
at 22 kHz

Figure 2   HALF SECTION THROUGH NOZZLE AND
COIL SHOWING COMPUTED
ELECTROMAGNETIC   FORCES

The distribution of forces is plotted in figure 2 which is a half section through a practical flow control nozzle. These forces are essentially radial and reach a maximum near the mid-plane of the coil. Conversely the induction $\underline{B}$ is essentially axial and also has a maximum on the mid-plane. The computed results confirm that at this frequency the 'skin' depth is small compared to the radius of the stream.

The effect of the radial forces is to increase the pressure over the central region of the stream by an amount $\int F_r \, dr$. This pressure increase will clearly be a maximum near the mid-plane

of the coil and it is this maximum value which governs the mass
flow through the nozzle. Now the electromagnetic force can be
written

$$\underline{F} = - \nabla \left(\frac{B^2}{2\mu} \right) + \frac{1}{\mu} \underline{B} \cdot \nabla \underline{B} \tag{2}$$

where μ is the magnetic permeability. On the mid-plane of the
coil $B_r \simeq 0$ and $\partial B_z / \partial z \simeq 0$ and therefore the second term in equation
(2) is very small. Consequently the maximum overpressure is given
by $B^2/2\mu$ where B is the value of the induction at the edge of
the stream on the mid-plane of the coil, the value at the centre
of the stream being zero. This quantity $B^2/2\mu$ is usually referred
to as the magnetic pressure.

The above argument shows that molten metal flowing through
the central portion of the nozzle flows into a region of increasing
static pressure as the mid-plane of the coil is approached thus
causing it to slow down. Axial gradients in the electromagnetic
forces will also introduce a rotational component of velocity.
Possible effects of this rotational component are discussed later,
but in order to proceed with a simple theory the flow is for the
moment assumed to be irrotational. With this assumption a mechanical
energy balance along a streamline, usually known as Bernoulli's
equation, can be written as

$$\frac{1}{2} q^2 + E + \frac{P}{\rho} + \psi = \text{Constant} \tag{3}$$

where q is the velocity, E the internal energy, P the static pressure,
ρ the density and ψ the potential due to gravitational and electromagnetic forces. For steady irrotational flow, with no change
in internal energy it can be shown that

$$\frac{1}{2} V^2 + \frac{B^2}{2\mu\rho} - gh = 0 \tag{4}$$

where V is the velocity corresponding to the maximum induction
B at the edge of the stream, h is the depth of metal above the
nozzle, and g is the acceleration due to gravity.

The velocity V_0 corresponding to zero induction is:

$$V_0 = \sqrt{2gh} \tag{5}$$

Combining (4) and (5)

$$\frac{V}{V_0} = \sqrt{1 - \frac{B^2}{2\mu\rho gh}} \tag{6}$$

Equation (6) identifies the ratio of the magnetic pressure
to the metallostatic pressure due to the depth of metal h as the
parameter governing the velocity through the nozzle. Clearly equation

(6) is invalid when the magnetic pressure $B^2/2\mu$ exceeds the metallo-static pressure ρgh, since partial levitation of the melt would then occur. In view of the sweeping assumptions made in deriving equation 6, it is unlikely that the velocity, and therefore the mass flow, will follow exactly the simple dependence on $B^2/2\mu\rho gh$ predicted. Therefore, experiments have been conducted with a range of nozzles and different molten metals to determine how the mass flow varies with field strength.

3. MEASURED FLOW RATES

To measure flow rates, a ladle with a nozzle mounted in its base, was supported on a cantilever system equipped with a sensitive load cell. A stopper fitted into the top of the nozzle to prevent the discharge of metal while the ladle was filled. Once the ladle was filled, the stopper was raised to release the metal through the valve, while the load cell output was continuously recorded. At any instant the weight of metal in the ladle could be used to determine the depth of metal above the nozzle, and the rate of change of weight gave the flow rate. An induction coil placed around the nozzle could be supplied with a range of medium frequency currents. Thus, flow rates could be measured for a wide range of both B and h.

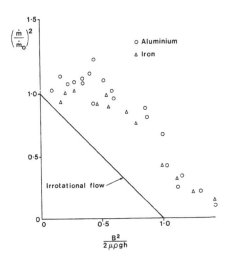

Figure 3 FLOW RATES FOR ALUMINIUM AND IRON

Flow rates were measured for aluminium in a 13.5mm diameter nozzle and iron in a 22mm diameter nozzle. Frequencies were between 21 and 22kHz. The measured rates $\dot{m}$ were non-dimensionalised by the flow rate $\dot{m}_0$ for zero induction and this ratio squared, $(\dot{m}/\dot{m}_0)^2$ was plotted against $B^2/2\mu\rho gh$ as shown in figure 3. The value of B should be that at the edge of the metal stream, but this is not easy to calculate without a computer program. However, if the value of B at the centre of the empty coil is used, the parameter $B^2/2\mu\rho gh$ is found to correlate (to within the accuracy of the measurements) the flow rates for both aluminium and iron over the full range of B and h. The advantage of calculating B in this way is that the data of figure 3 can then be used to predict flow rates without recourse to lengthy computer procedures. As the parameter $B^2/2\mu\rho gh$ increased from zero to 0.4, the flow rate increases by approximately 10% and then for larger values of the parameter up to 1.4 decreases approximately linearly to 30% of $\dot{m}_o$. However, for values of $B^2/2\mu\rho gh$ larger than 1.4 there is no further significant reduction in flow rate.

The observed behaviour of the flow rate differs considerably from that predicted by equation (6) for irrotational flow. Clearly the rotational part of the force field has a significant influence on the flow pattern in the nozzle. For these nozzles, with no applied magnetic field, the discharge coefficient is typically 0.65 and the vena-contracta just below the exit of the nozzle is therefore 80% of the actual exit diameter. When a weak magnetic field is applied there is a measurable increase in stream diameter corresponding to the increased mass flow. It appears that the rotational part of the force field changes the streamline shapes in the nozzle so as to improve the discharge coefficient. This effect at first more than compensates the influence of the magnetic pressure in slowing down the stream.

4.   ELECTROMAGNETIC VALVE

For values of $B^2/2\mu\rho gh$ greater than one, the magnetic pressure exceeds the metallostatic pressure and metal is levitated away from the walls of the nozzle. Under these conditions, as the field strength increases the separation point moves further up inside the nozzle, but the metal still continues to run through the centre portion of the nozzle. This suggests that if a centre body was placed in the nozzle, the flow could be cut off completely. Thus a nozzle with an off axis exit as shown in figure 4 could be used as a valve capable of controlling and stopping the stream. The simple mathematical model discussed above was therefore modified to allow the equilibrium position of the air metal interface to be computed by an iterative process. The model indicates that the free interface will take up a shape as sketched in figure 4. A simple Rayleigh-Taylor criteria shows that for the dimensions of the nozzle, surface tension should stabilise the interface, and the alternating magnetic field will also have a strong stabilising effect.

A nozzle was constructed to the shape shown in figure 4, with an exit slot sufficiently far off the axis that, according to the computer model, the metal would be levitated away from the slot at maximum field strength. This nozzle was mounted in the base of the ladle and flow rates measured as before. Values of $(\dot{m}/\dot{m}_o)^2$ plotted against $B^2/2\mu\rho gh$ in figure 5, follow quite closely the simple linear variation predicted by equation 6. The flow is cut

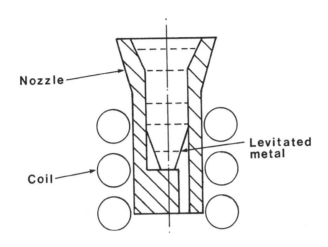

Figure 4   ELECTROMAGNETIC VALVE

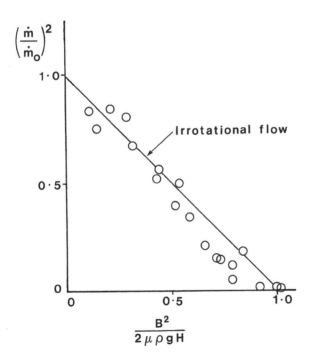

**Figure 5   FLOW RATES OF ALUMINIUM THROUGH
ELECTROMAGNETIC VALVE**

off entirely for values of $B^2/2\mu\rho gh$ slightly greater than 1. This simple linear characteristic is a big improvement on the highly non-linear characteristic obtained for a stopper and nozzle. In particular when the electromagnetic valve is used in a closed loop flow control system the control algorithm will be much simpler than that needed to control the movement of a stopper.

5. CONCLUSION

Techniques have been demonstrated for regulating the flow of molten metals through a nozzle by the application of medium frequency electromagnetic fields. For a nozzle with a uniform bore, flow rates can be regulated between 110% and 30% of the rate obtained without the magnetic field. By placing a centre body in the nozzle, so that the metal exit is off axis, the nozzle can be used as a valve capable of shutting off the stream entirely. This shut off is achieved when the magnetic pressure exceeds the metallostatic pressure due to the depth of metal above the valve. Thus, in those applications when the metal depth allows this condition to be satisfied for practically achievable magnetic field strengths, the electromagnetic valve offers a viable alternative to the stopper and nozzle.

6. REFERENCES

[1] Birlec Ltd., 1953, "A New or Improved Method of, and Apparatus for, Controlling or Preventing the Discharge of Molten Metal from Containers." British Patent Specification 777213.

[2] M. Garnier, 1982, "Une analyse des possibilités de contrôle des surfaces libres de métaux fondus", Thèse de Doctorat d'Etat, INPG.

[3] M. Garnier et R. Moreau, 1975, "Dispositif électromagnétique de confinement des métaux liquides", Brevet n° 75 21 075.

[4] M. Garnier et R. Moreau, 1977, "Dispositif électromagnétique de confinement des métaux liquides pour effectuer une régulation de débit", Brevet n° 77 21 121.

[5] J. Garnier, M. Garnier et R. Moreau, 1979, "Procédé et Dispositif pour réaliser le confinement des métaux liquides par mise en oeuvre d'un champ magnétique", Brevet n° 79 14 011.

Session G:
Measurements

DIAGNOSTICS OF LIQUID METAL FLOWS USING FIBRE-OPTIC VELOCITY SENSOR

V.G. Zhilin, K.V. Zvyagin, Yu.P. Ivochkin, A.A. Oksman
Institute of High Temperatures
USSR Academy of Sciences
127412, Moscow

ABSTRACT. The present work is connected with a theoretical and experimental study of the application of two-component fibre-optic velocity sensors for diagnostics of liquid metal turbulent flow. The above sensors have high spatial resolution ($10^3 mm^3$) and exhibit insignificant sensitivity to admixtures and the magnetic field; they could also be used for liquid metals in the presence of strong currents (≈ 10 A). The experimental data on the velocity fields and intensity of its pulsations in strongly nonuniform liquid metal magnetohydrodynamic flows are given. Dynamical errors of velocity measurements at a frequency of 200 Hz are negligible.

1 INTRODUCTION

Two main instruments are used at present for the local velocity measurements in the turbulent flows of liquid metals - the hot-wire anemometer and the electric potential probe. The use of these instruments is connected with several major problems caused by the properties of liquid metals and the investigated flow characteristics.

General disadvantage of the thermoanemometers used for liquid metals is the instability of the transfer function owing to the impurities that accumulate on surface of the sensing element [1]. In addition, the frequency range of the thermoanemometer is limited by the considerable thickness of the thermal boundary layer because of the low Prandtl number ($Pr \approx 10^{-2}$) and the thermal inertia of the wire electric insulation layer. For this reason the upper limit of the frequency measured without a distinct dynamical distortion is 50Hz [2].

The main drawbacks of the electric potential probe are the need for the uniformity of the flow in the direction of the magnetic field and a low spatial resolution determined by velocity distribution in the region, which has the range several times greater than the distance between the measuring electrodes [3]. Neither the thermoanemometer nor the electric potential probe are fit for the measurement of the velocity of electrovortex flows that arise under the interaction of the strong electric currents with its own magnetic field [4] These drawbacks were the

373

J. Lielpeteris and R. Moreau (eds.), Liquid Metal Magnetohydrodynamics, 373–379.
© *1989 by Kluwer Academic Publishers.*

reason for creation of the fibre-optic velocity sensor, which was free from the most of them [5]. This report presents the main metrological parameters of fibre-optic sensors for measuring in MHD-flows of liquid metals.

2. DESIGN AND OPERATION PRINCIPLE OF FIBRE-OPTIC VELOCITY SENSORS

The design of a two-component fibre-optic velocity sensor for measure-ments in the flows of liquid metals is shown in figure 1. The sensing element is formed as a thin-walled glass cone with a glass pointer sealed in its end. The free end of the pointer is blackened and placed in the air gap between the two pairs of light guides, two of which are optically connected to the source of light and the other two - to the photocells. The light guides in the air gap form two perpendicular beams of light. In the initial position (at zero velocity) the pointer inter-cepts partially the two beams. Under the flow action the sensing element deflecte and the pointer changes the amount of light reaching the photocells by changing its position. In such a way the values of electric signals of the photocells are functionally dependent on the values of the two velocity components The algorithm for the restora-tion of the two velocity components W_x and W_y with the values of output signals of the photocells U_x and U_y is based on the following simple relationships [6]:

$$U_x = a\, W_x'$$
$$U_y = b\, W_x'^{n-1}\, W_y' \qquad (1)$$

The constant n depends on the range of Reynolds numbers and the configuration of the sensing element. Its value, as well as the values of constants a and b are determined by calibration of the sensor in the apparatus adapted for independent assignment of two orthogonal velocity components W_x and W_y, which lie on the plane perpendicular to the axis of the sensor. The value n varies within the limits of 1,6-1,8 for all the sensors tested. The algorithm for calculations of W_x and W_y, their mean values, mean square values of velocity fluctuation W_x' and W_y' and Reynolds stress $\langle W_x'\, W_y'\rangle$ is computed by an analogue electronic device [5]. One can see the validity of the equation (1) and (2) from figure 2 where the calibration results of one of the sensors are shown (n=1,77).

The usual diameter of the sensing element end is 0,04 mm.Therefore, the magnetic interaction parameter N is small and the magnetic fields with flux density B<1T practically do not influence the calibration curve of the sensor [7],which makes the sensor convenient for the inves-tigation of MHD flows of liquid metals.

The fibre-optic sensor has low sensitivity to the fouling of the sensing element, since it only increases the diameter of the latter. For a cylindrical sensing element the value of the signal of the sensor in a first approximation is $V \approx d_s^m$ where m depends on the Reynolds number Re and varies from 0,2 for Re<1 to 1 for Re>10^3.

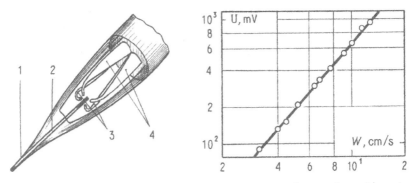

Figure 1. Design of the fibre-optic velocity sensor. 1-sensing element; 2-pointer; 3,4-light guides.

Figure 2a. The output signal of the sensor as a function of the velocity of the flow.

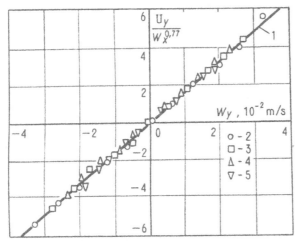

Figure 2b. Comparison of the experimental data with the equation (2). 1-equation (2); Experimental data-W_x, cm/s: 2-10,5; 3-8,7; 4-7,3; 5-6,0.

3. SPATIAL RESOLUTION

In measuring the two components of velocity the fibre-optic sensor reveals high spatial resolution. The volume over which the two velocity components are averaged is by several orders less than that of the thermoanemometer and the electric potential probe. This volume is $\pi d^2 \ell_{ef}/4$, where d is the diameter of the sensing element and ℓ_{ef} its effective length. The value ℓ_{ef} cannot be determined by immediate measuring since the sensing element diameter varies smoothly from 0,04 mm at its end to the diameter of the mount (about 0.55 mm, Fig.1). Therefore ℓ_{ef} was measured by the experiment [8]. The sensor was placed in a uniform flow of air at the nozzle cross section; it was partly covered by a hydrodynamic screen (hollow glass cone having a diameter of the hole in the narrow part about 0,5 mm, Fig.3). The sensor was displaced in the axial direction and the signal was measured to an accuracy of $\pm 1\,\mu$m as a function of the length of the projecting part. The measurements showed (see Fig.3) that ℓ_{ef} 0,6 mm for all the sensors tested. Taken as ℓ_{ef} was length for which the ratio of the signal of the sensor   to its maximum value U_m reached 0,95. The value of volume V does not exceed 10^3 mm^3 which is substantially less compared with other methods of measurements.

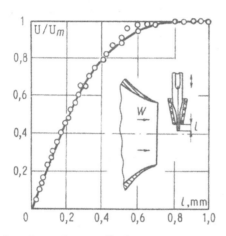

Figure 3. Dependence of the sensor output signal on the length of the projecting part.

4. DYNAMIC CHARACTERISTICS

As the velocity fluctuations are measured, the sensing element and the pointer take part in a complicated oscillating process, which parameters are determined not only by the frequency and amplitude of oscillations but also by the characteristics of the sensor itself. The most important of these characteristics are resonant frequencies of the pointer and the sensing element. It is known that measurements by means of mechanical sensors are possible only when the reference frequency of the parameters to be measured differs significantly from the proper frequency of the sensor In the opposite case the measurements will lead to dynamic errors that will grow infinitly when these frequencies coincide.

    Another important characteristic is the damping coefficient of oscillations that is connected with dissipative processes in the surrounding liquid. When the resonance frequency of the sensor ω_r and the damping coefficient α are known, it is possible to calculate the coefficient of dynamic amplification of amplitude λ which characterizes the dynamic error of the measurements:

$$\lambda = \left[\left(1 - \omega^2/\omega_\iota^2 \right)^2 + \alpha^2 \ \omega^2/\omega_\iota^2 \right]^{-1/2} \tag{3}$$

where ω is the frequency of the measured parameter. If $\lambda = 1$, there is no dynamic error.

The velocity sensor is a complicated oscillating system and its dynamic characteristics are determined by two oscillatory links: the sensing element and the pointer. Calculations and measurements show that the resonance frequency of the sensing element $\omega_{\iota se}$ is higher by an order than the resonance frequency of the pointer $\omega_{\iota p}$= 1000 Hz [9]. Therefore, one can estimate the possible dynamic errors by the determining parameter λ_p only for the pointer. Since the pointer oscillates in the air, one can assume $\lambda_p \approx 0$ [9] and

$$\lambda_p = \left| 1 - \overline{\omega}_p^2 \right|^{-1} \tag{4}$$

where $\omega_p = \omega / \omega_{\iota p}$.

As an example we define the relative dynamic error δ for measuring the mean-square velocity fluctuations in a turbulent flow at the upper frequency limit of energy spectrum of fluctuation ω_m=200 Hz. Assuming that $\omega_{\iota p}$ =1000 Hz and the spectral intencity is constant in the frequency range from 0 to ω_m we obtain an evaluation by the maximum dynamic error :

$$\delta = \frac{1}{\overline{\omega}_m} \int_0^{\omega_m} \frac{d\omega_p}{\left| 1 - \omega_p^2 \right|} - 1 = \frac{1}{2\,\overline{\omega}_m} \ln \frac{1 + \overline{\omega}_m}{\left| 1 - \overline{\omega}_m \right|} - 1$$

As ω_m=0,2, we get δ=0,014. Thus, the maximum dynamic error is $\delta <$1,5%, and so it is not reasonable to use any electronic correctors of the amplitude-frequency characteristics of the sensor in this case.

When measuring the turbulent velocity fluctuations, it is necessary to have in mind one principal limitation of the fibre-optic sensor. With the mean velocity of the flow $\langle W \rangle$ that corresponds to Reynolds number Re$\approx$50 the Karman vortex trail arises behind the sensing element. The Karman frequency f and, consequently, the drag force fluctuation frequency are estimated by the following relationship:

$$\frac{f d}{\langle W \rangle} \approx 0.2 \left(1 - \frac{20}{Re_d} \right)$$

Here Re$_d$ is Reynolds number determined by the diameter of the sensing element . The amplitude of the drag force fluctuations can be compared with the mean value of this force. Consequently, the sensor with a cylindrical sensing element cannot be used for the measurements of the turbulent fluctuations when $\langle W \rangle > 50 \ \nu/d$. The measurements of the mean velocity of the flow $\langle W \rangle$ are possible even if Re$_d >$50 but the frequency estimated according to (5) does not coincide with the resonant frequency of the pointer. If mercury is used as the working fluid ($\nu_\rho = 10^7 m^2/s$) and $d = 0,04$ mm, then $\langle W \rangle \leqslant 12$ cm/s.

5. THE EXAMPLES OF THE FIBRE-OPTIC SENSORS APPLICATION

Some results of the mean-square and mean velocity measurements in the trail immediately behind the cylinder and in the electrovortex flow are given below. Measurements in both these cases were possible only when fibre-optic sensor was used.

Shown in Fig.4 are the velocity longitudinal fluctuations profiles in the flow of mercury past a cylinder ($\frac{x}{d} = 2$) with the diameter $d = 5$ mm in the presence of the magnetic field directed along the cylinder axis [10]. One can see that the intensity of the fluctuations in the magnetic field with the induction $B = 0,6T$ is much higher than without the field.

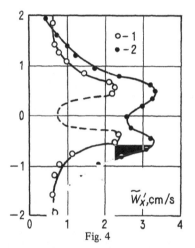 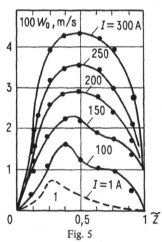

Figure .4. Velocity fluctuation profiles in the flow past a cylinder. 1-B=0; 2-B=0,6T.

Figure 5. Dependence of the axial velocity on the distance from the "point" eléctrode. 1-numerical calculation.

Also measured for the first time was the velocity of electrovortex flows in the cylindrical box 60 mm in diameter and length of 60 mm [11]. The current was taken to the edges of the box, the diameter of the first one was 60 mm and the second one ("point" electrode)-12 mm. The measurements of the velocity W_0 were made on the axis of the system with the different values of the current and distances $\tilde{Z}$ from "point" electrode. The results of the measurements are given in Fig. 5 with the results of numerical calculation for the current $I = 1A$. Today the fibre-optic sensors are used for the investigations of the structure of the electrovortex flows and grid turbulence in the presence of the magnetic field.

6. REFERENCES

[1] Malkolm D.G., 1970, 'Unsteady magnetohydrodynamic flow measurement in liquid metals using thermo-anemometry with applications of insulated platinum films in mercury', Magnitnaya Gidrodinamica, No. 2, 55.

[2] Markova L.G., 1974, 'A study of frequency characteristics of thermo-anemometer wire sensors', Magnitnaya Gidrodinamica, No. 3, 141.

[3] Korsunskii L.M., 1974, 'Space average range of conduction anemometer with the external magnetic field', Magnitnaya Gidrodinamika, No. 4, 148.

[4] Boyarevich V.V., Freiberg J., Shilova Ye.I., Shcherbinin E.V., 1985, Electrovortex Flows, Zinatne, Riga.

[5] Zhilin V.B., 1987, Fibre-Optic Measuring Transdusers of Velocity and Pressure, Energoatomizdat, Moscow.

[6] Zhilin V.G., Ivochkin Yu.P. & Oksman A.A., 1984, 'Method of two velocity components measurements in the turbulent flow of liquid metals using fibre-optic velocity sensor', Teplofiz. Vys. Temp., Vol. 22, No. 6, 1178.

[7] Zhilin V.G , Ivochkin Yu.P. & Oksman A.A., 1984, 'On an absence of a transverse magnetic field influence on the fibre-optic velocity sensor', Teplofiz. Vys. Temp., Vol. 22, No. 5, 1024.

[8] Zhilin V.G., Ivochkin Yu.P. & Oksman A.A., 1985, 'Spatial resolution of fibre-optic velocity sensors for liquid metals', Magnitnaya Gidrodinamika, No. 3 137.

[9] Zhilin V.G., Zvyagin K.V. & Ivochkin Yu.P., 1986, 'Dynamic response and spatial resolution in fibre-optic liquid-metal velocity sensors', Teplofiz. Vys. Temp.,Vol. 24, No. 5, 967.

[10] Zhilin V.G., et. al., 1986, 'The study of a flow near a circular cylinder in a magnetic field', Magnitnaya Gidrodinamica, No. 4, 130.

[11] Zhilin V.G., et. al., 1986, 'An experimental investigation of the axisymmetric electrovortex flow velocity field in a cylindrical container', Magnitnaya Gidrodinamica, No. 3, 110.

ON LOCAL MEASUREMENTS OF THE UP AND DOWNSTREAM MAGNETIC WAKE OF A CYLINDER AT LOW MAGNETIC REYNOLDS NUMBER

J. LAHJOMRI, Ph. CAPERAN & A. ALEMANY
Institut de Mécanique de Grenoble
B.P. 53 X - 38041   Grenoble Cedex - France
S. CRISTOLOVEANU
E.N.S.E.R.G.
28 Av. des Martyrs - Grenoble - France

ABSTRACT. We made experiments on the up and downstream MHD wakes of a non conducting cylinder translating with a velocity aligned to the magnetic field. Measurements of the magnetic field perturbation had been undertaken by means of magnetodiodes, which allow for local probing – a typical spacial resolution corresponds to $1,5$ mm^3 . In the upstream wake, longitudinal and transverse measured profiles of the magnetic perturbation correlate well with the available theory taking into account the confinement. In the downstream wake, we had been able to sense the magnetic perturbation associated with the Karman vortex street. Joined velocity and magnetic field measurements had been made, providing phase and correlation relationship. The Strouhal number S has been found independent of Re in the range considered. It decreases as the interaction parameter N increases, which means a downstream shift of the unsteady separation point. The disappearance of the vortex street is characterized by a critical curve Re(M) (M = Hartmann number).

1. INTRODUCTION

The phenomenon of MHD wakes has been widely studied for years, however the available theories were made for low Reynolds number Re (Chester |1|, Yosinobu |2|), or in the asymptotic regime for which Re and the Alfven number are infinite (Tamada |3|). Experiments with aligned magnetic and velocity fields had been driven on sphere (Maxworthy |4|, Yonas |5|) or Rankine body (Maxworthy |6|, Alhström |7|, Lake |8|), with an interaction parameter N ranging from 0 up to 70. In these studies, measurements of pressure had been made on the body while the magnetic field was sensed by coils of different diameter, providing integrated profiles of the magnetic perturbation. Velocity profiles had also been obtained in |8|.

We propose some new experimental results on the velocity and magnetic perturbation field in the configuration sketched on figure 1, where the obstacle is a cylinder, a situation only studied in |9| at these days. A theory which corresponds to the range of non dimensional

381

J. Lielpeteris and R. Moreau (eds.), Liquid Metal Magnetohydrodynamics, 381–390.
© *1989 by Kluwer Academic Publishers.*

parameters attainable in our facility is proposed, which fit fairly well our experimental data in the upstream part of the flow. New results are provided in the downstream part of it, particularly dealing with the domain of existence of the Karman vortices. We stress about the fact that all the magnetic field observations presented hereafter had been obtained by means of magnetodiodes, a new type of sensors of which use is very similar to hot film probes, and with a spacial resolution of about 1.5 mm^3.

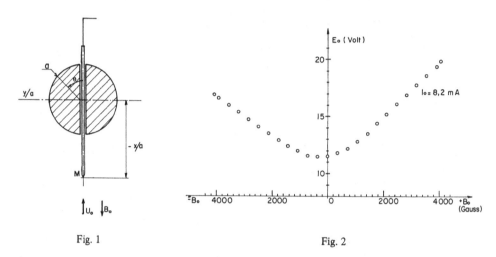

Fig. 1   Fig. 2

Fig. 1. Theoretical and experimental configuration probe traversing mechanism

Fig. 2. Typical calibration curves

2. EXPERIMENTAL FACILITY AND MEASURING DEVICE

The experimental facility is constituted by a vertical cylindrical reservoir of mercury (inner radius = 10 cm, length = 2 m) inserted in the core of a solenoid of which field ranges from 0 to 0.4 T. This facility is very much the same than in |10, 11|, except that the usual grids are replaced with horizontal cylinders (a = 6, 3 and 1,5 cm). We can set the velocity of descent of these obstacles in mercury over a range from 0 to 50 cm/s.

We use hot film probes for the measurements of the velocity perturbation, while the magnetic field ones are realized with magnetodiodes, the physical principle of which has been fully described by Cristoloveanu |12|. In this paper we will describe their metrological properties, and their possibilities on a test case. These probes are actually handmade, which hinder to reproduce identical sensors, but this trouble is largely counterbalanced by large advantages over other usual probes. The active part of the sensor is a piece of germanium of size L = 3 mm and l = 1 mm, and thickness 0,5mm.

It senses the component of magnetic field parallel to 1. When feeded with a constant current generator the electrical tension on the probe is related to field intensity. A sample of calibrating curve (fig. 2) shows that a typical static sensitivity is about 2 mV/Gauss, a very good value when compared with Hall sensor ones. We may note the possibility to dissymmetrize the calibration curves by degrading the surface state on one sensor face parallel to B. This causes a shifting of the minimum voltage point away from the origin, and allows measurements of fields on range including sign changes of it. The sensor spectral noise, of B dependent amplitude, follows a 1/f law. This permits a discrimination of 10 KHz of the mean field at low frequency. The flatness of the dynamic response of the magnetodiodes has been checked at least up to 10 KHz. As a matter of fact, these sensors seem very well fitted for use in mercury. No tests are available at high temperature, but the only limit of their use should come from the degradation of the p^+n n^+ junctions of the semiconductor, which should occur for such probes at around 250°C. The first attempt to use magnetodiodes in mercury had been made in |11|, they have been brought to a fully operational state in |13|, from which this paper is extracted.

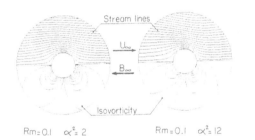

Fig. 3. Map of stream function and isovorticity

3. THEORETICAL ANALYSIS

We consider velocity and magnetic field perturbation as much smaller than their mean homogeneous component. This allows to linearize MHD equations and obtain : |7, 14|

$$\frac{\partial \omega}{\partial x} = - \alpha^2 \frac{\partial J}{\partial x} + \frac{1}{Re} \Delta \omega \qquad (1)$$

$$\frac{\partial J}{\partial x} = - \frac{\partial \omega}{\partial x} + \frac{1}{Rm} \Delta J \qquad (2)$$

where j and ω are non dimensional current and vorticity perturbations (The lengths are scaled with cylinder diameter a). Our laboratory conditions typically corresponds to low Rm ($\sim$ 0,05) high Re ($\sim$ 5 10^4)

and moderate value of the Alfven number α ($\alpha \sim 15$). Allowing for high value of Re in (1) gives :

$$\omega = -\alpha^2 \ J \tag{3}$$

$$2 \ k \ \frac{\partial J}{\partial x} = \Delta J \qquad \text{with } k = \frac{Rm}{2} (1 - \alpha^2) \ . \tag{4}$$

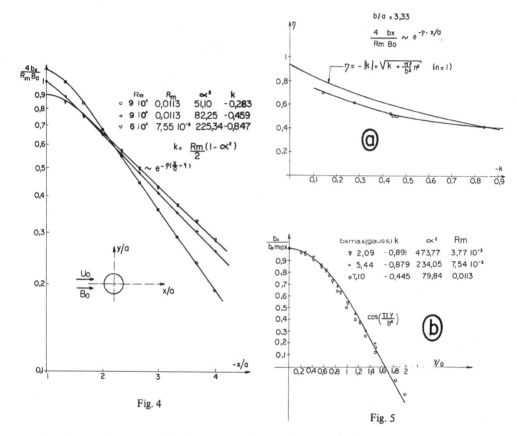

Fig. 4

Fig. 5

Fig. 4. x profiles of the x magnetic field perturbation

Fig. 5. a. Experimental decay rate compared with its value from equation 7.

b. y experimental profile of magnetic field b_x

This approximation is valid if no separation occurs in the boundary layer, which is a question to be discussed later. For now we consider that viscous effects are confined in a very thin layer on the positive x axis. The equation (3) shows that ω and j have the same topology. Vorticity and current propagates upstream or downstream according to the sign of the magnetic Oseen number k , in the Oseen equation (4). Using proper boundary conditions at infinity and on the body for the velocity and magnetic fields (the body is non conductive) gives us a development in series of Bessel functions of the magnetic and kinematic stream functions, which can be obtained as a Taylor expansion in the neighbourhood of the body. Maps of the kinematic stream function are given on fig. 4, together with the isovorticity (or current) lines for the subalfvenic regime, which is our exper- imental case. The expression of the density current is as follows :

$$J = - Rm \left\{ \frac{\sin 2\theta}{r^2} + Rm \frac{\sin 3\theta}{4 \ r^3} + Rm(\alpha^2 - 1) \right.$$
$$\left. \left[\frac{\sin \theta}{4 \ r} + \frac{\sin 3 \theta}{4 \ r} (1 + \frac{2}{r^2}) \right] \right\} \qquad (5)$$

The first two terms of (5) result from interaction of the potential velocity U_{pot} with B_0 and with its own induced field. The other terms of order $Rm \ k$ generate α^2 dependent assymmetry of the flow. As a matter of fact, they generate the MHD wake effect, and correspond to the rearrangment of the flow under the propagation of Alfven waves. We see on the flow map a deceleration of the fluid upstream of the body, which is a nascent slug region ahead of it, as observed in |4, 6, 8|. The expression for b_x in the upstream wake on x axis, which is the measured quantity, is of the form :

$$b_x = \frac{Rm}{.2} \frac{1}{x} (1 - \frac{1}{2 \ x^2}) + 0(Rm.k) \qquad (6)$$

From this we see that the maximum perturbation field at body surface is $Rm.Bo/4$ at first order, a value which is used as scaling of b in what follows. Moreover, the higher order terms tend to decrease this value as k increases. This fact is illustrated by the streamlines, which show an effect of deceleration of the fluid, due to MHD forces, as k increases.

4. EXPERIMENTAL STUDY

4.1. Upstream Wake

We made measurements of the magnetic field in the upstream part of the flow. The probe holder was whether fixed through the cylinder (fig. 1) or on the carriage downstream of it. This way it was possible to make x or y profiles by moving the holder and rotating the cylinder.

The x profiles of b on centerline show (fig. 4) a very good agreement with the scaling given by (6), and the expected decrease of b_x at body surface as k increases is observed. However, the experimental decay of b_x follows an experimental law of which rate depends only on k , instead of an algebraic one as one could expect from (6). This discrepancy from theory comes from a too high confinement ratio a/b, which is 0.3 in our experiment. This problem has been encountered by Yonas |5| or Alhström |7| in previous studies. To deal with, we model our experimental geometry by a cylinder between two plane walls at a distance 2 b. The solutions to this problem could be expressed as a sum of eigenfunctions with characteristic non dimensional wave numbers in the y direction given by :

$$k_y = n \cdot \frac{\pi a}{b} \qquad n \in \mathbb{N}^*$$

This introduced in (4), and with the condition of subalfvenic regime for which k is negative, gives us a decay rate :

$$\eta = k \left(1 - \sqrt{1 + \frac{k_y^2}{k^2}} \right) \tag{7}$$

for the upstream wake (for which x < 0). This expression, computed for the first mode, fits fairly well the experimental data (fig. 5a) and experimental transverse profiles, on fig. 5b, confirm that the first eigenfuntion contains most of the information on the phenomenon. Thus this part of our study confirms for cylinder the results obtained by previous authors with spheres. We had been able to measure the steady magnetic perturbation with an improved spacial resolution, and second order effects have been sensed. In order to complete this study, we follow on studying the downstream unsteady wake.

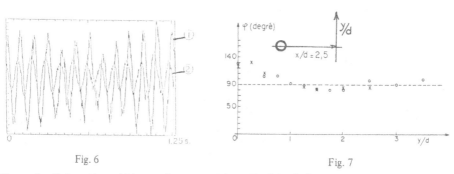

Fig. 6   Fig. 7

Fig. 6. Velocity (1) and magnetic field (2) perturbation in the downstream part of the flow

Fig. 7. y phase profile $2 \widehat{ub}/\pi$ for two different values of Re

4.2. Downstream Wake

The velocity and magnetic field perturbations have been simultaneously measured in the downstream part of the flow. The confinement ratio used for this study was equal to 0.05, corresponding to an aspect ratio of the cylinder of 20 which keeps us from unstability of the Karman street. The observed signals are given on figure 6 : the magnetodiode permits us to observe the "magnetic" vortices associated with the usual Karman street. We computed from these signals the different correlations and power spectral densities. From these, we obtain transverse profiles of phase shift between b_x and V_x on figure 7. A phase shift of $\pi/2$ is found for the outer region of the wake. This is in agreement with the linearized induction equation written in wavevector space, for a steady flow (we consider the own translational velocity of the vortex street as much smaller than the cylinder one) which gives us

$$\widehat{b} = e^{-1 \, \pi/2} \, \frac{Rm \cdot k_x}{k^2} \, \widehat{u}$$

where ^ stands for Fourier transform.

At this outer position, the induced field could be considered as mainly resulting from the interaction of the near field B with the potential velocity induced from the street. This is no more the case in the wake where non linearity takes place, that could explain the phase shift observed in this zone. However, it is worth to note that the active coaling velocity on the hot film probe in the wake, periodically changes its direction, which could induce some bias on the measurements.

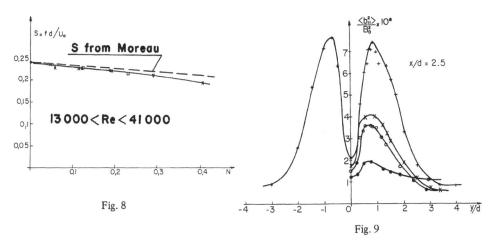

Fig. 8

Fig. 9

Fig. 8. Evolution of the Strouhal Number versus N comparison with the value expected from ref. |15|

Fig. 9. Profiles of magnetic energy perturbation

We found that the Strouhal number S is independent of Re but slowly decreasing as N increases (fig. 8). This could be interpreted as a downstream displacement of the separation point, as Moreau found out for the flow around a cylinder with a radial magnetic field |15|. The time of restructuration of the boundary layer after each shedding

388

could be evaluated by $a\theta_S/V_0$, where θ_S is the separation angle. If this time is related to the shedding frequency, it gives us a Strouhal number as directly proportional to θ_S^{-1}. The curve from Moreau prediction for θ_S is plotted on fig. 8. It tangents the experimental points and fit them fairly well for N < 0.1. Profiles of magnetic energy confirm the shifting of the separation point, as can be seen from the decrease of the distance between vortices (fig. 9). The maximum energy decreases with increasing M/Re (M is the Hartmann number), this is illustrated on figure 10 for different value of Re. This behaviour suggests a complete disappearance of the Karman street for some value of Re(M). This critical curve Re(M) is plotted on fig. 11. It is fitted by the relation :

$$Re = 0.53 \ M^{2.29} \quad \text{or} \quad N = 1.88 \ M^{-0.29}$$

under it, no vortices are sensed while the Karman street exists above it. At this time, we have not determined whether standing eddies exist for parameters under the critical curve, or the separation points completely disappear. However, some mean velocity profiles have been obtained which show a deepening of the wake as M increases. The W shaped profile, characteristic of standing eddies |16| does not appear on the M = 103 curve, but the uncertainties of mean velocity measurements do not permit to decide on this case. We may note that various authors found separated flow around sphere (|4|, 0 < N < 40) and cylinder (|17|, 0 < N < 4.3) despite higher values of N than in our experiment, which should confirm the possible existence of standing eddies in a domain of the (Re, M) plane to be determined.

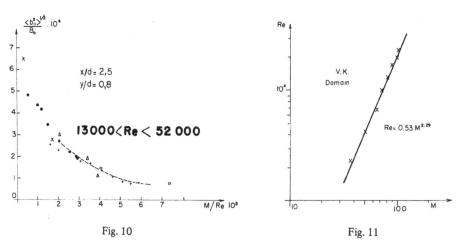

Fig. 10   Fig. 11

Fig.10. Magnetic energy for different Re, versus M/Re
Fig.11: Critical curve Re(M) for disappearance of the Von Karman
        Vortex street

5. CONCLUSION

We had been able to measure local variation of magnetic field by means of magnetodiodes. These sensors provided us a complete study of the upstream wake of a cylinder under subalfvenic regime. These results compare well with available theories, and they bear a strong resemblance with data obtained for spheres and Rankine body. The downstream part of the study correlate velocity and magnetic perturbation in and out of the wake. It leaves some open questions as whether or not standing eddies could subsist after the disappearance of the vortex street, while the interaction parameter is increasing. We expect to precise the condition of existence of the separation point, using the analytical value of the outer flow found in |13|. This work was supported by the C.N.R.S. A.T.P. "noyau terrestre" under contract n° 90.1639.

REFERENCES

|1| W. CHESTER, 'The effect of a magnetic field on Stokes flow in a conducting fluid'. J.F.M., Vol. 3, p. 304 (1957)

|2| H. YOSINOBU, 'A Linearized theory of Magnetohydrodynamic flow past a fixed body in a parallel magnetic field', J. of the phys. Soc. of Japan, vol. 15, n° 1, p. 175 (1960)

|3| K. TAMADA, 'Flow of a slightly conducting fluid past a circular cylinder with strong, aligned magnetic field', The physics of fluids, vol. 5, p. 817 (1962)

|4| T. MAXWORTHY, 'Experimental studies in magneto-fluid dynamics : pressure distribution measurements around a sphere', J.F.M., vol. 31, p. 801 (1968)

|5| G. YONAS, 'Measurements of drag in a conducting fluid with an aligned field and large interaction parameter', J.F.M., vol. 30, p. 813 (1967)

|6| T. MAXWORTHY, 'Experimental studies in magneto-fluid dynamics : flow over a sphere with a cylindrical afterbody', J.F.M. vol. 35, p. 411 (1969)

|7| H.G. ALHSTROM, 'Experiments on the upstream wake in magneto-fluid dynamics', J.F.M., vol. 15, p. 205

|8| B.M. LAKE, 'Velocity measurements in regions of upstream influence of a body in aligned-fields MHD flow', J.F.M., vol. 50, p. 209 (1971)

|9| B.P. BORISSOV, E. Yu. KRASSIL'NIKOV, 'Experimental determination of the condition for the absence of boundary layer separation in a flow around a cylinder', 1981, Proc. of the RIGA Sem. on M.H.D., p. 57 (in russian)

|10| A. ALEMANY, R. MOREAU, P.L. SULEM, U. FRISCH, 'Influence of an external magnetic field on homogeneous MHD turbulence', J. Mécanique, vol. 18, p. 277 (1979)

|11| Ph. CAPERAN, 'Contribution à l'étude expérimentale de la turbulence homogène MHD, première caractérisation de son

anisotropie', Thèse présentée à l'U.S.T.M. et l'I.N.P. de Grenoble, 7 décembre 1982

|12| S. CRISTOLOVEANU, 'Transport magnétoélectrique dans les semi-conducteurs en présence d'inhomogénéités naturelles ou induites par effets de recombinaison et d'injections', Doctorat d'Etat, U.S.M. Grenoble (1981)

|13| J. LAHJOMRI, 'Caractérisation de la structure des sillages MHD amont et aval d'un cylindre à petit nombre de Reynolds magnétique', Doctorat Univ. Joseph Fourier-Grenoble (1988)

|14| M.B. GLAUERT, 'Magnetohydrodynamic wakes', J.F.M., vol. 15, P. 1 (1963)

|15| R. MOREAU, 'L'effet d'un champ magnétique transversal sur le décollement', C.R. Acad. Sc. Paris, t 258, p. 1731, (1964)

|16| M. NISHIOKA, H. SATO, 'Measurements of velocity distributions in the wake of a circular cylinder at low Reynolds numbers', J.F.M., vol. 65, p. 97 (1974)

|17| A.E. GULOUBIEV, A.P. KACHOULIN, E.Yu. KRASSIL'NIKOV,V.G.LOUTCHIK, 'Flow of liquid metal around a cylinder in a longitudinal magnetic field', Proc. of the 1987 RIGA Sem. on M.H.D., p. 95. (in russian)

METALLURGICAL ASPECTS OF ELECTROMAGNETIC PROCESSING OF MATERIALS

Shigeo ASAI
Department of Iron & Steel Engineering
Nagoya University
Furo-cho, Chikusa-ku, Nagoya 464
JAPAN

ABSTRACT. Electromagnetic processing of materials is newly defined as the engineering field where electromagnetism is utilized for processing electrically conductive materials. The need for such processes is described. The existing processes and the proposed ones are classified and discussed form the view points of theoretical principle and applications to engineering.

Introduction

There is a real need to show that the study of magnetohydrodynamics is useful in the electromagnetic processing of materials as well as it has been useful in the original fields of astrophysics and plasma physics. This is because liquid metal is a typical electromagnetic material and the vast knowledge of MHD which has been accumulated over a long period of time in fusion research and in the field of astrophysics has a high potential for solving metallurgical problems.

Nowadays, electric current and magnetic field are being widely used in many processes. Their functions are partly to generate and control a flow field within liquid phase, partly to control the shape of solidifying specimens and partly to add heat to materials. A large amount of electric energy is used to produce high grade materials.

However, metallurgical engineers did not think to make use of MHD phenomena until the late 1970s.

As the first step, a research organization (MADYLAM) was established in 1978 in France. Then an IUTAM Symposium (The International Union of Theoretical and Applied Mechanics entitled "Metallurgical Applications of Magnetohydrodynamics" was held at Cambridge, UK, 1982. This timely event made a few metallurgists recognize the importance of MHD and was a good warning to metallurgical companies throughout the world. Especially in Japan, the Steelmaking Investigative Subcommittee of the Research Planning Committee in the ISIJ (The Iron and Steel Institute of Japan) discussed the application of MHD to metallurgy and selected it as one of the four major innovative subjects for future steelmaking technology in Japan. Indeed, the term "Electromagnetic Metallurgy" itself was first coined by this subcommittee. Upon the request of a report [2] from the subcommittee, the ISIJ established a committee on "Electromagnetic Metallurgy" in 1985, under the author's chairmanship, in which its definition was discussed and defined to cover not only the application of electromagnetic force, but also the electric and/or magnetic phenomena connected with metallurgy.

1. Background and Characteristics

The background where the study of electromagnetic processing of materials is needed and characteristics of electromagnetic processing are described here.

J. Lielpeteris and R. Moreau (eds.), Liquid Metal Magnetohydrodynamics, 391–400.

1.1. Background

1.1.1. Increase of electric energy consumption in metal industries.
Increase in the conseumption of electric energy has accompanied the strong demand for high grade materials. Especially in the process of steel-making the number of treatments to refine steel is increasing with the demand for clean steel. Therefore, the time needed for refining tends to be prolonged and the addition of thermal energy to molten steel is recognized to be an important technical subject. This thermal energy is mainly supplied by electric energy.

1.1.2. Development of the technology connected with this field.
Nowadays, the technology connected with electricity and magnetism is highly developed. For instance, the recently produced high energy permanent magnet (Fe-Nd-B system) provides a strong magnetic field at a cheap cost. Also, a superconducting magnet will be used in the related industries in the near future, assuming that the critical temperature of superconducting materials keeps increasing. Moreover, an electrode made of ZrB_2, (melting temperature : 3000° C and electric resistivity : 10^{-5} Ω.m), has been developed. This has made it possible to impose electric current directly on metals with a high melting point such as molten steel and seems to have a large impact on the metallurgic process. Until now the electric current induced by an alternating magnetic field has been used in metal industries (for example, linear electromagnetic stirring, electromagnetic pump, etc...) instead of directly imposing electric current on molten metal. With the direct imposition of current we can now expect alternative methods to transport and mix molten metal, which make more efficient use of electric power.

1.2. Characteristics

The characteristics of electromagnetic processing of materials are given as follows :

1.2.1. High density, cleanliness and controllability of electric energy.
Making use of electric current and magnetic field is the cheapest and the most convenient way to impose high density energy on metal. This energy source is extremely clean except for the contamination from electrode. Furthermore, the techniques for control of electric current and magnetic field have recently advanced a great deal.

1.2.2. Effective use of electric energy.
In the metal refining field, electric energy has been mainly used as the heat source. The heating rate calculated from the values of ordinary mixing power employed in steelmaking processes (1W/ton to 500 W/ton) is from about 1° C/one week to 1°C/30 min. On the other hand, the heating rate required as heat addition in the steelmaking processes is about 1 to 6 ° C/min. The amount of mixing power is extremely small in comparison with that for heat addition. Thus, a part of electric energy should be converted into thermal energy by a way which adds some functions of electromagnetism such as mixing, splashing, etc. (the details will be mentioned later) as shown in Fig. 1. The energy used for functions of electromagnetism is not wasted but is finally converted into thermal energy.

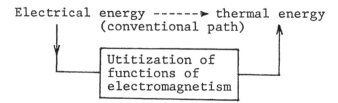

Figure 1. The recommended energy path (——) and the conventional one (---).

1.2.3. Application of magnetohydrodynamics to metallurgical problems.

The term magnetohydrodynamics coined in 1942 in Alfven, is the study of the dynamics of an electrically conductive fluid such as plasma, and can cover the dynamics of molten metal. The comparison between plasma and liquid metal is as follows :

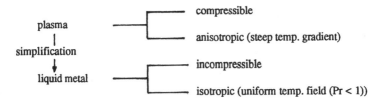

The vast knowledge stored in magnetohydrodynamics can be applied to solving the problems of liquid metal processing. The advantage of this application is the simplified procedure from the compressible anisotropic fluid to the incompressible and isotropic one.

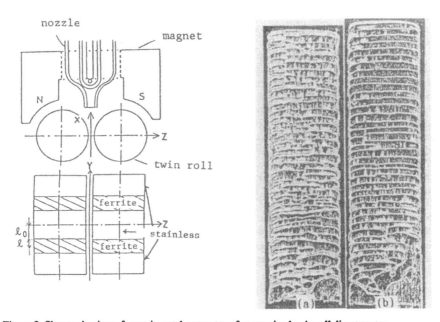

Figure 2. Shcematic view of experimental apparatus of magnetized twin roll direct caster.

Photo 1. Over view of Oscillation marks a) without magnetic field, b) with magnetic field.

2. Functions of Electromagnetism Applied to Processing of Materials

2.1. Function of shape control

In order to control the shape of molten or solid metal, the magnetic pressure, $P_m = - B^2/2\mu$, which is induced by a high frequency magnetic field is used in processes such as cold crucible, levitation melting, vertical electromagnetic casting and plastic deformation. Another type of shape control is a process utilizing the Lorentz force, $\mathbf{f} = \mathbf{J} \times \mathbf{B}$, which is caused by the imposed direct magnetic field, $\mathbf{B}$, and the current, $\mathbf{J} = \sigma \mathbf{v} \times \mathbf{B}$, induced by fluid motion, $\mathbf{v}$, under the magnetic field. The deformation of a liquid jet by use of a non-uniform direct magnetic field belongs to this type of shape control and its application to a process such as the production of thin plates can be expected [3].

2.2. Function of fluid driving

A travelling magnetic field induces a body force which drives molten metal along the direction of travelling and pinches it perpendicularly. The combined imposition of direct electric current and direct magnetic field induces a body force which drives molten metal to a direction perpendicular to those of current and field. Typical processes which employ the function of fluid driving are electromagnetic stirring of continuous casting and electromagnetic pumping.

2.3. Function of flow suppression

The Lorentz force, $\mathbf{f} = (\sigma \mathbf{v} \times \mathbf{B}) \times \mathbf{B}$, which was already explained, is used to suppress fluid flow motion. This force can suppress all of the fluid motion except the flow parallel to the direction of the imposed magnetic field. In the magnetic Czochralski method the natural convection due to the temperature difference is suppressed to control the oxygen content in the single crystal of silicon. In order to reduce nonmetallic inclusions in steel and to get a good surface quality of slab, direct magnetic field is applied to the liquid pool of continuous casting. It reduces the velocity of nozzle flow reduces wave motion in the meniscus. This is referred to as a magnetic brake [4].

In the production of thin sheet by a twin roll direct caster, a saw edge shape often appears on both sides. To get a smooth edge shape, direct magnetic field is applied to a pair of composite rolls made of a magnetic and a non-magnetic material, such as plain steel and stainless steel as shown in Fig. 2 [5]. A strong magnetic field appears at the plain steel part between the rolls and reduces the velocity of molten metal along the axis of the rolls. The quick reduction of the velocity at the vicinity of both edges gives comparatively smooth edges.

The wave motion existing at the meniscus of conventional continuous casting of steel causes irregular oscillation marks on the surface. Thus, the suppression of wave motion at the meniscus is crucial to get a good surface quality of the slab. It allows slabs to be be sent directly to the rolling stage without surface treatment and saves a large amount of thermal energy for reheating them. The effect of magnetic field on the appearance of oscillation marks was studied in a metal with a low melting point by Hayashida et al. [6]. As shown in Photo 1, the direct magnetic field has the function to change irregular oscillation marks to regular ones. The magnetic effect was obvious in the low frequency oscillation range.

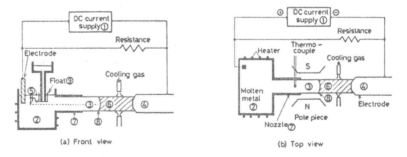

(a) Front view   (b) Top view

Figure 3. Schematic views of the experimental apparatus used in horizontal electromagnetic casting.

2.4. Function of levitation

Molten metal is levitated by balancing the gravity force with the electromagnetic force induced by direct electric current and direct magnetic field, $\rho\, \mathbf{g} = - \mathbf{J} \times \mathbf{B}$. On the basis of this principle, a new process of horizontal electromagnetic casting (HEMC) called nomold continuous casting was developed [7][8]. It not only enables the elimination of surface defects of the slab caused by contact with the wall of the mold, but also determines the near net shape casting. The additional advantage of this process is the favorable application to the casting of heavy metals, with less investment than for the ordinary electromagnetic casting, making use of high frequency magnetic field. Figure 3 shows a schematic view of the HEMC apparatus. A magnified view of the cast rod and thin plate and a cross-section of the thin plate are given in Photo 2. The surface appearance of products is very smooth without any defects and the cross sectional shape is slightly deformed by surface tension from the slit shpae of the nozzle.

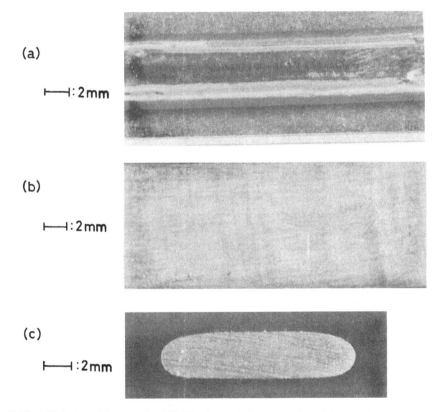

(a)

$\vdash\!\!\!\!-\!\!\!\dashv : 2\,mm$

(b)

$\vdash\!\!\!\!-\!\!\!\dashv : 2\,mm$

(c)

$\vdash\!\!\!\!-\!\!\!\dashv : 2\,mm$

Photo 2. Magnified view of the cast a) rod, b) thin plate and c) cross-section of the thin plate.

By imposing alternating electric current and direct magnetic field, the electromagnetic body forces which are exerted upon the molten metal to oppose gravity are induced periodically with the same switching frequency as that of the imposed current. Following this principle, frequency of bubble formation can be controlled as shown in Fig. 4 [9]. Bubbles are formed randomly when no electric current and no magnetic field are applied, but the frequency of bubble formation can be precisely controlled by their application.

Another example is the electromagnetic flotation of non-metallic inclusions where electromagnetic force induced by direct electric current and direct magnetic field is exerted upon the molten metal in the same direction as that of the gravity force. This force yield an apparent density heavier than that of molten metal and accelerates the separation rate of non-metallic inclusions [10].

2.5. Function of splashing

When the electromagnetic force induced by direct electric current and direct magnetic field is much larger than the gravity force, $|\rho g| < |J \times B|$, we observe a splashing phenomenon. By making use of this principle, a new atomization process, called electromagnetic atomization, has been developed. When electric voltage and magnetic field, perpendicular to each other, are imposed between a nozzle and an electrode, molten metal flowing out from the nozzle is splashed at the moment it contacts the electrode. The splashing phenomenon shuts down the current passing through the metal. Consequently, molten metal flowing out from the nozzle contacts with the electrode and is splashed again. By repeating this circulation, molten metal is thus atomized to produce particles with nearly the same size. Another advantage of this process is that it has no mechanical driving device and the splashing direction is easily controlled by changing the direction of the magnetic field. This process can be of great use for designing a new type of atomizer in spray casting or a gun in spray coating.

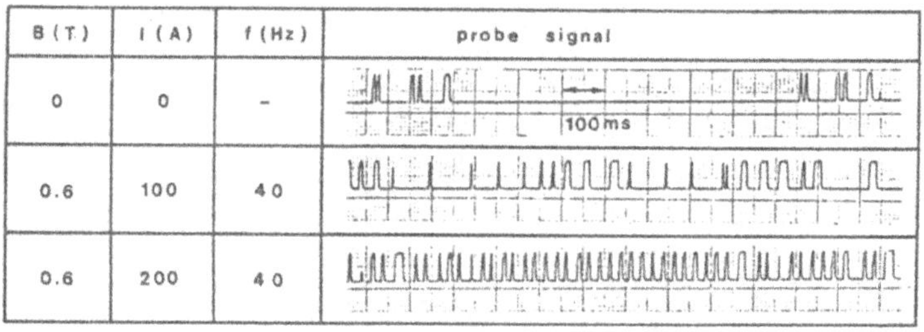

B (T)	I (A)	f (Hz)	probe signal
0	0	–	100 ms
0.6	100	40	
0.6	200	40	

Figure 4. Formation of bubbles under various electric and magnetic conditions.

2.6. Function of heat generation

There are two methods to heat up the molten metal by making use of Joule heat ($q = J^2/2\sigma$). One is the direct imposition of electric current on molten metal and the other is the indirect method in which electric current is induced in molten metal by imposing a high frequency magnetic field. The second one has mainly been employed by metal industries in spite of the inefficient conversion from electric to thermal energy because an electrode is not needed. The former is not practical for heating since it needs a current supply with high current capacity under low voltage due to the high electric conductivity and high melting point has recently been developed, the direct imposition of current should be noted as an important future technology to develop alternative figures for electromagnetic stirring and pumping [12].

2.7. Function of velocity detecting

A velocity sensor was proposed by Vives [13]. The principle of this sensor is based on Lenz's law ($E = - v \times B$), in which the velocity of molten metal is detected by observing E between two parallel lead wires set on both sides of a strong and small permanent magnet. This sensor is useful enough to observe fluctuation of turbulent flow. However, it is not usable over the Curie temperature of the magnet. Thus, it is necessary to develop a sensor able to work in high temperatures.

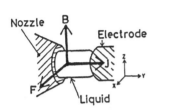 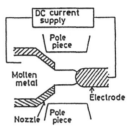

Figure 5. Electromagnetic atomization process.

2.8. Combined function

By taking advantage of the electromagnetic effects mentioned above, functions such as refining and solidification control can be combined.

2.8.1. Refining function.
An electromagnetic refining process is presented in which the driving force of molten metal caused by gas lifting in RH degassing process is replaced by electromagnetic force [14]. The advantage of this process is that the splashing of molten metal accompanied with arc spark in a vacuum chamber can accelerate mass transfer enormously and the capacity of gas ejector system can be reduced significantly due to the elimination of argon gas injection.

The electromagnetic separation of non-metallic inclusions described before also belongs to this function [10].

2.8.2. Control of solidification structure.
As a direct magnetic field suppresses natural convection in the vicinity of solidifying front by the induced body force, $f = J \times B = - \sigma (v \times B) \times B$, the nuclei multiplicated on the liquid-solid interface have no chance to be transported into the bulk liquid. This leads to a columnar structure [15]. On the other hand stirring fluid-driven causes multiplication due to the breaking of dendrite tips and transports nuclei into bulk liquid so that the solidified structure generally tends towards isotropy [15].

Recently, a levitation technique was applied to get a large amount of undercooling, as much as 200 K in steel [16]. This seems to be one of the promising methods for obtaining bulk amorphous metal.

Lielausis et al. [17] presented an interesting paper on the direct effect of electric current and magnetic field on solidification phenomenon. This paper seems to suggest the discover of a new function of electromagnetism connected with the solidification of metal.

3. Conclusion

The electromagnetic processing of materials has been newly defined as one of the promising engineering fields in which the functions of electromagnetism are utilized for processing electrically

Table 1 : Classification of processes on the basis of function of electromagnetism

Basic principle	Function of material processing	Processes
Electromagnetic body force $f = J \times B$	① Shape control $P_m = -B^2/2\mu$ $f = \sigma(v \times B) \times B$	cold crucible, levitation melting, vertical electromagnetic casting, magnetic shaping of metal film, plastic deformation, deformation by a non-uniform magnetic field
	② Driving fluid	stirring by direct electric and magnetic currents, electromagnetic stirring, electromagnetic pump
	③ Suppression of flow $f = \sigma(v \times B) \times B$	magnetic Czochralski method, magnetic brake, magnetic edge control of thin film, suppression of wave motion
	④ Levitation $|J \times B| \leq |\rho g|$	horizontal electromagnetic casting, control of timing of bubble formation, electromagnetic separation of non-metallic inclusion
	⑤ Splashing $|J \times B| > |\rho g|$	electromagnetic atomization
Joule heat	⑥ Heat generation $q = J^2/2\sigma$	cold crucible, levitation melting, heating by high frequency magnetic field , direct heating
Lenz's law	⑦ Detecting $E = -v \times B$	sensor of velocity
Combined function	⑧ Refining ②+④+⑤+⑥	electromagnetic refining, electromagnetic separation of non-metallic inclusion
	⑨ Control of solidification structure ②+③+④	grain refinement, grain coarsening growing single crystal, supercooling (amorphous)

conductive materials such as liquid metal, plasma and electrolyte. The background requiring this processing and the engineering characteristics supporting it were described. The existing processes and the proposed ones making use of electric current and magnetic field were classified on the basis of the functions of electromagnetism and were discussed from the view point of theoretical principles and engineering applications.

The existing and proposed processes are classified in Table 1 on the basis of the function of electromagnetism.

The electromagnetic processing of materials is located in a small valley between the two big mountains known as "Metallurgy" and "Electromagnetism". These mountains have provided very useful mines for a long time and supplied profound benefits to humankind. However, nobody knows whether the immense vein of the mines found in the two mountains meet under the valley floor or not. Sometimes, one found useful ores there by chance. In fact, most metallurgists have unconsciously made use of the phenomena which should belong to the electromagnetic processing of materials.

Let MHD and metallurgical engineers and scientists cooperate to find new functions which should exist on the boundary between magnetohydrodynamics and metallurgy. The discovery of new functions and those applications are indispensable for developing the electromagnetic processing of materials as a new engineering field.

REFERENCES

[1]   MOFFATT H.K. and PROCTOR M.R.E., 1984, "Metallurgical Application of Magnetohydrodynamics, Proceedings of a Symposium of the IUTAM, (The Metal Society).

[2]   KAWAKAMI K., 1984, "Avenues to Innovative Steelmaking Technologies in Japan", Trans. ISIJ, 24, 754.

[3]   OSHIMA S., YAMANE R., MOCHIMARU Y. and MATSUOKA Y., 1986, "The Shape of Liquid Metal Jets under a Non-Uniform Magnetic Field 1st Report and 2nd Report", Trans. of the Japan Society of Mechanical Engineers (series B), 52, 2888, 2897.

[4]   NAGAI J., SUZUKI K., KOJIMA S. and KOLLBERG S., 1984, "Steel Flow Control in a High-Speed Continuous Slab Caster Using an Electromagnetic Brake", Iron and Steel Engineer, 61, 41.

[5]   YUHARA T., KOZUKA T., ASAI S., and MUCHI I., 1988, "Effect of Direct Magnetic Field on the Edge Shape of Thin Plate Cast by Twin Roll Process", CAMP-ISIJ, 1, 389.

[6]   HAYASHIDA M., OHNO T., ONO H., and TSUTSUMI K., "The simulation of Relationship between Surface Waves in Mold and the Quality of Strands Surface Using Sn-Pb Alloy", Tetsu-to Hagane, 73, S686.

[7]   ASAI S., KOZUKA T., and MUCHI I., 1986, "Process Development and stability Analysis of Horizontal Electromagnetic Casting Method" Tetsu-to Hagane, 72, 2218.

[8]   KOZUKA T., ASAI S., and MUCHI I., 1988, "Horizontal Electromagnetic Casting Process of Thin Plate and its Stability Analysis", Tetsu-to Hagane, 74, in print.

[9]   TAKEDA K., NAKAMURA M., OHNO H., KUWABARA K. and OHASHI T., 1987, "Development of Electromagnetic Valve for Continuous Casting Tundish", Tetsu-to Hagane, 73, S1449.

[10]   MARTY P. and ALEMANY A., 1984, "Theoretical and Experimental Aspects of Electromagnetic Separation", Proceedings of a Symposium of the IUTAM, (The Metal Society), 245.

[11]   SASSA K., AGATA N., KOZUKA T., and ASAI S., 1988, "Atomization of Molten Metal by Use of Electromagnetic Force", CAMP-ISIJ, 1, 390.

[12]   WADA K., TAKEUCHI E., ANDO K., KITAMINE S., 1987, "Direct Electric Charging to Molten Steel", Tetsu-to Hagane, 73, S687.

[13] VIVES Ch. and RICOU R., 1985, "Experimental Study of Continuous Electromagnetic Casting of Aluminum Alloys", Met. Trans., 16B, 377.

[14] ASAI S., Feb. 1986, Report of the Committee on Electromagnetic Metallurgy in ISIJ.

[15] ASAI S., YASUI K., and MUCHI I., 1978, "Effects of Electromagnetic Forces on Solidified Structure of Metal", Trans. ISIJ, 18, 754.

[16] NAKATA M., OZEKI A., and MORI K., 1988, "Solidification Behavior of Undercooled Molten Steel by Levitation Method", CAMP-ISIJ, 1, 264.

[17] LIELAUSIS O., MIKELSON A., SHCHERBININ E., and GELFGAT Yu., 1984, "Electric Currents in Molten Metals and Their Interaction with a Magnetic Field", Proceedings of a Symposium of the IUTAM, (The Metal Society), 234.

Session H:
Dynamo Theory

LIQUID METAL MHD AND THE GEODYNAMO

H.K. MOFFATT
Department of Applied Mathematics and Theoretical Physics
Silver Street
Cambridge CB3 9EW
U.K.

ABSTRACT. The magnetic field of the Earth is generated by dynamo action associated with the upwelling of buoyant material in the liquid outer core. It is argued that this upwelling occurs in the form of mushroom-shaped blobs of material released from the mushy zone at the inner core boundary (ICB), and having a very small density defect $\delta\rho/\rho$. The rise of buoyant material with velocity w is compensated by the slow rate of growth of the solid inner core. The resulting mass balance, combined with approximate geostrophic force balance in the core leads to estimates

$$\delta\rho/\rho \sim 3 \times 10^{-9} \quad , \quad w \sim 2 \times 10^{-4}\text{m/s}.$$

Each rising blob drives a Taylor column, and the helicity and α-effect associated with this flow is estimated. A mean-field dynamo driven by this α-effect in conjunction with differential rotation generates a magnetic field whose strength is determined in order of magnitude by the plausible assumption of magnetostrophic equilibrium.

1. Introduction

The most plausible scenario for the dynamics of the geodynamo is that first proposed by Braginskii [1], in which gravitational energy is released through the process of solidification at the inner core boundary (ICB) of the iron alloy of which the liquid outer core is composed. Pure iron solidifies, leaving a layer of molten alloy slightly richer in the lighter ingredient (which we shall assume to be sulphur, although this is not the only possibility). Dendrites grow and collapse under their own weight, forming a 'mushy zone' [2] of depth of the order of a few kilometers, this layer providing a continuous source of the lighter fluid material.

This layer of lighter fluid is gravitationally unstable by the Rayleigh-Taylor mechanism. The character and growth-rate of the instability are controlled partly by conditions within the mushy zone, which may be most simply treated as a porous medium with a Darcy law of resistance, and partly by the dynamics of the overlying liquid zone in which both Coriolis forces and Lorentz forces play an important, and probably dominant, role.

Whatever the nature of the instability, the resulting rise of lighter material is compensated by the descent of heavier material and by the gravitational condensation through solidification onto the solid inner core. In this process, gravitational

J. Lielpeteris and R. Moreau (eds.), Liquid Metal Magnetohydrodynamics, 403–412.
© *1989 by Kluwer Academic Publishers.*

energy is released at a rate of order 10^{12}W[3], and is converted partly to kinetic energy of the rising blobs, and partly to heat. If there were no dynamo action and therefore no magnetic field, all of this power would have to be dissipated by viscosity; the flow would be turbulent with energy dissipation rate

$$\epsilon \sim u^3/\ell \sim 10^{-8}\text{W}/\text{m}^3$$

where u and ℓ are velocity and length-scales of the turbulence. Thus, for example, if $\ell \sim 10$ km, then $u \sim 5$ cm/s., a hundred times larger than the value that is inferred from studies of the secular variation of the Earth's magnetic field extrapolated down to the core-mantle boundary (CMB) [4].

Dynamo action *does* however occur, because strong Coriolis forces associated with the Earth's rotation ensure that the buoyancy-driven convection has a helical character, with helicity distribution [5] that is antisymmetric about the equatorial plane. Provided the typical scale ℓ of the convection is small compared with the core radius (and this is the essence of the process proposed in this paper), a mean-field approach [5-8] is relevant, and dynamo action occurs through the α-effect that is an inevitable concomitant of the non-zero helicity. Thus the kinetic energy is immediately converted to magnetic energy which is in turn converted to heat by Joule dissipation. The rate of Joule dissipation associated with a predominantly toroidal field of order 20G is of order 4×10^8W. The actual Joule dissipation will be greater than this, because of the dissipation associated with the small-scale field; and it may be supposed that the system adjusts itself so that the net rate of dissipation of energy into heat is just equal to the gravitational power supplied.

This description of the dynamo process is set out schematically in figure 1, in which some of the important subsidiary mechanisms are also indicated. Broadly speaking, these involve interactions between small-scale and large-scale fields, and between their poloidal and toroidal ingredients. A number of feedback processes are identified, which can all play a part in the delicate balance that is established. Ideally, the dynamo theoretician would like to explain not only the existence and present behaviour of the Earth's magnetic field, but also its long-term behaviour, including the phenomenon of reversals, over geological time. This target is still some way off, but it is evident that the complexity of the nonlinear effects represented in figure 1 is quite sufficient to incorporate such phenomena as reversals, without seeking more exotic physical mechanisms.

2. Bubble Convection

Following the above discussion, we postulate that Rayleigh-Taylor instability causes intermittent release of buoyant blobs (or 'bubbles') from the dendritic layer where the density defect is continuously created (figure 2). The molecular diffusivity D of sulphur in iron just above the melting point is estimated [9] to be

$$D \sim 10^{-8}\text{m}^2/\text{s}$$

and is very small compared with the product $w\ell$ of the expected scale ℓ and rise velocity w (see below). In a first approximation, we may neglect this molecular

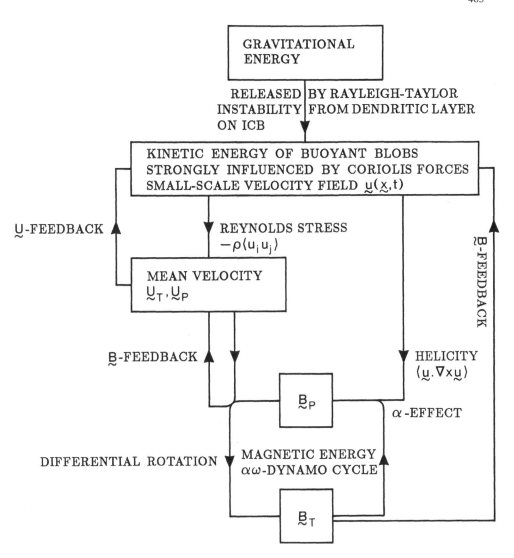

Figure 1. Energetics and feedback cycles of the Geodynamo

diffusion altogether. The bubbles, once formed, then rise through the outer core, and spread out horizontally as they approach the CMB. [This picture of bubble convection has some affinity with the bubble convection proposed by Howard [10] as a model for thermal convection at very high Rayleigh number.]

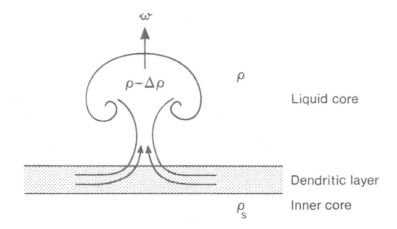

Figure 2. Release of buoyant blob from dendritic layer on ICB

Consider first the mass balance associated with this process (figure 3). Let w represent the mean upward velocity of blobs near the ICB, carrying density defect $\Delta\rho$; this is compensated by a downward mass flux

$$\frac{dM}{dt} = 4\pi R_I^2 \dot{R}_I (\rho_s - \rho)$$

required to provide the growth of the inner core; here R_I is the inner core radius, ρ_s the density of the solid inner core, and ρ the density of the liquid core, both taken near the ICB, and we assume $\rho_s - \rho \approx \rho/20$. This mass balance then gives

$$w \, \Delta\rho/\rho \approx \dot{R}_I/20. \tag{2.1}$$

Secondly, consider the force balance which determines the rate of rise of the bubble. In a non-rotating environment, this rate of rise would be given (on dimensional grounds) by

$$w \sim (ga \, \Delta\rho/\rho)^{1/2}, \tag{2.2}$$

where a is the radius of the bubble (or the radius of curvature at its upper surface); this merely reflects a balance between the buoyancy force $\Delta\rho \, g$ and the convective acceleration term $\rho \underline{u}.\nabla \underline{u}$ of the Navier Stokes equation. In a rotating medium with

$$a\Omega/w \gg 1, \tag{2.3}$$

the Coriolis acceleration $\rho \underline{\Omega} \wedge \underline{u}$ replaces the convective acceleration in the force balance, so that (2.2) is replaced by

$$w \sim (g/\Omega)\Delta\rho/\rho, \tag{2.4}$$

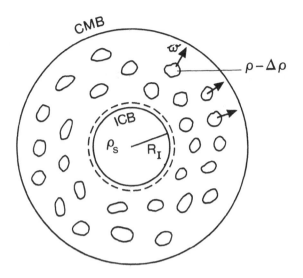

Figure 3. The rise of blobs with density defect $\Delta\rho$ implies a downward mass flux just sufficient to account for the slow growth of the inner core.

independent of the size of the bubble.

Combining (2.1) and (2.3), we find

$$\frac{\Delta\rho}{\rho} \sim \left(\frac{\Omega\dot{R}_I}{20\,g}\right)^{1/2} \quad , \quad w \sim \left(\frac{g\,\dot{R}_I}{20\,\Omega}\right)^{1/2}. \tag{2.5}$$

With $\Omega \sim 7 \times 10^{-5}\mathrm{s}^{-1}$, $g \sim 3\ m/s^2$ (near the ICB) and $\dot{R}_I \sim 10^{-11}\mathrm{m/s}$ (on the assumption that the inner core has been growing steadily over the whole lifetime of the Earth), (2.5) gives

$$\frac{\Delta\rho}{\rho} \sim 3 \times 10^{-9} \quad , \quad w \sim 2 \times 10^{-4}\mathrm{m/s}. \tag{2.6}$$

This estimate of w is entirely plausible, being of the same order of magnitude as velocities at the CMB inferred from the secular variation studies referred to previously. Note that $w/\Omega \sim 3$m so that any blob of scale a much greater than a few meters will easily satisfy the condition (2.3).

3. Helicity and Dynamo Action Induced by Bubble Rise

Under the same condition (2.3), the rising bubble may be expected to drive flow in a Taylor column [11] consisting of the cylinder of fluid circumscribing the bubble and parallel to the rotation vector $\underset{\sim}{\Omega}$. The effect is difficult to analyse in the spherical annulus geometry, but may be understood at least qualitatively with reference to the plane geometry indicated in figure 4. As the bubble rises, it pushes the column of fluid ahead of it; this column must spread out in a layer near the upper boundary where convective acceleration associated with the change of direction of the flow is not negligible; and the fluid must then descend and flow in to feed the Taylor column at the lower boundary.

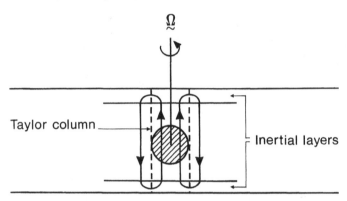

Figure 4. Taylor column associated with a rising bubble in a plane geometry; inertial layers are needed at top and bottom to satisfy the condition of zero normal velocity on the boundaries.

The upward flux in the column is $Q = \pi a^2 w$, and this is compensated by an equal and opposite downward flux outside the column. The downward moving fluid rotates more slowly than the upward moving fluid on account of the tendency to conserve angular momentum in the inertial layers at the top and bottom, this reduction in angular velocity being proportional to Ω. Hence the total helicity of the flow, defined by

$$\mathcal{H} = \int \underset{\sim}{u} \cdot \underset{\sim}{\omega} \, dV \tag{3.1}$$

is given in order of magnitude by

$$\mathcal{H} \sim V_c w \Omega, \tag{3.2}$$

where V_c is the volume of the Taylor column. More generally, we may expect that if there are a number of such columns randomly distributed, then the local mean helicity density will be

$$\bar{\mathcal{H}} \sim c \, w \Omega \tag{3.3}$$

where c is the proportion of the volume occupied by columns. [A similar mechanism for the production of helicity has been found by Hunt and Hussain [12] in the opposite limit of a weakly rotating fluid.]

The situation is more complicated in the spherical annulus geometry, as can be seen from figure 5. Outside the cylinder circumscribing the inner core and parallel to $\underset{\sim}{\Omega}$, the helicity induced by blobs originating from northern and southern hemispheres at the ICB will tend to cancel; no such cancellation occurs however *inside* this circumscribing cylinder, and the mean helicity in the fluid part of this region will be given by $\bar{\mathcal{H}} \sim \pm cw\Omega$ in the northern and southern hemispheres respectively.

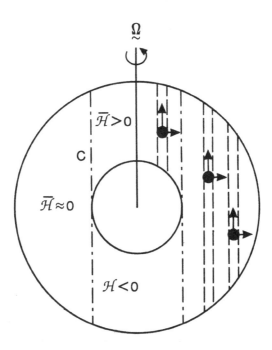

Figure 5. Taylor columns associated with rising blobs in spherical geometry. Helicities associated with blobs outside the cylinder C tend to cancel.

We may now estimate the α-effect [6,7] associated with this helicity. To be specific, let us suppose that the scale of the bubbles is of order $a \sim 10$km, which seems quite plausible in relation to the scale of the dendritic layer from which the bubbles originate. The magnetic diffusivity of the liquid core, η , is of order $3\mathrm{m}^2/\mathrm{s}$, so that the magnetic Reynolds number associated with the bubble rise is

$$R_m \cong \frac{wa}{\eta} \sim 1 . \tag{3.4}$$

Under this condition, α may be determined by first-order smoothing theory [7], and is given in order of magnitude by

$$\alpha \sim -\bar{\mathcal{H}}a^2/\eta \sim c(\Omega a)R_m . \tag{3.5}$$

The magnetic Reynolds number based on α and the core radius R_c is then

$$R_\alpha = \left| \frac{a\, R_c}{\eta} \right| \sim c \left(\frac{\Omega R_c}{w} \right) R_m^2 \tag{3.6}$$

so that, for example, with $c \sim 0.01$, we find

$$R_\alpha \sim 3 \times 10^3, \tag{3.7}$$

a value that is amply sufficient to drive a dynamo of either α^2 or $\alpha\omega$-type [5,8].

The above discussion of course neglects the back-reaction of the magnetic field on the flow, via the Lorentz force. It is to be expected that the magnetic field will saturate at a level at which the three forces of buoyancy, Coriolis and Lorentz are comparable:

$$\underline{g}\Delta\rho \sim 2\rho\underline{\Omega} \wedge \underline{u} \sim \underline{j} \wedge \underline{B}. \tag{3.8}$$

With $j \sim \sigma w B$, this leads to the estimate

$$B \sim (2\rho\Omega\mu_o\eta)^{1/2} \sim 20G. \tag{3.9}$$

Since the poloidal field in the core is of order 1G, this suggests that the dynamo is of $\alpha\omega$-type with a dominant toroidal field of order 20G.

4. Discussion

We have adopted the view [1] that the source of energy for dynamo action in the Earth's core is the gravitational potential energy that is released when iron alloy solidifies at the ICB. We have argued that this process leads to the release of bubbles of material with density defect $\Delta\rho/\rho \sim 3 \times 10^{-9}$, rising with velocity $w \sim 2 \times 10^{-4}$m/s; that these rising blobs generate helicity and so an α-effect in the rotating environment; and that this, in conjunction with differential rotation, drives an $\alpha\omega$-dynamo.

The description appears self-consistent as far as orders of magnitude are concerned. A number of problems arise, however, which need closer investigation: (i) the problem of Rayleigh-Taylor instability in a dendritic layer, taking due account of Coriolis forces and Lorentz forces in the overlying liquid; (ii) the problem of the rise of a blob of slightly buoyant material in a spherical annulus, the dynamics being dominated by Coriolis and Lorentz forces; (iii) the horizontal spread of such a blob when it reaches the CMB, and the associated transient coupling between core and mantle that this may induce.

There may also be a longer period cyclical behaviour associated with the gradual growth of the layer of more buoyant material near the CMB. The rate of growth of this layer is such that it will fill the spherical annulus in a time of approximately 4000 years. At this stage the density would be $\rho - \Delta\rho$ throughout the liquid core, and the process would in effect start again. Over the whole lifetime ($\sim 4 \times 10^9$

years) of the Earth, the mean density of the liquid core has decreased by an amount $\Delta\rho$ about 10^6 times if this picture is correct, the total decrease being of order $.002\rho$, just sufficient to compensate the growth of the denser inner core. It is tempting to speculate that the oscillations of the Earth's dipole moment on time-scales in the range $10^3 - 10^4$ years may be caused by this cyclical process in the core, and even that field reversals may be triggered by this mechanism.

Acknowledgements. The work described here was started during the tenure of a Green Scholarship at IGPP, La Jolla, April-August 1987. I am grateful to Professor John Miles and others at IGPP for providing such an excellent opportunity for geophysical research. The warm and friendly hospitality of the Latvian SSR Academy of Sciences during the IUTAM Symposium in Riga is greatly appreciated.

This work has been presented also at the First SEDI (Study of Earth's Deep Interior) Symposium, held in Blanes, Spain, 23-25 June 1988.

References

[1] BRAGINSKII S.I., 1963 'Structure of the F-layer and reasons for convection in the Earth's core', Dokl. Akad. Nank SSSR **149** 8.

[2] LOPER D.E. and ROBERTS P.H., 1978 'On the motion of an iron-alloy core, containing a slurry I General theory', Geophys. Astrophys. Fluid Dyn. **9** 289.

[3] GUBBINS D., 1976 'Observational constraints on the generation process of the Earth's magnetic field', Geophys.J. **47** 19.

[4] GUBBINS D., 1987 'Mechanisms for geomagnetic polarity reversals', Nature **326**, 167.

[5] MOFFATT H.K., 1978 *Magnetic Field Generation in Electrically Conducting Fluids*, Cambridge University Press, 47 et seq.

[6] STEENBECK, M., KRAUSE, F., and RADLER K-H, 1966 'A calculation of the mean electromotive force in an electrically conducting fluid in turbulent motion, under the influence of Coriolis forces'. Z. Naturforsch. **21a**, 369 [in German].

[7] MOFFATT H.K., 1970 'Turbulent dynamo action at low magnetic Reynolds number', J.Fluid Mech. **41**, 435.

[8] KRAUSE F. and RADLER K-H, 1979 *Mean-field Magnetohydrodynamics and Dynamo Theory*. Pergamon Press.

[9] LOPER D.E. and ROBERTS P.H., 1981 'A study of conditions at the inner core boundary of the earth'. Phys. Earth Planet. Inter. **24**, 302.

[10] HOWARD L.N., 1966 'Convection at high Rayleigh number' in Applied Mechanics (Proc. XIth Int. Cong. Appl. Mech. Munich 1964, Ed. H. Görtler)

1109.

[11] GREENSPAN H.P., 1968 *The Theory of Rotating Fluids*, Cambridge University Press.

[12] HUNT J.C.R. and HUSSAIN A.K.M.F., 1988 'A note on velocity, vorticity and helicity of fluid elements and fluid volumes'. Unpublished.

[13] FEARN D.R., LOPER D.E. and ROBERTS P.H., 1981 'Structure of the Earth's inner core'. Nature 292-232.

[14] LOPER D.E. 1983 'Structure of the inner core boundary', Geoph. Astrophys. Fluid Dyn. **25**, 139.

THE HELICAL MHD DYNAMO

A. GAILITIS, O. LIELAUSIS
Institute of Physics, Latvian SSR Academy of Sciences
229021 Riga, Salaspils
USSR

B.G. KARASEV, I.R. KIRILLOV, A.P. OGORODNIKOV
Efremov Institute of Electrophysical Apparatus
188631 Leningrad
USSR

ABSTRACT. The uniform MHD dynamo is often presented working on helicity occuring in multi-eddy turbulence. Such turbulence consumes energy, which is unacceptable in the laboratory. For the purpose of experimentation, we are studying helical models with one or two moving parts. The models can be semi-analytically calculated and are selected in increasing complexity to show the level of information each contains. The current status of the experiment is also reported.

1. Formulation

Let us examine an electrically uniform (σ = const) fluid with a velocity v(r) given inside an infinite cylinder. We consider three versions : i) in the simplest case [1, 2] the cylinder moves as a solid one in a steady helical motion - rotation + axial translation (Fig. 1) ; ii) two (or even more) coaxial solid cylinders are moving one within the other ; iii) instead of solid motion velocity, v(r) having a polynomial dependence on r.

Due to symmetry the generated field behaves on t, $\varnothing$, z exponentially :

$$\exp (pt + ikz + im\varnothing) . \tag{1}$$

In both bodies, the solid cylinder I and the immobile cylinder surrounding II, the fields radial behaviour can be presented by Bessel functions of complex argument - an exact solution of Maxwell equations. Matching conditions leads to a closed but a bit tedious secular equation in the form

$$F(Rm, k_R + ik_I , p_R + ip_I , \text{geom. param.}) = 0 . \tag{2}$$

Magnetic Reynolds number $Rm = \mu_0 \sigma v_{max} R$ is determined by maximum velocity v_{max} within the inner cylinder and its radius R (below R = 1).

The complex equation (2) is exact but transcendent (Bessel functions) and without any exact solution. To get two of the five arguments named in (2) for given combinations of the remarking three, we use numerical iteration.

With the polynomial velocity profile the equation (2) cannot be presented in a closed form using standard functions. Nevertheless, it causes only minor changes in the computer code, for the field inside the cylinder the Bessel expansion must be replaced by a not very complicated expansion evaluated directly from the Maxwell equations.

413

J. Lielpeteris and R. Moreau (eds.), Liquid Metal Magnetohydrodynamics, 413–419.

2. Models with one Moving Part

2.1. The Ponomarenko model

In Fig. 2, disconnected neutral curves ($k_I = p_R = 0$) look like neutral mechanical stability loops for laminar boundary layer with an inflected profile. Inside, the field grows, while outside it decays. An AC (rotating) field is generated, the frequency p_I varies along the loop from high on the upper branch to low on the lower one. At $Rm \gg 1$, on the upper branch, the field is almost frozen within the cylinder, on the lower branch - in an immobile surrounding. The left most point on each loop marks the critical Rm for particular m. It depends on $x = v_z/wR$. Without rotation ($x = \infty$) or translation ($x = 0$) no field is generated ($Rm = \infty$). Confirming Cowling's theorem, no axisymmetric field ($m = 0$) can be generated. The curves for $m = 1$ [2] are presented in Fig. 2 to see higher modes placed at much higher Rm. Hence, below we are dealing with the main mode only : $m = 1$.

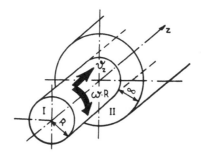

Figure 1. Ponomarenko dynamo

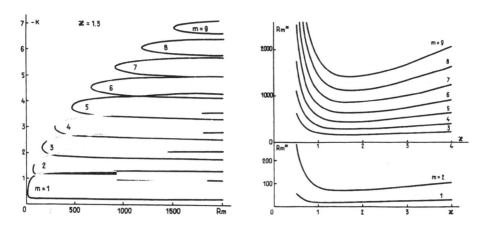

Figure 2. Ponomarenko dynamo. Neutral curves (left) and critical Rm versus x (right).

2.2. Polynomial profile of velocity

The Ponomarenko model has a velocity jump on the surface of the cylinder. For comparison we calculated neutral curves for radial profiles with no jumps and breaks : $v_z = (1 - r^2)^2, r < 1 ; v_z = 0, r > 1$ (Figs. 3 and 4). Near the threshold there is no serious difference with the Ponomarenko model ; a smaller effective radius can explain higher critical Rm. Asymptotes at Rm >> 1 differ in decline.

The polynomial form was used to search for the most stable velocity profile leading to the lowest possible critical Rm. We get the lowest Rm = 12.2 for /v/ = const and a ratio rotation/translation rising outwards.

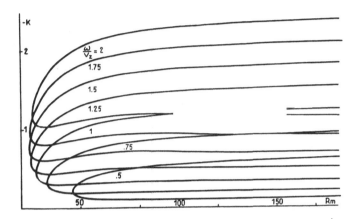

Figure 3. Neutral curves for smooth velocity profiles. Rotation similar to translation.

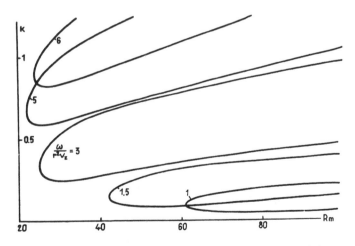

Figure 4. Neutral curves for smooth velocity profiles. Rotation different from translation.

416

3. Models with two Moving Parts

3.1. More then one critical Rm :

With one moving cylinder we have only one critical Rm. The solid type helical stream separated from the surrounding cylinder by some moving coaxial layer can have two (Fig. 5) or even three critical Rm (Fig. 6).

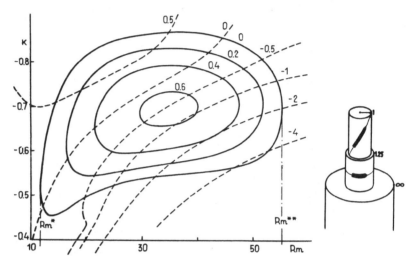

Figure 5. Layer rotating together with the central stream. Generation starts at Rm*, passes a maximum and ends at Rm**. Solid lines : p_R = const Dashed lines : p_I = const. x = 1.3.

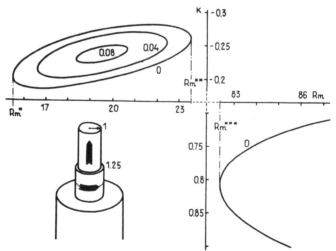

Figure 6. Layer rotating round axial central stream. Generation starts at Rm* and ends at Rm**. Restart at Rm*** happens at the inner surface.

3.2. Model with flow reversal

The degree of asymmetry with respect to the sign of z modifies the correspondence of an infinite model to a long but finite one. For instance, a helical stream surrounded by an axial flow reversing the whole central flow rate obtains a new type of critical Rm (Fig. 7). The first critical Rm (= 14.28) remains as before at the leftmost point of the neutral curve $k_I = p_R = 0$. The new critical Rm (= 15.74) is at the leftmost point of a member of the family of curves $k_I = const$, $p_R = 0$.

Between the two critical Rm the generation is of the convective type like fluid instabilities in boundary layers, channels etc. The width of the neutral loop indicates an interval of wave number k_R giving temporal growth ($p_R > 0$). The corresponding frequency pi interval gives spatial growth ($k_I < 0$). The growing field is moving axially with a nonzero group velocity $v_{gr} = dp_I / dp_R$. Hence the model is only valid for a limited period [3]. The temporal growth ends with the last premordial perturbation leaving the model. Any further field must be supported by some AC perturbation at the entrance all the time. The longer the model is the exponentially smaller the support may be.

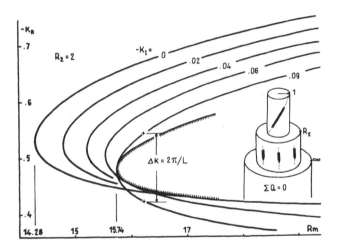

Figure 7. Evolvent of $k_I = const$ indicates an absolute generation. $x = 1$.

At the leftmost point of the stability boundary $dRm/dk = - (dF/dk)/(dF/dRm) = 0$. Because of (2) there is zero for the complex group velocity too : $dp/dk = -(dF/dk) / (dF/dp) = 0$. A long model starts to generate without any entrance support. The Ponomarenko model has no evolvent and hence no absolute generation.

Limited models start to generate at some higher Rm. For any such Rm one curve $k_I = const$ has equal frequencies on the lower and the upper branches as marked by crosses in fig. 7. Both solutions form a standing wave. The node distance $L = 2 \pi/(k_1 - k_2)$ is an estimate for a model generating length (Fig. 8).

4. Helical MHD Dynamo Model Experiment

Due to the lowest known critical Rm we are using the reverse schematics for a laboratory MHD experiment [4].

418

The model is all-welded from stainless steel and is filled, except for an open measuring channel 1, with liquid sodium (Fig. 9). Sodium passes through the model from entrance 2 to exit 10. In the helical labyrinth 3, the sodium stream is accelerated and made helical : axial motion + rotation round the axis. When passing through the main channel 4, the sodium rotation is maintained by inertia only. Here, both the calculation and the water test indicate less than 20 % loss of angular momentum. Rotation ends in the twelve-wall labyrinth 5 of the reverse system ; hence the flow is axial in the reverse channel 6.

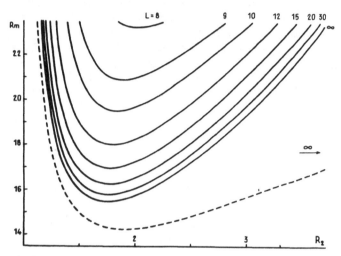

Figure 8. Generation Rm depends on model length L and outer radius of reverse flow R_2. Dashed line - a convective generation margin.

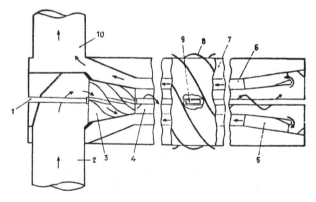

Figure 9. Experimental model. Length 3 m, ⌀ 0.5 m, empty mass 400 kg.

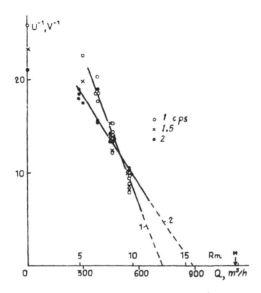

Figure 10. Models response to external excitation.

The reverse channel is separated from the main channel and from an immobile sodium volume 7 by two thin (1 mm) coaxial cylinders, providing good electrical connection in sodium. Electrodynamics of the field generation is calculated accounting for the real electrical resistance of both inner cylindrical walls and for insulator outside the model and inside the measuring channel too. All dimensions are set at optimum to have the calculated critical Rm as low as possible.

To start the experiment at subcritical flowrates, the whole model was located in a helical three-phase excitation winding 8 linked with a low-frequency (0-10 CPS) generator. In the channel 1, we placed a coil 9 sensitive to the transverse magnetic field and connected to a voltmeter. To find the maximum signal U, we tuned the three-phase generator frequency. The inverse signal (1/U) versus flowrate for three frequencies is plotted in Fig. 10. The critical flowrate is actually the point leftmost where 1/U curves cross the abscissa. In Fig. 10, the crossing point is not reached as it seems to be even lower than theoretical value (*).

The experiment was carried out at 200° C ; the flowrate of the pump used can be up to 1200 cmph. The experiment was interrupted at 660 cmph before reaching self-excitation when our model was damaged by unexpected mechanical vibration.

At subcritical conditions model response to external excitation agrees with theory. We are looking forward to reaching self-excitation after eliminating the mechanical vibration.

REFERENCES

[1] PONOMARENKO Y.B., 1973, "On the theory of hydrodynamic dynamos", Zh. Prikl. Mech. Tech. Fiz. (USSR), vol. 6, 47.
[2] GAILITIS A. & FREIBERG J., 1976, "On the theory of helical MHD dynamo", Magn. Gidrodin., n° 2, 3 (Magnetohydrodynam. Vol. 12, 127).
[3] LANDAU L.D. & LIFSHITZ E.M., Fluid Mechanics, 1959, Pergamon Press.
[4] GAILITIS A.K., KARASEV B.G., KIRILLOV I.R., LIELAUSIS O.A., LUZHANSKII S.M., OGORODNIKOV A.P., PRESLITSKII G.V. 1987, "Liquid metal MHD dynamo model experiment", Magn. Gidrodin., n° 4, 3.

THE PONOMARENKO DYNAMO

Andrew D. Gilbert,
D.A.M.T.P.,
Silver Street,
Cambridge,
United Kingdom.

ABSTRACT. This is a very simple kinematic dynamo which can be solved exactly. In the limit of high magnetic Reynolds number dynamo action occurs with the magnetic field concentrated in a thin boundary layer. An asymptotic analysis of the unstable modes has shown that the dynamo is fast; magnetic fields on sufficiently small scales grow on the convective time-scale, in the limit of large magnetic Reynolds number. We also consider a generalisation of the Ponomarenko dynamo in which the flow field is oscillatory.

1. Dynamo Theory

Dynamo theory is the study of the generation and maintenance of magnetic fields by the motion of conducting fluids (see [h,m] for reviews). Because of the difficulty of solving the full non-linear magnetohydrodynamic equations, attention has been largely focussed on the more tractable kinematic dynamo problem. In this a velocity field, $u(x)$, is prescribed and the magnetic induction equation,

$$\partial_t b = \nabla \times (u \times b) + \eta \nabla^2 b,$$

is solved. If dynamo action occurs there will be magnetic field growth:

$$b \propto \exp(\lambda t) \quad \text{with} \quad \text{Re } \lambda > 0$$

(for the case of a time-independent flow). We also introduce the magnetic Reynolds number, $R_m \equiv |u|l/\eta$, where l is a typical length scale of the flow field. The aim of kinematic dynamo theory is to understand the structure of velocity fields capable of dynamo action and the nature of the magnetic fields that are generated.

Dynamo theory has a number of applications. Experiments are underway to demonstrate dynamo action in the laboratory [e]; the principal difficulty is attaining magnetic Reynolds numbers large enough for magnetic field excitation. It is, however, possible that sufficiently large values might be achieved in the liquid sodium cooling circuits of some nuclear reactors, a possibility that in general must be avoided. The main application of dynamo theory is in explaining the maintenance of magnetic fields within cosmical bodies, such as the earth, sun and the galaxy. In

J. Lielpeteris and R. Moreau (eds.), Liquid Metal Magnetohydrodynamics, 421–424.
© *1989 by Kluwer Academic Publishers.*

these geophysical and astrophysical contexts the magnetic Reynolds numbers are typically very large [m], and this motivates the study of the singular limit, $R_m \rightarrow \infty$. Thus in the kinematic dynamo problem we are interested in prescribing a velocity field, and examining dynamo action in the limit $\eta \rightarrow 0$ or, equivalently, $R_m \rightarrow \infty$. If the growth rate of magnetic fields tends to zero in this limit, the flow is called a slow dynamo; however if the growth rate remains positive and finite, the flow is said to be a fast dynamo [m]. Thus fast dynamos regenerate magnetic energy more efficiently than slow dynamos; field growth occurs on the convective time scale, $l/|u|$, of the fluid motion, in the limit of large R_m.

2. The Ponomarenko Dynamo

Ponomarenko [j] introduced a simple flow field and showed that it is capable of dynamo action. In cylindrical polar coordinates, (r, θ, z), the velocity field takes the form:

$$\mathbf{u} = \begin{cases} (0, r\omega, U) & r < a, \\ \mathbf{0} & r > a. \end{cases}$$

Inside the cylinder, $r = a$, the flow is helical solid body motion; the magnetic field diffuses and is advected, but not stretched. Outside the cylinder the flow is zero, and the magnetic field simply diffuses. The flow is discontinuous across the cylinder, where the field is sheared and stretched. The magnetic Reynolds number is defined by $R_m \equiv a(a^2\omega^2 + U^2)^{1/2}/\eta$. The magnetic field may be decomposed into normal modes,

$$\mathbf{b}(r, \theta, z, t) = \mathrm{Re}\left[\mathbf{B}(r)\exp(im\theta + ikz + \lambda t)\right],$$

and the induction equation solved with the boundary conditions that the magnetic and tangential electric fields be continuous across the cylinder [j]. The resulting dispersion relation has been analysed numerically, and the critical magnetic Reynolds number of $R_m \simeq 17$ obtained [b,c,d,j].

In the astrophysical limit, $R_m \rightarrow \infty$, boundary layer techniques may also be used [k]. The case when $m, k = O(1)$ and $m\omega + kU \simeq 0$ as $R_m \rightarrow \infty$ was analysed in [k,m]. These magnetic fields become localised in a boundary layer about the cylinder of width $O(R_m^{-1/3})$ as $R_m \rightarrow \infty$, and are approximately aligned with the motion just inside the cylinder. Their growth rates fall off as $O(R_m^{-1/3})$; so these magnetic field modes do not indicate fast dynamo action [m]. However these are not the most unstable magnetic modes [a]. In [f,g] magnetic modes with $m, k = O(R_m^{1/2})$ and $m\omega + kU \simeq 0$ were analysed. These small-scale magnetic modes are localised in thin layers about the cylinder of width $O(R_m^{-1/2})$; in [i] it is argued that fast growing magnetic fields in fast dynamos necessarily have structure on these scales. The growth rates of these modes are given to leading order by:

$$\mathrm{Re}\,\lambda = \frac{\eta^{1/3}}{2}\left(\frac{|m\omega|}{2a}\right)^{2/3} - \eta\left(\frac{m^2}{a^2} + k^2\right).$$

The growth rates are $O(1)$ and can be positive. In other words growth of these small-scale magnetic fields occurs on the convective time-scale and so the Ponomarenko dynamo is a fast dynamo [a,f,g]. Numerical evidence has also been presented to support these asymptotic analyses and the conclusion that the dynamo is fast.

The physical mechanism of the dynamo has been elucidated in [1]. A radial magnetic field is stretched by the shear across the cylinder to give axial and azimuthal field; this is an omega effect. Now in cylindrical geometry the radial and axial components of a vector field do not diffuse independently, and in the Ponomarenko dynamo the diffusion of axial field regenerates the radial field and closes the cycle.

3. An Oscillatory Ponomarenko Dynamo

We have also considered a generalisation of the Ponomarenko dynamo in which the velocity field is oscillatory,

$$\mathbf{u} = \begin{cases} (0, r\omega, U)\cos(\gamma t) & r < a \\ 0 & r > a, \end{cases}$$

with frequency, γ; in this flow the motion of each fluid element is bounded. We assume a magnetic field of the form:

$$\mathbf{b}(r, \theta, z, t) = \text{Re}\left[\exp(\lambda t)\exp(im\theta + ikz)\sum_{j=-\infty}^{j=\infty}\mathbf{B}_{(j)}(r)\exp(ij\gamma t)\right],$$

and take $m\omega + kU = 0$ for simplicity. A difference equation results when this is substituted into the induction equation and the boundary conditions are applied. Solving the difference equation numerically and demanding that $|\mathbf{B}_{(j)}| \to 0$ as $j \to \pm\infty$, yields growth rates, λ. We have shown that this oscillatory dynamo gives fast dynamo action for any value of the frequency of oscillation, γ. We have also calculated critical magnetic Reynolds numbers (for magnetic modes with $m\omega + kU = 0$), when γ/ω is between 0.1 and 0.6. The oscillatory dynamo gives the greatest growth rates and lowest critical magnetic Reynolds numbers when the oscillations have a long period.

I am grateful to the sponsors and organisers of the Woods Hole Oceanographic Institution 1987 Geophysical Fluid Dynamics Program, during which the analysis in [f,g] was carried out. I should also like to thank S. Childress and A. M. Soward who initiated and supervised my research there.

References

[a] Childress, S., 1988, 'Order and disorder in planetary dynamos,' to appear in Woods Hole Oceanographic Institution Technical Report, WHOI-88-16.

[b] Gailitis, A.K. & Freiberg, Ya.G., 1976, 'Theory of a helical MHD dynamo,' Magn. Gidrodin., No. 2, 3, [Magnetohydrodynam., Vol. 12, 127 (1977)].

[c] Gailitis, A.K. & Freiberg, Ya.G., 1980, 'Non-uniform model of a helical dynamo,' Magn. Gidrodin., No. 1, 15, [Magnetohydrodynam., Vol. 16, 11 (1980)].

424

[d] Gailitis, A.K. & Freiberg, Ya.G., 1980, 'Nature of the instability of a turbulent dynamo,' Magn. Gidrodin., No. 2, 10, [Magnetohydrodynam., Vol. 16, 116 (1980)].

[e] Gailitis, A.K., Karasev, B.G., Kirillov, I.R., Lielausis, O.A., Luzhanskij, S.M., Ogorodnikov, A.P. & Preslitskij, G.V., 1987, 'Liquid metal MHD dynamo model,' preprint, Physics Institute, Latvian SSR Academy of Sciences.

[f] Gilbert, A.D., 1988, 'The Ponomarenko dynamo,' to appear in Woods Hole Oceanographic Institution Technical Report, WHOI-88-16.

[g] Gilbert, A.D., 1988, 'Fast dynamo action in the Ponomarenko dynamo,' to appear in Geophys. Astrophys. Fluid Dyn.

[h] Moffatt, H.K., 1978, Magnetic Field Generation in Electrically Conducting Fluids, Cambridge University Press.

[i] Moffatt, H.K. & Proctor, M.R.E., 1985, 'Topological constraints associated with fast dynamo action,' J. Fluid Mech, Vol. 154, 493.

[j] Ponomarenko, Y.B., 1973, 'On the theory of hydromagnetic dynamos,' Zh. Prikl. Mech. Tech. Fiz. (USSR), Vol. 6, 47.

[k] Roberts, P.H., 1987, 'Dynamo theory,' in Irreversible Phenomena and Dynamical Systems Analysis in Geosciences, ed. C. Nicolis & G. Nicolis, page 73, D. Reidel.

[l] Ruzmaikin, A.A. & Sokoloff, D.D., 1980, 'Helicity, linkage and dynamo action,' Geophys. Astrophys. Fluid Dyn., Vol. 16, 73.

[m] Zeldovich, Ya.B., Ruzmaikin, A.A. & Sokoloff, D.D., 1983, Magnetic Fields in Astrophysics, Gordon and Breach.

MHD PHENOMENA IN FAST NEUTRON REACTORS WITH A LIQUID METAL HEAT CARRIER

G.E. KIRKO
Politechnical Institute
614600 Perm
USSR

Present day atomic-power engineering provides scientists with a very interesting object for investigating processes in large liquid-metal masses, the liquid-metal fast breeder reactor.

The fast breeder reactor BN-600 at the Beloyarsk atomic power station has been investigated. The main experimental results have been published in the papers [1-5].

Two phenomena out of many have been chosen for this paper. The first one is the increase of the amplitude of oscillations of the magnetic field strength component in the pump's rotor area of the first circuit (called main circulating pump-MCP). It was called magnetic field generation [4]. The second phenomenon under discussion is magnetic field self-excitation in the central part of the fast breeder reactor, which is caused by processes in the pressure chamber under the reactor's core [5].

Both of them are characterized by critical values of the revolution per minute (RPM) and of the reactor's heat power. In other words, the emergence of magnetic field generation and magnetic field self-excitation is defined by the critical values of magnetic Reynolds number and Lu number :

$$Lu = \Delta t \cdot \alpha_T \cdot \mu_0 \, \sigma^2 Ro \, (\mu_0/\rho)^{0,5},$$

where Δt is the temperature difference between the reactor's core input and its output ; Ro - pressure chamber radius ; μ_0 - magnetic constant ; σ and ρ - liquid sodium conductivity and density respectively ; α_T - thermoelectric power of the conducting medium between sodium and steel. The emergence points of generation and self-excitation are depicted with the coordinates Re_m and Lu in Fig. 1*.

Assume that the physical nature of the processes in question is the same. Then a line can be drawn through these emergence points (form of the curve having not yet been found). The points in the shaded part of Fig. 1 describe either proper magnetic field self-excitation in the pressure chamber or the operation of the mechanism of self-excitation. Both of them are related to $Re_m \geq Re_{mc}$. Proper self-excitation takes place when Lu = 4 (see Fig. 2).

Several hypotheses can be put forward to explain the observe phenomena, which confirm the experimental results.

1. Magnetic field generation in pump's rotor area

This phenomenon is observed only while the reactor is put into operation [4]. Parallel circuits of thermoelectric currents generally caused by a "gigantic thermocouple" - reactor's core and heat exchanger - are present inside the reactor [3]. Some of these currents pass through the pumps as well (240 A through each pump).

---

* Prof. H.K. MOFFATT (Cambridge University, UK) also suggested the possibility of the existence of the analoguous relationship in a personal letter.

J. Lielpeteris and R. Moreau (eds.), Liquid Metal Magnetohydrodynamics, 425–429.
© *1989 by Kluwer Academic Publishers.*

426

The nature of the phenomenon observed, viz. magnetic field generation, is stated as follows : the amplitude of H_r component oscillations is equal to 1/2 of the direct component, the latter changes while the power increases. It takes place when heat power equals 30 % of the nominal power under the values $Re_m > Re_{mc}$. Other magnetic field strength components are oscillating as well but there is no such amplitude increase.

Magnetic Reynolds number is very big in the rotor pump's area, $Re_m = 120$ (rotor pump's radius and linear velocity on the rim are characteristic numbers). That is why one can speak about the "freezing-in" of the magnetic field and state that H_ϕ is carried by the liquid and deformed with it, and the magnetic line force configuration has to be close to the rotor pump's guide configuration. If it is true, the ratio H_r/H_ϕ has to be equal to the tangent of β, the twist angle of the moving liquid ($\beta \approx 25°$).

In fact, $H_r = 0,17$ Oe, $H_\phi = 0,43$ Oe when $\Delta t = 175°$ C. Hence, $\beta \approx 22°$. The way the H_ϕ component changes is very closely connected with the processes taking place in the pressure chamber.

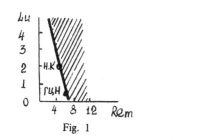

Fig. 1

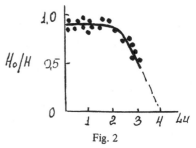

Fig. 2

2. Magnetic field self-excitation in pressure chamber

In the case when Re_m is greater than Re_{mc}, the transducers under the bottom indicate the emergence of the self-excitation mechanism in the pressure chamber. Pump's rotor RPM ≥ 650 rpm.

Relationships of $H_r = f(W)$ and $H_z = f(W)$ are shown in Fig. 3.When the reactor's heat power is changed from 0 to 70 % of nominal the pump's rotor RPM keeps constant, viz. 720 nominal pump's rotor RPM changes from 720 to 930 rpm. When the reactor's heat power is 45 % from nominal break of curve is observed on the diagram of functional dependence $H_r = f(W)$ and $H_z = f(W)$. It corresponds to Lu = 2. The magnetic fields ϕ component is not changed when the heat power increases. But when MCP operates nonsynchronically ϕ component changes its value.

The magnetic field self-excitation mechanisms under discussion are based on liquid metals hydrodynamics in the pressure chamber. Both mechanisms were discussed in detail by the author [6], with one of them treated by means of the so-called theory of laminar dynamo, vortical liquid movement apears there and interacts with the magnetic field. The theoretical results of this analysis are qualitatively conform to the experiment.

Statistical, or turbulent, theories can be used to explain the emergence of the MHD-dynamo operation. Nowadays meanfield magnetohydrodynamics is part of these theories [7].*

Liquid metal motion in the pressure chamber conforms to the requirements of the theory mentioned. It is non-miror symmetric and gives rise to left-handed helical motions in the left-hand of the chamber. The chamber's structure is such that the liquid metal flow inside each of the vortices falls into main small vortices twisted principally into the same direction as the main vortex is twisted. Due to this the

* Prof. M. Steenbeck pointed out the possibility of the magnetic field self-excitation presence in BN-600 and important role α - effect in this process in one of his letters to Soviet scientists.

helicity (v' curl v') is not equal to zero. In its turn it causes the appearance of the current parallel to some local mean field B_0. The corresponding contribution into the mean current density can be calculated in the form of :

$$j_\alpha = \sigma\alpha \cdot B_0.$$

The main difficulty of the theoretical approach to the description of the processes in the pressure chamber with the help of the mean-field magnetohydrodynamic notions is concerned with the calculation of the α coefficient. It is still impossible to calculate the α coefficient value because there are no detailed experimental data about the liquid sodium motion in the pressure chamber.

The analysis of the possibilities of magnetic field generation in the MCP-4 area and magnetic field self-excitation in the pressure chamber has proved the significance of the thermoelectric currents, caused by the reactor's heat power, for these processes. It has given the opportunity to introduce one more dimensionless parameter Lu for pointing out the emergence of the generation mechanism as well as the self-excitation one alongside with the critical values of the magnetic Reynolds number Re_m.

Frequency spectra of measured temporal dependencies of the magnetic field strength components were received by means of Fourier analysis method, frequency spectrum of vibration in the MCP-region being shifted towards the long wavelength side as compared to that of the magnetic field strength in the pressure chamber region [6] (Fig. 4).

The different parts of the magnetic vibration spectrum observed are of different origin. Out of all possible mechanisms Alfven waves arising in the pressure chamber and surface waves on the free surface of fluid sodium in the upper part of the apparatus make the greatest contribution.

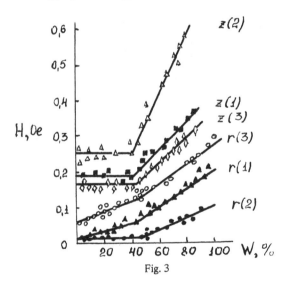

Fig. 3

As has already been shown the whole volume of fluid metal filling up the region above the reactor core, active zone and pressure chamber is traversed by the vertical thermoelectric current, the mean density value of which we shall designate j_0. It results in the appearance of the magnetic field, the force lines of which will be depicted with ring lines. The magnetic induction of this field may be evaluated as $B_\varphi = 0{,}5$ Uo $j_0 r$. Transversal Alfven waves $\lambda = aT = 0{,}5$ $j_0 r T (\mu_0\rho^{-1})^{0,5}$ can propagate along these force lines, where T is the period of vibration caused by hydrodynamic effects. In case the wave number

428

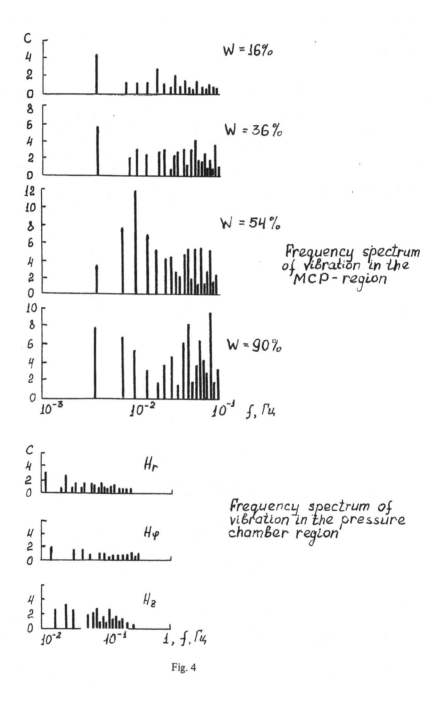

Frequency spectrum of vibration in the MCP-region

Frequency spectrum of vibration in the pressure chamber region

Fig. 4

on the length of the magnetic line of force $2\pi r$ is an integer the whole system wil turn into a resonance structure, which due to peculiarities of the active zone consisting of a great number of vertical canals is characterised by magnetic field vibrations h_z laid on the main field B_φ. If it is accepted that $j_0 = 950$ Am^{-2} then resonance periods $T_p = 4\pi/j_0 K \quad \rho/\mu_0$ are equal to 332,5s. 116,2 s., 110s., 58, 1s. and so on. The distance at which an Alfven wave fades out in 1 times can be evaluated as $\delta = T(\sigma\mu_0 \; l)$ then for T_{p1} - $\delta \approx 4m$ and for T_{p2} - $\delta \approx 2{,}3$ m and so on. Never before have Alfven waves with so little fading been observed in laboratory experiments with fluid metals. The magnetic Reynolds number in the central part of the apparatus reaches 7. In the same part of the apparatus the intensive upward metal motion occurs. All these taken together allow us to consider the motion of Alfven wave resonance structures taken by the vertical metal flow. They are carried through the main circulating pumps to the periphery of the reactor tank, the turn of the magnetic field vibration vector h from the vertical direction into the radial one being inevitable. This accounts for the existence of the especially strong vibration H_r of the magnetic field component in the MCP region. The carrying-out of Alfven wave resonance structures from the pressure chamber region into the MCP region results in shifting magnetic field vibration spectra in the "long wavelength" part of spectrum (characteristic period is about 100 s), this phenomenon being observed experimentally.

"The short wavelength" part of the spectrum (characteristic period is about 10 s) is accounted for by the existence of surface waves in the upper part of the apparatus.

References

[1] Kirko I.M., Kirko G.Ye., 1979, O vozmozhnosti nablyudeniya MGD - yavleniy v obyeme zhidkogo metalla pervogo kontura bloka BN-600 Beloyarskoy electrostantsii, DAN USSR, v. 246, n° 3, 593-596.

[2] Kirko I.M., Mitenkov F.M., Barannikov V.A., Gaylitis A.K., Lielausis O.A., Mukhametshin A.M., Ponomarev V.N., Telichlo M.T., Sheynkman A.G., Nablyudenie MGD - yavleniy v obyeme zhidkogo metalla pervogo kontura reaktora na bystrykh neytronakh BN-600 Beloyarskoy atomnoy electrostantsii, DAN USSR, v. 257, n° 4, 861-863.

[3] Kirko I.M., Kirko G.Ye., 1982, Telichko M.T., Sheynkman A.G. O sushchestvovanii termoelectricheskikh tokov v reactore BN-600 Beloyarskoy atomnoy electrostantsii, DAN USSR, v. 266, n° 4, 854-856.

[4] Kirko I.M., Kirko G.Ye., 1982, Telichko M.T., Sheynkman A.G., 1982, Eksperimentalnoe obnaruzhenie generatsii magnitnogo polya zhidkom metalle pri znachenii magnitnogo chisla Reynoldsa mnogo bolshe edinitsy, DAN USSR, v. 266, n° 6, 1384-1386.

[5] Kirko G.Ye., Telichko M.T., Sheynkman A.G., 1983, Samovozbyzhdenie magnitnogo polya /MGD-dinamo/ v tsentralnoy oblasti yadernogo reactora s zhidrometallicheskim okhlazhdeniem, DAN USSR, v. 270, n° 4, 874-876.

[6] Kirko G. Ye., 1985, Generatsiya i samovozbuzhdenie magnitnogo polya v tekhn. ustroystvakh - M. : Nauka, 88.

[7] Steenbek M., Krauze F., 1967, Vozniknovenie magnitnykh poley zvezd i planet v resultate turbulentnogo dvizheniya ikh veshchestva, Magnitnaya gidrodinamica, n° 3, 19-44.

Session I:
Turbulence

THE EFFECT OF INITIAL AND BOUNDARY CONDITIONS UPON THE FORMATION AND DEVELOPMENT OF MHD TURBULENCE STRUCTURE

I.PLATNIEKS, S.F.SELUTO
Institute of Physics
Latvian SSR Academy of Sciences
229021, Riga, Salaspils
USSR

Abstract. Hot-wire measurements of turbulent velocities in the flows behind various grids show that the process of development of turbulence structure in flows with the transverse magnetic field is determined by initial and boundary conditions. Moreover, the magnetic field is able to perform great influence on both structure of generated disturbances and process of developing them into turbulence right up to arising the phenomenon of coherent wakes generation from spatially separated bodies. The process of development of regular quasi two-dimensional structures into a turbulent motion is decelerated. When the essentially three-dimensional turbulence decays in the transverse magnetic field, the deformation of turbulence structure proceeds in the presence of effective Joule dissipation. The relationship between the two processes depends greatly on boundary conditions, orientation and location of turbulent eddies in the flow, as well as on the relation between integral scale of turbulence and the mean flow scale in the magnetic field direction. If in the near wall regions the structure of turbulence in a relatively short time acquires two-dimensional properties then in the bulk of the flow the decay displays many features of the decay of grid turbulence in the axial magnetic field, the distinctions being connected with the orientation of generated wakes regarding to the field direction.

Introduction. There are convincing proofs now for the possibility to realize two-dimensional MHD-turbulence [1-3]. Besides, it has been shown [4] that the properties of this MHD-turbulence observed experimentally can under certain conditions be compatible with the dynamic ones of the ordinary two-dimensional turbulence. Though, the transformation of the initially three-dimensional turbulence structure under the influence of magnetic field is treated controversially [5-6]. This can be attributed to the fact that boundary and especially initial conditions are accounted for insufficiently in consideration of the MHD-turbulence structure formation and development.

The paper is concerned with the development of turbulence structure behind various grids under the influence of a uniform transverse magnetic field at low magnetic Reynolds numbers. All measurements have been performed at constant Reynolds number $Re_M=10^4$.

J. Lielpeteris and R. Moreau (eds.), Liquid Metal Magnetohydrodynamics, 433–439.
© *1989 by Kluwer Academic Publishers.*

Experimental equipment. Experiments were carried out in the horizontal closed mercury loop. The voltage and frequency of generator supplying the induction pump, the current through the electromagnet and the
temperature (stability was higher than 0.05°C) of mercury in the channel
have been stabilized. The formation of M-shaped velocity profiles was
prevented by installing the honeycombs at the entrance and exit of flow
into/out of the magnetic field. A scheme of test section and the grids
configurations are shown in Fig.1.

The X-array and single hot wire probes have been manufactured. The
insulated (SiO) sensors were coated additionally with a thin layer of
gold [7]. As a result, a good stability of hot wires calibration curves,
independently of pressure changes, were achieved during up to two weeks.

Conditioned signals from DISA 55M01 thermal anemometers were recorded on the FM tape recorder and, later, digitized and processed.

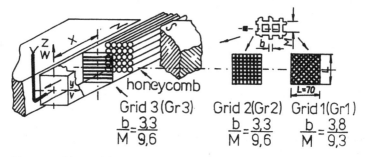

Figure 1. Flow configuration.

Initial conditions. Figure 2 shows the energies of u- and v- components of velocity fluctuations aligned accordingly with the mean velocity and magnetic field directions at x/M=7.5. If at weak magnetic fields
the behaviour of these fluctuations behind Gr.2 is close to those past
Gr.1, then as long as the field increases the properties of turbulence
behind Gr.2 approach those past Gr.3. High correlation in the field direction of u-component of velocity just behind Gr.2 (Fig.3) demonstrates
that there are generated quasi two-dimensional wakes at stronger magnetic fields. The appearance of such disturbances is the consequence of
long-range nature of electromagnetic forces which synchronizes the vortices breaking away the aligned with a magnetic field elements of Gr.2.

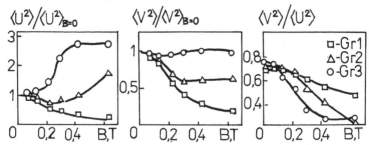

Figure 2. Effect of the field on velocity fluctuations behind the grids.

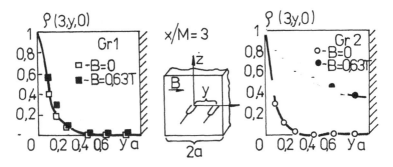

Figure 3. The coherent wake generation.

Further, this coherent wakes, due to the electromagnetic diffusion [4] of vorticity, joins forming close to two-dimensional eddies. As for Gr.1, there is formed essentially three-dimensional turbulence.

Decrease of the relative portion of energy of regular disturbances downstream the grids is presented in Fig.4a. These data obtained from measured autocorrelations and spectra show the deceleration of the process of development of quasi two-dimensional disturbances generated behind Gr.3 with a growth of magnetic field. The deceleration is displayed even at weak magnetic field B=0.2T when three-dimensional turbulence is developed downstream this grid. The process of this deceleration is stopped after the magnetic field reaches values sufficient for the supporting two-dimensional vortices structure near after Gr.3. For comparison, in Fig.4b there is shown the corresponding values of virtual origin - x_0/M, obtained when approximating measured turbulence decay laws (see Fig.8).

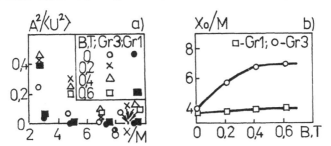

Figure 4. Development of regular disturbances into turbulence.

Boundary conditions. To estimate the role of electrically insulated, surfaces perpendicular to magnetic field lines in development of initially three-dimensional homogeneous turbulence, the measurements of velocity fluctuations components behind Gr.1 at various distances from this walls have been perfomed.It has been proved that there is an appreciable increase of turbulence anisotropy near such surfaces with the growth of magnetic field (Fig.5), the effect being more pronounced for large scale eddies. Moreover, at the initial state of decay not only expected decrease of pulsations aligned with the field,but the increasing of those in the flow direction took place there in spite of high degree of uniformi-

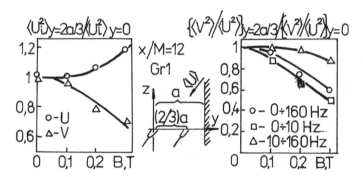

Figure 5. The anisotropy of MHD turbulence near insulated walls.

ty of mean velocity profile in the transverse magnetic field. The phenomenon may be explained by comparing the distributions of currents induced by vortices motions in the near wall regions with those in a free
volume. For eddies with the prevailing vorticity component in the field
direction (quasi two-dimensional disturbances) the induced currents are
closed in the planes orthogonal to the magnetic field lines. In the near
wall regions those eddies electromagnetically diffused in a short time
to the surface with the very thin Hartmann layer formed. The Joule dissipation and correspondingly the rate of decay of these quasi two-dimensional eddies decreases compared with the ones oriented in the same way in
the bulk of the flow (Fig.6). The rate of spreading of the near wall eddies along the magnetic field is also reduced. As for the disturbances
having the prevailing components of vorticity in directions orthogonal
to the field lines, the formation of the Hartmann layers does not take
place. In free volumes the relative rate of deceleration (both the Joule
and electromagnetic diffusion effects) of those vortices is decreased
owing to their stretching along the magnetic field. As the considered
surfaces prevent further stretching of the eddies, so their suppresion
proceeds in the near wall region at a relatively greater rate. Thus, in
the regions with magnetic field perpendicular to conductivity disconti-
nuities (e.g. solid or free boundaries, phase interfaces) MHD-turbulence
should be expected to have quasi two-dimensional features including the
reduced turbulent transfer coefficients along the field direction.

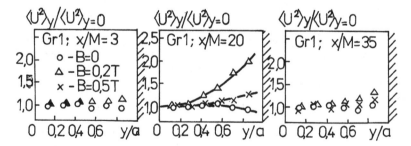

Figure 6. The profiles of directed with mean flow velocity fluctuations.

Turbulence decay. The main feature of MHD decay of initially three-
-dimensional turbulence behind Gr.1 as compared with the development of
quasi two-dimensional wakes past Gr.3 is that the increasing of anisotropy (Fig.7) is accompanied by higher dissipation rates (Fig.8) at equal

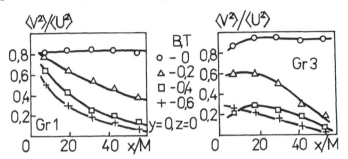

Figure 7. Development of anisotropy downstream the grids.

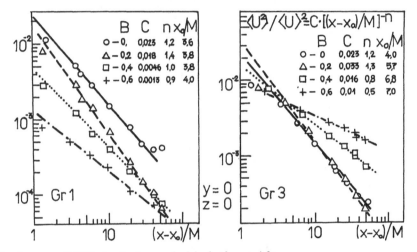

Figure 8. Decay of MHD-turbulence behind the grids.

values of magnetic field.It should be noted that the observable high decay rates are not only the consequence of the Joule dissipation but also
the electromagnetic diffusion, which elongates turbulent eddies along
the magnetic field direction.The discrepancies in the decay behind those
grids are also well illustrated by measured spectra (Fig.9) and cross-
correlations (Fig.10). As at the strongest field B=0.6T the level of turbulence downstream Gr.1 are compatible with the background, the reliable
measurement of the correlations was not possible. An interesting feature
of spectra evolution in magnetic field behind Gr.3 is the inverse energy
cascade, when the regular quasi two-dimensional disturbances are still
present. Two-dimensional turbulence has been realized (Fig.9, upper curve) only past Gr.3 when the field is strong to prevent the three-dimensional instability of generated two-dimensional wakes (Fig.7, lower curve).

438

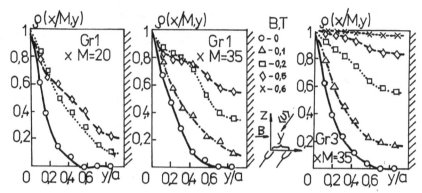

Figure 9. Variation of correlation factor with magnetic field.

Discussion. The comparison of the presented results with those obtained earlier allows us to give the following remarks.Behavioural patterns of the decay of turbulence behind the grid (the array of cylinders parallel and orthogonal to magnetic field), employed in [5] and treated there as the reformation of the structure of turbulence from three- to two-dimensional one, are the consequence of formation past such grids in strong magnetic fields close to two-dimensional vortices. The coherent generation of quasi two-dimensional wakes could take place also in the flow past the grid [3],which is a square net of holes. The obtained there at small x/M and B=0.68 T decay rate with n=1 characterizes the process of development of generated regular structures into two-dimensional turbulence but not a decay of turbulence. Moreover, as seen from [8] also, it is not sufficient, when studying the decay of MHD-turbulence, to use the equality of intensities of orthogonal components of velocity fluctualtions as a criterium of initial turbulence isotropy, since the character of MHD-turbulence in this case is determined first by the angular distribution of vorticity of disturbances generated in the flow when a magnetic field is strong enough.

References.
[1] PLATNIEKS I.A., 1972, 'Correlation study of turbulent velocity perturbations in a MHD-channel', Proc. Riga 7 MHD-Meeting, part 1, 31.
[2] KOLESNIKOV Yu.B., TSINOBER A.B., 1974, 'Experim. Invest. of Two-dimens. turbulence behind a Grid', Izv. AN SSSR, Mech.Zh.i Gaza, Vol.4,146
[3] SOMMERIA J., 1983, 'Two-dimensional behavior of MHD fully developed turbulence (Rm<<1)', J.de Mecanique, numero special, 169.
[4]SOMMERIA J., MOREAU R., 1982, 'Why, how and when MHD turbulence becomes two-dimensional',J.Fluid Mech., Vol.118, 507.
[5] VOTSISH A.D., KOLESNIKOV Yu.B., 1976, 'Transition from three-dim. to two-dim. turbulence in a magn. field', Magn. Gidrodin., Vol.3, 141.
[6] ALEMANY A., MOREAU R., SULEM P.L., FRISCH U., 1979, 'Influence of an external magn. field on homog. turbulence', J.de Mecanique, Vol.18, 277.
[7] HOFF M., 1969, 'Hot-film anemometry in liquid mercury',Instr. and Control Systems, Vol.42, 83.
[8] KLJUKIN A., KOLESNIKOV Yu.B., 1988, 'MHD turbulence decay behind grids', IUTAM Symp. on Liquid Metal MHD, Riga.

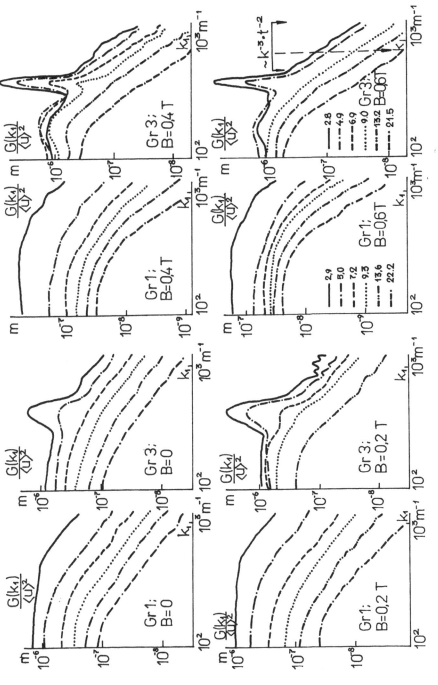

Figure 10. Evalution of one dimensional velocity spectra.

TWO-DIMENSIONAL M.H.D. TURBULENCE

Joël SOMMERIA*, Jean-Michel NGUYEN DUC*, Philippe CAPERAN**
*MADYLAM, ENSHMG, BP 95, 38402 ST MARTIN D'HERES Cedex
**I.M.G. BP68, 38402 ST MARTIN D'HERES Cedex, FRANCE

ABSTRACT. We have studied two-dimensional turbulence in a horizontal layer of mercury subjected to a vertical magnetic field. The flow is visualized on the free surface and measured from its associated electric potential. A lattice of small vortices is electrically driven by a corresponding lattice of electrodes inserted in the bottom of the box. When the vortex lattice has a square mesh, it undergoes a strong instability that leads to a developed two-dimensional turbulence with inverse energy cascade toward large scales. By contrast a hexagonal lattice undergoes a continuous instability that can generate robust non-linear traveling waves rather than turbulence.

I.   M.H.D. flows and two-dimensional turbulence:

Since the pioneering results obtained in Riga [1, 2], MHD turbulence in rectangular electrically insulating ducts is known to have the general aspects of two-dimensional turbulence. In particular, turbulent fluctuations are carried along by the mean flow with little dissipation, and energy is transferred toward large scales. In a purely two-dimensional flow, the ponderomotive force vxB is exactly balanced by potential gradients, so that the eddy currents vanish and there are no electromagnetic forces. Such a two-dimensional flow can develop three-dimensional inertial instabilities, but these are strongly damped by eddy currents: The resulting dynamics has been a subject of controversy. It has been recognized that the confinement of the flow by insulating walls perpendicular to the magnetic field is essential to suppress most three-dimensional perturbations so as to get two-dimensional dynamics. By contrast turbulence produced behind a grid moving in a deep vertical container subjected to a vertical magnetic field is strongly damped, an indication of a three-dimensional dynamics[3,4]. Sommeria and Moreau[5] argued that a turbulent eddy of scale L in the direction perpendicular to the magnetic field Bo has a natural length $L_z = N^{1/2} L$ in the direction of the field. (N is the interaction parameter $N = sB_o^2L^2/\rho v$, where σ, ρ and v are the conductivity, density and kinematic viscosity of the fluid respectively). Such an eddy would be damped by the electromagnetic effects in a time of the order of its turnover time. The eddies whose length $LN^{1/2}$ is larger than the channel width a, are strongly influenced by the boundaries. The boundary condition at a solid wall is a no-slip condition, but if we restrict

441

J. Lielpeteris and R. Moreau (eds.), Liquid Metal Magnetohydrodynamics, 441–448.
© 1989 by Kluwer Academic Publishers.

the study to the flow outside the thin Hartmann boundary layer, a new boundary condition can be established: near an insulating wall the normal derivative of the tangential velocity nearly vanishes. This condition of nearly free surface constraines the eddies with scale $LN^{1/2} < a$ to a two-dimensional structure [5]. The only dissipative effect is then a bulk friction of the two-dimensional flow, attributable to the Hartmann layer. Its characteristic time t_H is given in next section. Notice that this is valid only if the walls perpendicular to the magnetic field are insulating; if these walls are conducting the eddies are strongly damped, as shown experimentally by Platnieks and Freiberg [6].

The electromagnetic dissipation in different cases can be understood in a simple way by considering a vortex aligned with the magnetic field as an electric dynamo that produces a voltage of the order of UB_0L (U is a typical velocity of the vortex). If the vortex is limited by a conducting wall, a strong current circulates through this wall, and the kinetic energy of the vortex is transferred to electric energy and dissipated by Joule effect (Fig.1a). In the case of an insulating wall the current can only circulate in the Hartmann layer, therefore with much higher resistance than through a conducting wall (Fig.1b), and the dissipation occurs in a much longer time t_H. In the ideal case of a vortex spanning the whole gap between two free surfaces, the dynamo would be in a open circuit and no dissipation would occur (Fig.1c). By contrast, a vortex that does not span the whole width, which represents a flow that depends on z (i.e. a three-dimensional flow), is damped in a similar way as with a conducting wall(Fig.1d).

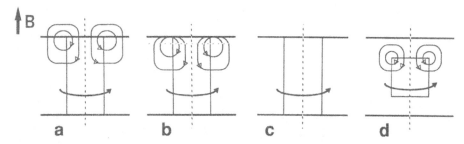

Fig.1 Eddy currents in a fluid eddy, represented by a cylinder spinning around an axis aligned with the magnetic field a) limited by conducting walls b) limited by insulating walls c) Ideal case with frictionless walls d) eddy that does not span the whole width.

2.Electrically driven two-dimensional flows:

In channel flows, turbulence is generated at the entrance in the magnetic field or behind a grid and this process is complex and difficult to control[7,8]. Such turbulence must be considered as decaying, in the reference frame moving with the mean flow. The time of observation is limited by the length of the duct and the size of the electromagnet. We have proposed instead an experimental device with a well-controlled electric forcing, which allows us to study two-dimensional turbulence in a permanently forced regime, without mean translation.

A layer of mercury with a free upper surface is enclosed in a box with a flat horizontal bottom. The layer is located in the uniform vertical magnetic field of a solenoid. The velocity is sufficiently small so that the vertical deformations of the free surface are negligible. The flow is electrically driven by a set of electrodes embedded flush in the bottom plate (Fig.2). In some of our experiments, the upper surface is free, so that the flow can be visualized by small floating particles. Flow measurements are made by digital processing of photographs of the particle streaks, or from the record of the electric potential at a set of electrodes inserted in the bottom plate. The voltage between two electrodes is proportional to the transverse velocity component between these electrodes.

The velocity virtually does not depend on the vertical position, above the thin Hartmann boundary layer of the bottom plate. To a good approximation, this velocity field $v(x,y,t)$ satisfies the two-dimensional Navier-Stokes equation with a forcing term $f(x,y)$, and a linear friction term [9].

$$\partial v / \partial t + v.\nabla v = -\nabla P + v \Delta v - v/t_H + f(x,y) \tag{1}$$
$$\nabla \times f = j_z$$

The production of vorticity is limited to the electrodes and is proportional to the injected current density j_z at the surface of the electrode. The friction time $t_H = s(a/B_0) (\rho/\sigma v)^{1/2}$ can be obtained from a simple calculation of the Hartmann boundary layer. The coefficient s is equal to 1 with a free upper surface, or to 1/2 with a rigid lid (in this case friction is twice larger due to the second Hartmann layer). The choice of the depth a results from a compromise: we need a sufficiently shallow layer, so that most eddies have a two-dimensional structure, except possibly the smallest ones. However if the depth is too small, the dynamics is dominated by the Hartmann friction rather than inertial effects. A depth of 2cm appears to be a good compromise, leading to a two-dimensional dynamics in a magnetic field of 0.25 Tesla with a ratio of the friction time to inertial time typically 20.

This ratio, called here Rh, is the non-dimensional parameter that characterizes the dynamical regime. The ordinary Reynolds number Re is much less relevant than Rh since horizontal viscous effects are negligible in comparison to this friction (Re is typically 10^4). In our experiments, the velocity is not directly set by the experimental parameters, so we estimate a typical inertial time from the acceleration produced by the forcing. This yields [9].

$$Rh = s (aI / B_0 \sigma v)^{1/2} D^{-1} \tag{2}$$

which increases with the forcing total current I (D is the width of the box).

A variety of two-dimensional flows can be produced by such a system, using different configurations of electrodes and different partitions of current through these electrodes. The simplest example is a single electrode at the center of a circular box, with a steady radial current between this electrode and the outer conducting frame[10]. In the axisymmetric vortex produced in this way, the non-linear advective effects vanish and the velocity profile results from a balance between forcing and friction. The vorticity is confined to a core above the electrode, and a 1/r irrotational velocity profile is obtained outside this core. When the current is switched off, the vortex decays in a time t_H in good agreement with eq.1. A vortex couple can be produced if a current pulse is injected through one electrode and returns through a second electrode, and the structure and evolution of such a couple is in good agreement with the two-dimensional Euler equation [10,11], except for the overall decay in a time t_H.

3.The inverse energy cascade of two-dimensional turbulence:

This technique is appropriate to force two-dimensional turbulence at small scale and to study the inverse energy cascade from the point of view of Kraichnan[12]. In Kraichnan's statistical theory the forcing is random, but we chose instead a steady forcing. We make this choice for simplicity, and also to study the instabilities of simple cellular flows that lead to turbulence.

In the first experiments a square box with a square lattice of 36 electrodes was used[9]. The current was equally dispatched through 18

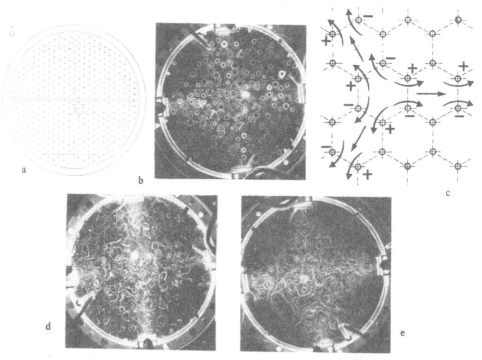

Fig. 2 Experimental device and streak photos. a) Top view of the box showing the hexagonal lattice of 282 electrodes and the double line of small measurement electrodes. The fluid boundary can be made hexagonal along the dashed line by using a copper ring. b) Steady hexagonal lattice, Rh = 1.3. c) Transport of perturbations along the three directions by the basic velocity field of the vortex lattice d) State with traveling instability waves (Rh = 1.6). e) Turbulence with inverse energy cascade.(Rh = 11.3).

electrodes and returned through the 18 other electrodes, generating a square vortex lattice of alternating sign. This vortex pattern is stable for weak forcing, but it becomes unstable beyond a threshold Rh=1.82, and yields directly a turbulent flow. This transition occurs with hysteresis: it is a subcritical bifurcation.

For low values of Rh the dominant turbulent scale is close to the forcing scale, equal to the mesh of the electrode lattice. For higher values of Rh, the dominant scale increases, and an inertial range with $k^{-5/3}$ energy spectrum appears for eddies ranging between the forcing scale and the dominant scale. Beyond a new threshold of about Rh = 40, most of the energy is condensed in the lowest available mode, which corresponds to a global rotation of the fluid. This rotation can have the two possible directions with equal probability, and random infrequent reversals are observed near the threshold Rh=40.

These results are a clear indication of the importance of inverse energy transfers. The inertial effects at the origin of these transfers are in competition with the bottom friction, which dissipates energy at all scales, and thus limits the range of the inverse cascade. For low Rh the inverse cascade is strongly limited by friction, but the cascade extends to lower and lower wave-numbers for increasing Rh. Finally, for very high Rh, energy accumulates in the lowest mode of the system, and the mean rotation appears. The approximation of isotropic turbulence then breaks down.

The two-dimensional structure of the flow has not been checked by direct measurements at different points along the vertical. However the dynamics is found[9] to be mostly determined by the parameter Rh, which supports eq.1. A three-dimensional dynamics would depend on two parameters instead, the Hartman number and the interaction parameter. Furthermore a good agreement is found with the results[13] of direct numerical simulations of the two-dimensional equation 1.

4.Waves and turbulence in a hexagonal vortex lattice:

In order to get a wider range of scales we have designed a new box with 282 electrodes (Fig.2). We use now an hexagonal lattice instead of a square one to explore the influence of the forcing geometry. The outer boundary of the domain is either circular or hexagonal. The forcing in the hexagonal domain is equivalent, in Fourier space, to a forcing on a discrete infinite hexagonal lattice of wave-numbers; while in a circular domain, a part of the forcing takes place also on a continuous range of low wave numbers. These differences in the forcing lead to different dynamics at high Rh.

The steady flow obtained for low forcing is unstable above a threshold Rh =1.4, but the nature of the instability is quite different than in the square lattice. We observe the emergence of an oscillation (Fig.3), whose amplitude increases continuously with Rh without hysteresis: it is a supercritical instability. The frequency is constant near the threshold, which characterizes a Hopf bifurcation. The spatial structure of the oscillation can be studied using space-time spectra: the square of the two-dimensional Fourier transform of the velocity in time and along the line of measurement. We observe that the oscillations correspond to a well-defined wave-number $k_I/3$, where k_I is the forcing wave-number.

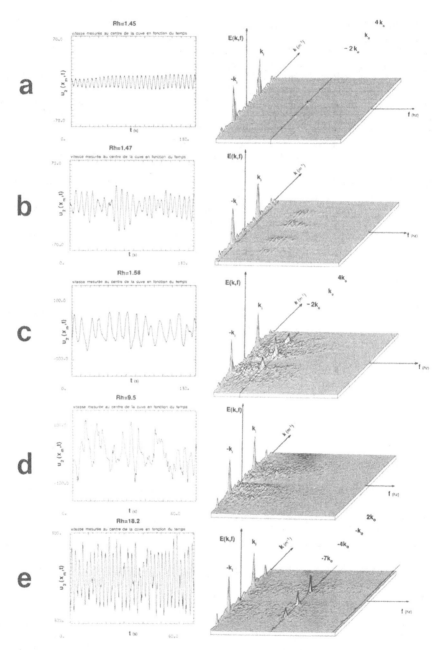

Fig. 3 Time records (of the transverse velocity at the box center) and space-time spectra. The two peaks at zero frequency correspond to the steady forcing. The two signs for the wave-number correspond to the two directions of propagation a) periodic oscillation near onset of instability. b) chaotic state. c) coherent waves emerge from chaos. d) New chaotic regime. e) Coherent waves at high Rh.

There is no symmetry between positive and negative wave-numbers, which means that the wave propagates only in one direction along the line of measurement (but harmonics propagate in the opposite direction).

This propagation can be understood from the basic flow pattern. Indeed a mean velocity exists along lines between hexagones (Fig.2c), like the line of measurement, and it is not surprising that perturbations move along these lines. Three different directions of propagation are possible, in relation with the symmetry of the system.

For higher forcing, further instabilities lead to turbulent regimes, with broad spectra (Fig.3b,d). However new regimes with coherent waves appear at higher Rh (Fig.3c, e) and the wave-number is still equal to $k_I/3$. This ratio 1/3 can be explained by a resonance between the three waves and the basic lattice both in space and time. This resonance is remarkably strong since it prevents the appearance of strong turbulence and inverse energy cascade. The behaviour at low Rh is independent of the outer boundary. However the regime with coherent waves at large Rh (Fig.3e) appears only in an hexagonal domain, while a strong turbulence always occurs at large Rh in the circular domain.

In this latter case the turbulent fluctuations are fairly homogeneous and isotropic. This has been shown by comparing the correlation functions of the two components of velocity along the line of measurement. The cross-correlation, that should vanish for isotropic turbulence, is less than 0.1, and the two auto-correlations satisfy the theoretical relation for isotropic turbulence[13]. The energy spectra for the fluctuations are smooth, and the slope is close to $k^{-5/3}$ (Fig.4a). (The smaller slope at low wave-number on Fig.4a could be due to the friction that drains energy along the cascade). A corresponding Kolmogoroff constant equal to 3 can

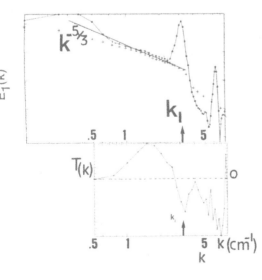

Fig. 4 a) One-dimensional spatial spectra of the longitudinal velocity component. o whole signal, + fluctuating part only. The two peaks correspond to the forcing wave-number and its harmonic.

b) Bispectrum calculated[15] as the Fourier transform of the triple correlation of the longitudinal velocity component $<u1(x)\ u1(x)\ u1(x+r)>$. A positive value of $T(k)$ indicates that energy is brought to the wave-number k by energy transfers. Energy is then removed from the injection wave-number and transferred to smaller wave-numbers.

be deduced from the level of the spectra[13]. Beside these quasi-isotropic turbulent fluctuations, a mean flow exists, but it is essentially confined to the scale of the forcing, and to very large scales, and it is weak in the inertial range. This can be seen by a comparing the spectrum of the whole signal to the spectrum of its fluctuating part (Fig.4a). The mean flow at large scale is at least partly due to the forcing, which has some component at low wave-number in a circular domain. This component disappears with an hexagonal boundary or in a very large domain. The sign of the energy transfers can be directly obtained from bispectra, calculated by a Fourier transform of the triple correlation function[15]. Fig. 4b indicates that energy is indeed transferred from the injection wave-number to smaller wave-numbers.

In conclusion, our method provides inertial flows with a two-dimensional dynamics within a good approximation. A steady forcing at small scale generally produces a quasi homogeneous isotropic turbulence with inverse energy cascade toward large scales. However in the specific case of a forcing in an hexagonal lattice, traveling waves are produced instead. These waves are locked with the basic lattice, and their presence can prevent the initiation of inverse energy cascade.

References

[1]   KIT L.G. & TSINOBER A.B. "Possibility of creating and investigating two-dimensional turbulence in a strong magnetic field", Magnitnaya Gidrodinamica (1971) 3, 27.

[2]   KOLESNIKOV Y.B. & TSINOBER A.B. "Experimental investigation of two-dimensional turbulence behind a grid" Isv. Akad. Nauk. SSSR Mech. Zhid. i Gaza (1974) 4, 146.

[3]   CAPERAN Ph. & ALEMANY A. "Turbulence homogène MHD à faible nombre de Reynolds magnétique..." J.M.T.A. (Paris) (1985) 4 (2) , 175-200.

[4]   IEVLEV V.M., KOROTEEV A.S. & LEVIN V.B. The effect of a uniform magnetic field upon stability transition and turbulence as a control means for liquid metal flow mixing" in this book.

[5]   SOMMERIA J. & MOREAU R. "Why , how, and when MHD turbulence becomes two-dimensional" J. Fluid Mech. (1982) 118, 507-518.

[6]   PLATNIEKS I.A. & FREIBERGS Yu. Zh. Magnetnaia Gidrodinamica (1972) 2, 29.

[7]   KLJUKIN A.A. & KOLESNIKOV Ju.B. "MHD Turbulence decay behind spatial grids" in this book.

[8]   PLATNIEKS I. & SELJUTO S.F. "The effect of initial and boundary conditions upon the formation and development of MHD turbulence structure" in this book.

[9]   SOMMERIA J. "Experimental study of the two-dimensional inverse energy cascade in a square box" J. Fluid Mech. (1986) 170, 139-168.

[10]   SOMMERIA J. "Electrically driven vortices in a strong magnetic field" J. Fluid Mech. (1988) 189, 553-569.

[11]   NGUYEN DUC J.M. & SOMMERIA J. "Experimental characterization of steady two-dimensional vortex couples", J. Fluid Mech.(1988) 192, 175-192.

[12] KRAICHNAN R.H. "Inertial ranges in two-dimensional turbulence" Phys. Fluids (1967) 10, 1417.

[13] VERRON J. & SOMMERIA J. "Numerical simulation of a two-dimensional turbulence experiment in magnetohydrodynamics" Phys. Fluids (1987) 30 (3)

[14] NGUYEN DUC J.M., CAPERAN Ph. & SOMMERIA J. "Experimental investigation of the two-dimensional inverse energy cascade" 5th Beer Sheva seminar, march 87, Israël. To appear in AIAA

[15] NGUYEN DUC J.M. "Instabilités et turbulence dans les écoulements bidimensionnels MHD" thèse INPG, Grenoble, France.

MHD INSTABILITIES AND TURBULENCE IN LIQUID METAL SHEAR FLOWS

A. A. KLJUKIN, Ju. B. KOLESNIKOV
Inst. Phys. Latv. SSR Acad. Sci., 229021, Riga, USSR
H. KALIS
Latv. State Univ., 226050, Riga, USSR

ABSTRACT. The stability of MHD flows which have free shear layers extending along the magnetic field has been investigated for the case of electrical analogue known as Lehnert flow. Experimental methods are based on velocity measurement using conduction probes. It was proved that critical Reynolds numbers depend on a parameter, which represents the ratio of dissipative electromagnetic forces and viscous forces. Nonsingle-valued nature and hysteresis state of perturbed flow with changing of critical parameters were discovered. Flow bifurcation sequence was observed. It was concluded that electromagnetic braking of perturbations plays the main role in affecting the stability of flow class being investigated and that in the investigated flow instability development takes place at a narrow spectral band of wave numbers.

Introduction

The most significant influence of a strong magnetic field on conductive medium flows is the formation of two-dimensional velocity structures with free shear layers in the plane perpendicular to the field. These are flows with M-shaped velocity profiles in channels and diffusers, as well as in channels with abrupt changes of cross-section or of wall-conductivity. Such "two-dimensional" flows with inflection points in velocity profiles are unstable with respect to perturbations of vorticity in the direction of the magnetic field [1].

MHD-flows with free shear layers often occur in practice, and the instability of such flows has been proved. However the investigation of the stability of these flows in the presence of a strong magnetic field has received insufficient attention until now.

1. Experimental Device

We studied rotating flow, which occurs in a cylinder due to the interaction of an axial uniform magnetic field with an electric current flowing through the liquid from two coaxial circular electrodes, mounted into one of the end planes. This is the electrical analogue of the well-known Lehnert flow.

The cylinder used in the experiment was made of plexiglass and had an inner diameter 100 mm and a height of 72 mm. Circular electrodes with inner diameters of 60 mm and 76 mm and with a width of 2 mm were made of copper. The cylinder was filled with the eutectic alloy In-Ga-Sn and was placed into an axially uniform magnetic field. The strength of the magnetic field varried from 0.06 T up to 1.5 T and the magnitude of applied electric current was up to 30 A. Azimuthal and radial velocity were measured with conduction probes for varying height and radius of the cylinder.

2. Experimental Results

Azimuthal velocity (U) profiles along r and z showed that in a region of the liquid pool, sufficiently

449

J. Lielpeteris and R. Moreau (eds.), Liquid Metal Magnetohydrodynamics, 449–454.
© *1989 by Kluwer Academic Publishers.*

uniform circular jet flow arises in the direction of the magnetic field. When the magnetic field increases, the flow is concentrated in the area between the electrodes and its uniformity along z increases sufficiently. Above the electrodes a free shear layer is formed parallel to the magnetic field lines (Fig. 1).

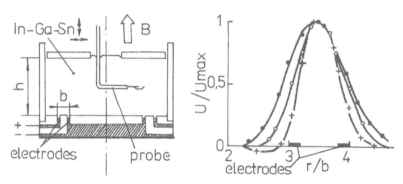

Figure 1. Experimental set and velocity profiles at different Hartmann numbers Ha=Bb/$\sqrt{(\sigma/\mu)}$: ● - 47; ○ - 100; + - 194.

Stability boundaries of flow were determined as a relationship I_{cr} = f(B) at different depths [2]. These experimental results are represented in Fig. 2.

Fig. 3 demonstrates the dependence of the velocity pulsation frequency on the magnetic field strength. As seen from the figure, after crossing the stability boundary in the flow there are some bifurcations accompanied by a decrease in the auto-oscillations frequency. The decrease of the non-linearity parameter leads to an opposite bifurcation sequence and in this situation we observe hysteresis of states [3].

Such frequency behaviour permits us to conclude that flow regimes, existing at the same values of parameters differ by the number of vortices in the vortex trail. In Fig. 3, experimentally determined regions of flow with different number of vortices in the vortex trails are shown.

We observed the evolution of sprectral energy density for radial velocity pulsations as the parameter Re was increased above its critical value up to the order Re/Re_{cr} = 18. Measurements were carried out at a constant value of the magnetic field. In Fig. 4, the sequence of resulting spectra is shown.

It should be emphasized that the range of wave numbers is restricted at the side of K and that in spectra there exists high order resonances of following type :

$$mK_1 + nK_2 = pK_3.$$

A characteristic feature of spectra is the absence of harmonics.

When investigating the general characteristics of instability development in the flow, one may separate the different steps : 1) -"stability changing" of auto-oscillations, 2) - occurence of quasiperiodic flow regime, 3) - synchronization, 4) - if possible, a bifurcation to a strange attractor (at Re/Re_{cr} = 4.2 exponential damping of autocorrelation appears), 5) - a slow evolution of spectrum, including increasing random components energy concentration in the small wave numbers range, establishement of equidistant spectrum at large wave numbers.

3. Discussion

The effect of jet profile formation in MHD flow is well-known. Qualitatively it can be represented as

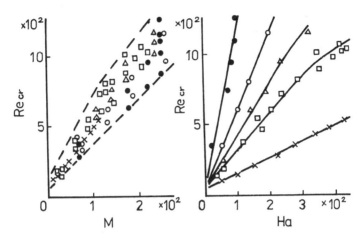

Figure 2. Critical stability parameter dependence on modified Hartmann number M and Hartmann number Ha at different ratios h/b: ● - 0.8; ○ - 1.5; △ - 2,7; □ - 4.0; ✕ - 7.2.

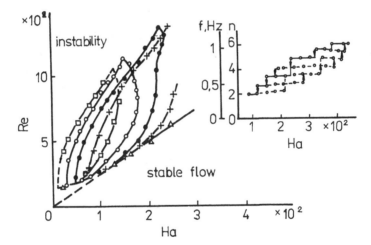

Figure 3. The hysteresis effect in auto-oscillation frequency dependence on Ha. The Re-Ha regions of existence of auto-oscillations with different number of vortices. $n = f/f_0$; $f_0 = V_0/2\pi R$; n: □ - 2.1; ○ - 2.7; ● - 3.5; + - 4.2.

follows : at the moment of switching, electrical current in the area between circular electrodes at radius R establishes a radial current component distribution with average current density height, of $\langle j_0 \rangle = I/2\pi Rh$. The resulting electromagnetic force

$$F_0 = \langle j_0 \rangle B \approx IB/2\pi Rh \qquad (1)$$

leads to a rotating movement of the liquid between the electrodes. Applying the induced potential difference UB across the Hartmann layers, we can estimate induced current density in the flow core $<j_c> \approx 2 \sqrt{(\sigma\mu)U/h}$ which is valid at Ha >> 1. The direction of these currents is opposite to that of current $<j_o>$. The electromagnetic force induced in the flow core

$$F_{el.m} = <j_c>B \approx 2 \quad (\sigma\mu) \; UB/h \qquad (2)$$

acts as a braking force.

Viscous forces in the flow core are of the order :

$$F_{visc} \approx \mu U/b^2$$

where b is a characteristic dimension of a formed jet profile. The ratio of induced electromagnetic forces and viscous forces is characterised by parameter :

$$F_{el.m}/F_{visc} \approx 2b^2 \; B \sqrt{(\sigma/\mu)}/h = M.$$

In accordance with this at M >> 1 in the flow core, viscous forces become negligible in comparison with electromagnetic braking forces. From the balance of (1) and (2) wa can get the evaluation of flow velocity in the region between electrodes :

$$V_0 = I/4\pi R \; \sqrt{(\sigma\mu)},$$

In the region outside electrodes, where average current density $<j_o>$ from external source is zero, the force moving the liquid is a viscous force $F_{visc} \approx \mu U/\delta^2$, where δ is characteristic radial velocity shift width. The opposing action of induced electromagnetic force is also valid for this region. From the balance we get the well known expression for the shift value

$$\delta \approx b/\sqrt{M} = h \sqrt{(2Hah)}.$$

Analytical and numerical considerations [4] prove these main relationships and demonstrate a good agreement with experimental data.

So, the main factor affecting the flow is electromagnetic braking. As an induced electromagnetic force is dissipative and may exceeed viscous forces, jet profile instability must occur, when inertia forces $F'_{in} \approx V_0 v'/b$, acting on perturbations become comparable with electromagnetic $F'_{el.m} \approx \sqrt{(\sigma\mu)}v'B/h$ [5] :

$$F_{in}/F'_{el.m} \approx Re/M \approx 0(1), \quad Re = \rho V_0 b/\mu.$$

This gives the approximate criterion $Re_{cr} \approx M$.

On the basis of energy consideration, the vortex interaction with the average profile can be described as follows. Energy pumping into vortices occurs due to Reynolds stresses while its dissipation takes place in the Hartmann layer on the walls perpendicular to the field. Such spatial separation of zones of generation and zones of dissipation leads to a situation in which the magnitude of the magnetic field and the perturbations length in the direction of the field become two competing factors. This fact can explain the dependence of Re_{cr} on the parameter M, in which B and h enter as their ratio. The strategy of the stability experiments was based on the verification of Re_{cr} dependence on ratio B/h [2]. Data shown in Fig. 2, demonstrate the above considerations. As one can see from the figure, experimental

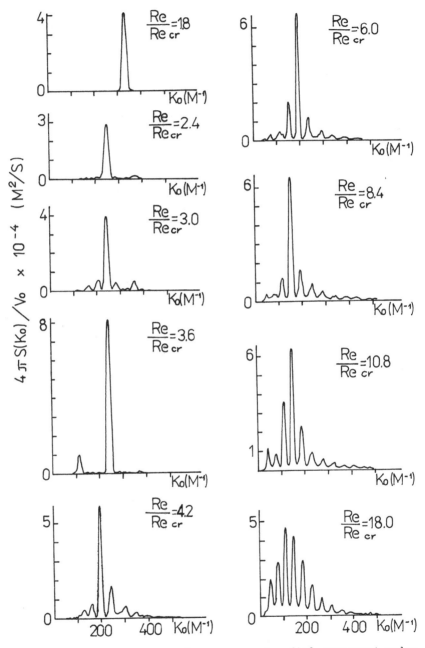

Figure 4. The energy density spectral sequence of radial component velocity pulsations for increasing supercritical parameter Re/Recr.

points fall into one curve, while the dependency of Re_{cr} on Ha has a strong dependance on the parameter B/h.

The bifurcation sequence observed in the near supercritical parameter Re/Re_{cr} range and accompanied by the changing of the number of vortices and hysteresis, can be attributed to a secondary instability of initial flow. Provided that the second unstable mode increases rapidly and the strong non linear absorption of the first mode takes place due to its interaction with the second one, the first mode is damped and in the flow a single mode regime is established.

As the supercriticality parameter changes, a series of bifurcations takes place. This sequence corresponds closely to the well-known Ruelle-Takens-Newhouse scenario [6]. Bifurcation at $Re/Re_{cr} = 4.2$, apparently, is related to the existence of a strange attractor, which is characterized by the appearance of new frequencies, by a slow increasing of spectral density in a broad band and by exponential correlation of the damping.

It should be emphasized that in the spectra sequence, shown in Fig. 4, expansion is made not in terms of frequencies, but in terms of "wave numbers" $Ko = 4\pi f/Vo$. Wave numbers imply an absence of dispersion and waves propagating with equal phase velocity 0.5 Vo. Taking into account the results of the paper [7] one may expect that in the flows under investigation dispersion is small for $Ko > (2b)^{-1}$. The absence of harmonics in spectra, which was empasized above, contradicts the assumption about small dispersion. But this fact could be explained by the limited range of wave numbers K. Restrictions on wave numbers and on perturbation scales may arise from the requirements that the diffusion time of the perturbations in the direction of field [8] must be smaller that the time of their dissipation due to electromagnetic braking. This requirement is a condition of the existence of perturbations as two-dimensional formations. From this we get the following evaluation

$$R^{-1} < K < \delta^{-1}.$$

Under some conditions when the initial harmonics generated by flow has a wave number close to boundary value, as it is in our case, Fig. 4, the generation of higher harmonics is impossible. As a result of harmonics generation with the remaining possibility of three-wave interaction, one can observe a generation of high order resonances.

Apparently, supercriticality increases the contribution of overlapping non linear resonances in the development of stochasticity.

4. References

[1] LEHNERT B., 1955, "Instability of laminar flow of mercury caused by an external magnetic field", Proc. Roy. Soc. Vol. A233, 299.

[2] KLJUKIN A.A. & LEVIN V.B., 1984, "Stability of free submerged rotating shear layer arising in the conducting fluid in axial magnetic field", Izv. A.N.S.S.S.R. Mekh. Zh. i Gaza, vol. 5, 37.

[3] KLJUKIN A.A. & KOLESNIKOV Ju, B., 1980, "Experimental study of a rotating MHD flow stability", Magn. Gidrodin, vol. 2, 140.

[4] KALIS H., KLJUKIN A.A., KOLESNIKOV Ju. B., 1978, "Strong magnetic field influence on shift flows in viscous incompressible conduction fluid", Proc. VI Int. conf. numerical meth. fluid dynamics, vol. 2, 115.

[5] KLJUKIN A.A., 1981, "On the one possible model of MHD turbulence", X Riga Conference on MHD, vol. I, 55.

[6] ECKMANN J.P., 1981, "Roads to turbulence in dissipative dynamical systems", Rev. Mod. Phys., vol. 53, n° 4, Part 1, 643.

[7] LEVIN V.B. & SHTERN V.N., 1985, "Stability of jet MHD flow between insulating plates in transversal magnetic field", Magn. Gidrodin., vol. 2, 23.

[8] SOMMERIA J. & MOREAU R., 1982, "Why, how and when, MHD turbulence becomes two-dimensional". J. Fluid Mech., vol. 118, 507.

Session J:
Stability with Uniform Field

DISPERSION AND CHAOS IN LINEAR ARRAYS OF VORTICES

P. TABELING, O. CARDOSO
Groupe de Physique des Solides de l'ENS
24, rue Lhomond
75231 Paris
France

ABSTRACT . An experimental study of the behaviour of linear arrays of vortices is presented. The vortices are produced by an electromagnetic forcing in a layer of electrolyte. Two problems are considered : (i) the onset of chaos and (ii) the dispersion of a passive contaminant. Concerning problem (i), we show that the system behaves like a linear chain of non-linearly coupled oscillators, each oscillator being sustained by a pair of vortices. About problem (ii), we reveal the existence of an anomalous diffusion regime, in excellent agreement with recent theoretical predictions.

1. Introduction

The idea of studying the onset of turbulence in forced periodic flows traces back to Kolmogorov (1960) (see (1)). His basic idea was to impose the scale of injection of the energy and study, as the Reynolds number is increased, how energy is transferred throughout a continuous band of wave-numbers. The model considered by Kolmogorov and further studied by Meshalkin and Sinai (2) , Greene (3) and Sivashinsky (4) was a plane periodic parallel shear flow. These authors have shown that above a certain treshold, large scale instabilities develop.In finite size systems, and after a transient, the system is driven towards a state where only two length scales survive, the lattice period and the system size (She (5), D'Humières (6)). One interesting feature of this model is that the process involved during the transient state seems to be an inverse cascade (Green (3), She (5)); the geometry of this cascade, which has some link with 2D turbulent cascades, is still the subject of theoretical studies.

Experimentally, Bondarenko et al (7) found that the first instability of the periodic flow drives the system towards a stable stationary supercritical state. The higher order instabilities were studied only qualitatively. Later on, very interesting results were obtained on two-dimensional periodic flows by Sommeria (8), and Nguyen Duc and Sommeria (9). They studied the case of a regular two-dimensional lattice of magnetohydrodynamically forced vortices. Depending on the lattice symmetry, and the boundary conditions, the system evolves towards either a fully developed turbulent state or a spatio-temporal chaos.

In this paper, we present an experimental study of two basic problems - (i)the

457

J. Lielpeteris and R. Moreau (eds.), Liquid Metal Magnetohydrodynamics, 457–463.

onset of weak turbulence and (ii) the dispersion of a contaminant - in a particular periodic flow, which consists in a linear array of vortices. We will present results concerning problems (i) and (ii) and compare with recent theoretical approaches.

2 - The experimental arrangement

The experimental arrangement, which has been described elsewhere (10), is a cell of 350 mm long, 35 mm large, machined out of Plexi-Glass, and filled partially with an electrolyte. The working fluid is limited laterally by the walls of the cell and horizontally by its free surface and the bottom of the cell ; the values of the thickness of the fluid layer (denoted by b) range between 1 and 3.5 mm. In order to force a steady flow, we apply a longitudinal electric current I through the electrolyte and we impose a vertical, stationary, spatially periodic magnetic field : such a field is produced by using permanent magnets, put together just below the fluid to form a line of alternated poles. The half period of the magnetic field, denoted by λ_0, is 11.2 mm. Due to the interaction of the magnetic field with the electric current, a flow is driven with a spatial structure imposed by the electromagnetic body forces.

The flow is visualized by using a sheet of light produced either by a He-Ne 30 mW laser (for chaos experiments) or by an air-cooled Argon laser (for dispersion experiments). The beam enters laterally just below the free surface with an almost razing incidence and is diffused by particules lying on it. The dynamics of the system is thus studied by measuring with a photodiode the local intensity of the laser beam reflected under the free surface.

For chaos experiments, we use a dilute solution of sulfuric acid as the electrolyte whereas for the dispersion experiments, we use a salt solution as the electrolyte and a mixture of fluorescein with salt water as the contaminant. We follow the tracer by using a CCD camera ; the frames are digitized at a resolution of 512 by 512 pixels.The accuracy of the local concentration measurement , obtained by using the video camera, is typically a few percent.

3 . Onset of chaos in linear arrays of vortices

3.1. THE FIRST BIFURCATION TOWARDS "STATE +"

The evolution of the system as the electric current I is increased from zero has been studied in Tabeling et al (10), and we summarize herein only the main results. At low current, the line is composed of steady counter-rotating vortices of size λ_0 ; as I is increased, some of the vortices decrease whereas the others increase ; finally, for larger values of I, the flow evolves towards a state where the vortices are corotating and have a size $2\lambda_0$(see Picture 1). We denote this new steady state by "state +".

3 - 2. ONSET OF TEMPORAL MODES

State + is characterized by the presence of high shear between each vortex pair. As the electric current is increased, such regions will become subjected to a shear instability, leading to the onset of temporal modes.

When we work with a single pair of corotating vortices, we observe that this

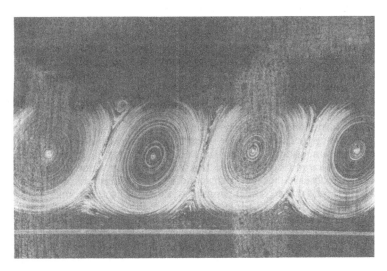

Picture 1 : State +, for I = 14 mA, b = 2.5 mm, in the case of 8 magnets.

instability gives rise to a monoperiodic regime, characterized by a well defined frequency f_0 and a treshold I_c.This is the basic oscillator of the system. In the general situation, each vortex pair sustains an oscillator. The oscillators interact nonlinearly to produce various states of flow, such as synchronized motion, quasi-periodicicity, chaos,..(see Picture 2 which shows a synchronized mode in the 8 magnets system).

Picture 2 : Optical mode in the 8 magnets configuration.

We present herein results obtained in the 16 magnets case, for which state + is a linear array of 8 corotating vortices ; the states of flow are represented in the phase diagram shown in Figure 1. Surprisingly, we find a sensitive dependence of the behaviour of the system with the depth of the fluid layer b, so that the phase diagram includes both I and b as parameters. There are two regions in the phase diagram. When b is larger than 2.9 mm (and below 4 mm), we observe the following sequence of events as the control parameter is increased from state + : chaos, quasi-periodic with two frequencies,monoperiodic, and then intermittency leading to chaos. The observation of chaos just above state + can be interpreted as related to the presence of many oscillators interacting nonlinearly. Such oscillators lock-in "partially" in the form of a quasi-periodic state. The "complete" lock-in state is observed at larger values of I, in the form of a monoperiodic state, with a well defined frequency, and whose domain of stability is quite large (see the phase diagram). By using video camera, and phase measurements, we observe that the corresponding spatial structure of this regime is an optical mode,i.e each vortex oscillates out of phase with its neighbour. As I is further increased, we observe the onset of intermittency, which persists and increases at larger values of the control parameter. Outside the boundary of the optical "tongue" of the system, the system shows intermittent behaviour, which turns out to be the particular form of chaos in this region of the phase diagram.

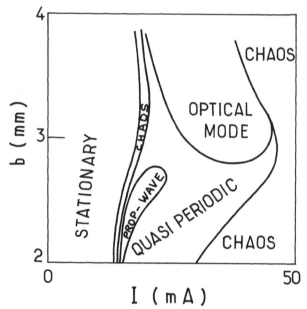

Figure 1 : Phase diagram in the 16 magnets configuration

.Another sequence of events is observed for small values of the thickness of the fluid layer : for b < 2.75 mm, we observe the following succession of dynamical events as I is increased from "state +": chaotic, quasi-periodic, monoperiodic, quasi-periodic and finally chaotic. In contrast with the regimes described above, the range of stability of the complete locking state is very narrow : it extends from Ic to typically 1.05 Ic. The transition from this state to the quasi-periodic regime is continuous. Further in the phase diagram, for much larger values of the control parameter, the system becomes chaotic. We thus observe at small values of b, a phenomenon of desynchronization of the chain of oscillators. Such an evolution may suggest that the linear chain of oscillators is structurally unstable in this region of the phase diagram.

The two regions of the phase diagram are separated by a "channel" where no "complete" lock-in state is observed : the system evolves from chaos to quasiperiodicity and from there to chaos without encountering monoperiodic regimes.

4 . Dispersion in linear arrays of vortices

For dispersion experiments, the procedure consists in injecting, at a given point in the lattice, a small quantity of fluorescent tracer and further following the invasion of the system. The lattice is such that there are only counter-rotating vortices of size λ_0. Experiments are performed for a Peclet number equal to $1.6 \ 10^5$. One remarkable fact - typical of large Peclet number situations -is that the tracer is found to be homogeneous along the streamlines but not within each vortex. We observe the presence of concentration boundary layers, on the separatrix between two adjacent vortices and close to the walls. As time is running, thee boundary layers thicken out and we end up, after a few hours, with a complete homogeneization of the tracer within each vortex. In the period of time which we consider, the tracer is confined in boundary layers on the edges of the vortices.

It is interesting to compare the thickening process for the boundary layers with boundary layer theory (11) which predicts that the local concentration close to the separatrix scales with reduced variable $xt^{-2/3}$ (where x is measured from the separatrix).

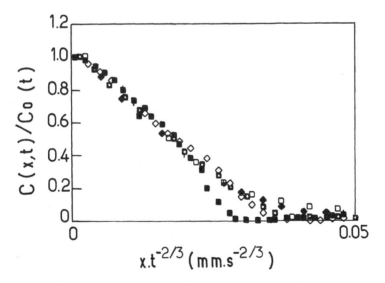

Figure 2 : Concentration curves close to one separatrix, at various times, plotted as a function of variable $xt^{-2/3}$; symbols correspond to different times.

We thus plot concentration curves obtained around one separatrix, renormalized in amplitude, against the reduced variable $xt^{-2/3}$, where x is the distance from the separatrix (see Figure 2).

For a long period (up to 25 mn, which represents 400 turn-over times), all these curves reduce to a single one, and therefore the process is self similar. The self-similarity is accurate except at low values of the concentration, where experimental errors become significant. At larger times, the boundary layer also thickens out but the concentration profiles cease to be self-similar ; this loss of self-similarity is due to three-dimensional effects.

Moreover, we have studied the invasion of the lattice by the tracer over scales much larger than the vortex sizes. Figure 3 shows the evolution of the position of the contamination front, or the "number" of invaded vortices, with time. We obtain a first regime, characterized by a power law with an exponent close to 1/3, in excellent agreement with theory. This law defines the so-called anomalous diffusion regime since normal diffusion would be characterized by a 1/2 exponent. This regime breaks at times larger than 35 mn, where a saturation effect is observed. As was pointed out above, the existence of this saturation is presumably a consequence of the influence of three-dimensional effects.

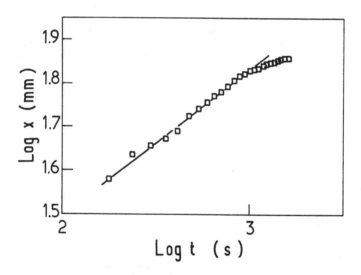

Figure 3 : Size of the contaminated region as a function of time ; the straight line has a slope 1/3, which is the prediction of Pomeau, Pumir, Young (11).

The same exponents have been obtained also for other values of the Peclet number. The only changes concern the duration of the anomalous diffusion regime and the width of the universal curves.

5 - Conclusion

Concerning the chaos experiments, it appears that the regimes of transition to weak turbulence are sensitive to the size of the system. For reasonably large lattices, the system has a complex behaviour, and several types of chaos are observed. It would be interesting to figure out the general laws which govern periodic system of large sizes.

Concerning dispersion experiments, our results are in excellent agreement with the theory of Pomeau, Pumir and Young (11) : anomalous diffusion exists and the exponents predicted by theory are observed. The duration of anomalous diffusion regime is shorter than that predicted by theory presumably on account of three-dimensionnal effects, which actually are, in our case, strongly inhibited by the vertical confinement; it would be interesting now to test the other aspects of the theory, concerning the scaling laws with the Peclet number, and the long time evolution of the system towards the normal diffusion regime.

References

(1) Arnol'd, V.I, Meshalkin, L.D., 1960 'The seminar of A.N. Kolmogorov on selected topics in analysis (1958-1959)', in *Usp. Mat. Nauk.* **15**, 247.

(2) Meshalkin, L.D., Sinai, Ya,G. 1961 'Investigation of the stability of the stationary solution of a system of plane motion equations of a viscous incompressible fluid'. *Prikl. Matem. Mekhan.*, **25**.

(3) Green, J.S.A. 1974, 'Two-dimensional turbulence near the viscous limit,' *J. Fluid. Mech,* **62**,273.

(4) Sivashinsky, G.I., 1983, 'Negative viscosity effect in large-scale turbulence ; long-wave instability of a periodic system of eddies', *Phys. Lett,* **95**A,3, 152.

(5) She, Z, 1987, 'Instabilités et dynamique à grande échelle en turbulence', Thèse, Univ. Paris7, Paris, France.

(6) D'Humières, D. Private communication.

(7) Bondarenko, N., Gak, M., Dolzhanskii, F., 1979 *Izv. Akad. Nauk SSSR Ser. Fiz. Atmosfer. i Okeana,* **15**, 1017.

(8) Sommeria, J., 1986, 'Experimental study of the two-dimensional inverse energy cascade in a square box', *J. Fluid. Mech.* **170**, 139.

(9) Nguyen Duc J.M., 1988, 'Instabilité et turbulence dans des écoulements bidimensionnels MHD', Thèse, INPG, Grenoble.

(10) Tabeling, P., Fauve, S., Perrin, B.,1987 'Instability of a linear array of forced vortices', *Europhys. Lett.*,**4**, 555-560

(11) Y. Pomeau, A. Pumir and W. Young, to appear in *Physics of Fluids*

STABILITY OF MAGNETOHYDRODYNAMIC FLOW OVER A STRETCHING SHEET

H.S. TAKHAR   AND   M. ALI
Dept. of Engg., Manchester Univ., U.K.

A.S. GUPTA
Mathematics Department,
Indian Institute of Technology,
Kharagpur, India.

ABSTRACT. The linear stability of two-dimensional flow of a viscous electrically conducting fluid permeated by a uniform transverse magnetic field over a flat porous deformable sheet is investigated when the sheet is stretched in its own plane with an outward velocity proportional to the distance from a point on it. Since the flow has curved streamlines as in a stagnation point flow, its stability is examined with respect to three-dimensional disturbances in the form of Taylor-Gortler vortices. Using a numerical method, the differential equations governing stability are integrated for various values of the suction(R) and magnetic(M) parameter. It is shown that the flow is stable for all values of R, M and non-zero values of the dimensionless wave number $\bar{a}$. Further for fixed $\bar{a}$ and R, the decay rate of the disturbances increases with M. This study is likely to be relevant to certain metallurgical processes where a magnetic field is used to control the rate of cooling of a stretching sheet by drawing it through a coolant.

INTRODUCTION

Flow of an incompressible viscous fluid over a stretching surface has an important bearing on several technological processes. For example in the extrusion of a polymer in a melt-spinning process, the extrudate from the die is generally drawn and simultaneously stretched into a sheet which is then solidified through cooling by direct contact with water. Further, the study of magnetohydrodynamic (MHD) flow of an electrically conducting fluid due to the deformation of the walls of the vessel containing this fluid is of interest in modern metallurgical and metal-working processes. In all these cases the properties of the final product depend to a great extent on the rate of cooling. Neglecting the induced magnetic field, Pavlov [1] gave an exact similarity solution of the steady two-dimensional MHD boundary layer equations for an incompressible fluid in the presence   of   a uniform transverse magnetic field, the flow being caused solely by the stretching of an elastic flat sheet in its own plane with a velocity varying linearly with distance from a fixed point. Chakrabarti and Gupta [2] found the temperature distribution in this flow when a uniform suction is applied

465

at the surface.

However the stability of the above flows does not seem to have received adequate attention despite its importance in the various technological processes mentioned above. This is due to the fact that unless the above flows are stable, products of desired physical characteristics cannot be achieved. The aim of this paper is to examine the linear stability of the MHD flow discussed in [2] with a view to studying the influence of the magnetic field on the stability characteristics.

STABILITY ANALYSIS

Let an infinite elastic surface y=0 experience in its plane a steady uniform stress which results in the movement of its points in the direction of x-axis with velocity Cx, where C > 0. The steady two-dimensional boundary layer flow in a conducting fluid occupying the half-space y > 0 and permeated by a uniform magnetic field B_o along y-axis (Figure 1) is governed by the equations

$$u_o \frac{\partial u_o}{\partial x} + v_o \frac{\partial u_o}{\partial y} = -\frac{1}{\rho} \frac{\partial p_o}{\partial x} + \nu \frac{\partial^2 u}{\partial y^2} - \frac{j B_o}{\rho}, \tag{1}$$

$$\frac{\partial u_o}{\partial x} + \frac{\partial v_o}{\partial y} = 0. \tag{2}$$

In writing (1), the magnetic Reynolds number is assumed small so that the induced magnetic field is neglected. The current density j along z-axis (normal to xy plane) is given by Ohm's law as $\sigma (E_z + u_o B_o)$, E_z being the electric field along z-axis. Since the pressure field in the boundary layer is obtained from that outside the layer, it follows from (1) and the fact that there is no flow outside the boundary layer

$$-\sigma E_z B_o - \frac{\partial p_o}{\partial z} = 0. \tag{3}$$

This gives from (1)

$$u_o \frac{\partial u_o}{\partial x} + v_o \frac{\partial u_o}{\partial y} = \nu \frac{\partial^2 u}{\partial y^2} - \frac{\sigma B_o^2}{\rho} u_o. \tag{4}$$

The similarity solution of (2) and (4) satisfying the boundary conditions

$$u_o = Cx, \quad v_o = -V_o \quad \text{at } y = 0; \quad u_o \to 0 \quad \text{as} \quad y \to \infty \tag{5}$$

was given in [2] as

$$u_o = Cx F'(\eta), \quad v_o = -(\nu C)^{1/2} F(\eta); \quad \eta = (C/\nu)^{1/2} y. \tag{6}$$

Here a prime denotes derivative with respect to η and

$$F(\eta) = A + B\,e^{-\alpha\eta}, \qquad \alpha = \frac{1}{2}\left[A + (A^2 + 4M^2)^{1/2}\right], \tag{7a}$$

$$A = \left[R(1+2M) + \left\{R^2 + 4(1+M)^{1/2}\right\}\right] \Big/ 2(1+M),$$

$$R = V_0/(\nu C)^{1/2}, \qquad M = \sigma B_0^2/\rho C, \tag{7b}$$

$$B = -2\Big/\left[R + \left\{R^2 + 4(1+M)\right\}^{1/2}\right], \tag{7c}$$

where R and M denote the suction Reynolds number and the magnetic parameter, respectively.

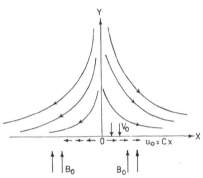

Figure 1: A sketch of the physical problem.

We shall now study the stability of the solution given by (6) and (7). Since the form of the solution is similar to that for a two-dimensional stagnation point flow, we examine following Görtler [3], the stability of this flow with respect to disturbances periodic in a direction normal to the plane of the basic flow. Hammerlein's [4] detailed study of the disturbance differential equations derived by Görtler suggests that instability can occur in the form of Taylor-Görtler vortices.

We take the perturbed state as

$$u = u_0 + \overline{u} = u_0 + C\,x\,u_1(\eta)\cdot(\cos\alpha z)\cdot e^{\beta t}, \tag{8a}$$

$$v = v_0 + \overline{v} = v_0 - (C\nu)^{1/2}\,v_1(\eta)\cdot(\cos\alpha z)\cdot e^{\beta t}, \tag{8b}$$

$$w = \overline{w} = \nu\alpha\,w_1(\eta)\cdot(\sin\alpha z)\cdot e^{\beta t}, \tag{8c}$$

$$p = p_0 + \overline{p} = p_0 + \rho\nu C\,p_1(\eta)\cdot(\cos\alpha z)\cdot e^{\beta t}, \tag{8d}$$

where u_0 and v_0 are given by (6), α is the wave number in the spanwise direction and β is the growth (or decay) rate. The linearized three-dimensional perturbation equations are given by

$$\partial \bar{u}/\partial x \; + \; \partial \bar{v}/\partial y \; + \; \partial \bar{w}/\partial z \; = \; 0 \tag{9}$$

$$\partial \bar{u}/\partial t \; + \; u_0 \, \partial \bar{u}/\partial x \; + \; \bar{u} \, \partial u_0/\partial x \; + \; v_0 \, \partial \bar{u}/\partial y \; + \; \bar{v} \, \partial u_0/\partial y$$
$$= \; - \left(1/\rho \right) \partial \bar{p}/\partial x \; + \; \nu \, \nabla^2 \bar{u} \; - \; \left(\sigma B_0^2/\rho \right) \bar{u}, \tag{10}$$

$$\partial \bar{v}/\partial t \; + \; u_0 \, \partial \bar{v}/\partial x \; + \; \bar{u} \, \partial v_0/\partial x \; + \; v_0 \, \partial \bar{v}/\partial y \; + \; \bar{v} \, \partial v_0/\partial y$$
$$= \; - \left(1/\rho \right) \partial \bar{p}/\partial y \; + \; \nu \, \nabla^2 \bar{v}, \tag{11}$$

$$\partial \bar{w}/\partial t \; + \; u_0 \, \partial \bar{w}/\partial x \; + \; v_0 \, \partial \bar{w}/\partial y \; = \; - \left(1/\rho \right) \partial \bar{p}/\partial z \; +$$
$$+ \; \nu \, \nabla^2 \bar{w} \; - \; \left(\sigma B_0^2/\rho \right) \bar{w}, \tag{12}$$

where ∇^2 is the three-dimensional Laplace operator. In writing these equations, we neglect the electric field due to polarization of charges. This is similar to the assumption made by Pavlov [1] who took the electric field to be zero (corresponding to current return through a short-circuit) even for the undisturbed flow (u_0, v_0). Substitution of (8) in (9)-(12) gives

$$u_1 - v_1' + \bar{\alpha}^2 w_1 = 0, \tag{13}$$

$$u_1'' + F u_1' - \left(\bar{\beta} + \bar{\alpha}^2 + M + 2 F' \right) u_1 = F'' v_1, \tag{14}$$

$$v_1'' + F v_1' - \left(\bar{\beta} + \bar{\alpha}^2 - F' \right) v_1 = - p_1', \tag{15}$$

$$w_1'' + F w_1' - \left(\bar{\beta} + \bar{\alpha}^2 + M \right) w_1 = - p_1, \tag{16}$$

$$\bar{\alpha}^2 = \nu \alpha^2/c, \qquad \bar{\beta} = \beta/c. \tag{17}$$

The no-slip boundary conditions are

$$u_1 = v_1 = w_1 = 0 \qquad \text{at} \quad \eta = 0 \tag{18}$$

and since the perturbations vanish at infinity, we have

$$u_1 = v_1 = w_1 = 0 \qquad \text{as} \quad \eta \to \infty. \tag{19}$$

Elimination of w_1 and p_1 from (13), (15) and (16) gives on using (14)

$$v_1^{IV} + F v_1''' + \left(F' - \bar{\beta} - 2 \bar{\alpha}^2 - M \right) v_1'' + \left(F'' - \bar{\alpha}^2 F \right) v_1' +$$
$$+ \left[\bar{\alpha}^2 \left(\bar{\beta} + \bar{\alpha}^2 - F' \right) + F''' \right] v_1 = 2 F' u_1' + 2 F'' u_1. \tag{20}$$

The boundary conditions (18) and (19) become on using (13)

$$u_1 = v_1 = v_1' = 0 \qquad \text{at} \quad \eta = 0 \quad \text{and} \quad \eta \to \infty. \tag{21}$$

Eqns.(14) and (20) subject to (21) constitute the eigenvalue problem for stability. For given R, M and $\bar{\alpha}$, we are required to calculate $\bar{\beta}$ from this eigenvalue problem.

The method due to Harris and Reid [5] is employed to solve the above system. Since (5) and (6) demand $F'(\infty) = 0$, in our calculations $\eta = \infty$ is replaced by $\eta = \delta$ where $|F'(\delta)| \leq 10^{-4}$. Putting

$$u_1 = \bar{A}, \quad u_1' = \bar{B}, \quad v_1 = \bar{C}, \quad v_1' = \bar{D}, \quad v_1'' = \bar{E}, \quad v_1''' = \bar{G}, \tag{22}$$

we write (14) and (20) as a system of first-order equations as

$$d\bar{A}/d\eta = \bar{B}, \quad d\bar{B}/d\eta = -F\bar{B} + (\bar{\beta} + \bar{\alpha}^2 + M + 2F')\bar{A} - F''\bar{C}, \tag{23}$$

$$d\bar{C}/d\eta = \bar{D}, \quad d\bar{D}/d\eta = \bar{E}, \quad d\bar{E}/d\eta = \bar{G}, \tag{24}$$

$$d\bar{G}/d\eta = -F\bar{G} - (F' - \bar{\beta} - 2\bar{\alpha}^2 - M)\,E - (F'' - \bar{\alpha}^2 F)\,\bar{D} -$$
$$- [\bar{\alpha}^2 (\bar{\beta} + \bar{\alpha}^2 - F') + F''']\,\bar{C} + 2F'\bar{B} + 2F''\bar{A}. \tag{25}$$

The boundary conditions (21) become

$$\bar{A} = \bar{C} = \bar{D} = 0 \quad \text{at } \eta = 0 \text{ and } \delta. \tag{26}$$

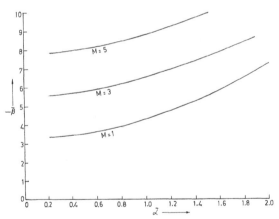

Figure 2. Variation of growth rate with M for R=2

We define three linearly independent solutions $\bar{A}_i$ and $\bar{C}_i$ (i=1,2,3) of (14) and (20) by imposing the initial conditions

$$\bar{A}_i = \bar{C}_i = d\bar{C}_i/d\eta = 0 \qquad (i=1,2,3) \tag{27}$$

$$[\,d\bar{A}_i/d\eta, \ d^2\bar{C}_i/d\eta^2, \ d^3\bar{C}_i/d\eta^3\,] = \begin{cases} (1, 0, 0), & i=1 \\ (0, 1, 0), & i=2 \\ (0, 0, 1), & i=3 \end{cases} \tag{28}$$

at $\eta = 0$ A solution of (14) and (20) satisfying (27) at $\eta = 0$ can therefore be written as

$$u_1 = \sum_{i=1}^{3} P_i \overline{A}_i , \qquad v_1 = \sum_{i=1}^{3} Q_i \overline{C}_i , \tag{29}$$

where P_i and Q_i are constants. The boundary conditions (21) at infinity are

$$\sum_{i=1}^{3} P_i \overline{A}_i = 0, \quad \sum_{i=1}^{3} Q_i \overline{C}_i = 0, \quad \sum_{i=1}^{3} Q_i \frac{d\overline{C}_i}{d\eta} = 0 \quad \text{at } \eta = \delta. \tag{30}$$

If P_i and Q_i do not vanish identically, we have upon introducing $\overline{B}_i = \overline{A}'_i$, $\overline{D}_i = \overline{C}'_i$, $\overline{E}_i = \overline{D}'_i$ and $\overline{G}_i = \overline{E}'_i$, the following relation

$$\begin{vmatrix} \overline{A}_1 & \overline{A}_2 & \overline{A}_3 \\ \overline{C}_1 & \overline{C}_2 & \overline{C}_3 \\ \overline{D}_1 & \overline{D}_2 & \overline{D}_3 \end{vmatrix} = 0 \quad \text{at } \eta = \delta. \tag{31}$$

This is the characteristic equation from which $\overline{\beta}$ can be determined for given values of R,M and $\overline{\alpha}$ by using the bisection method as follows :
(i) Start with a large value of $\overline{\beta}$ and integrate the eqns.(23)-(25) numerically using the Runge-Kutta-Merson method. This involves the integration of 18 equations for $\overline{A}_i, \overline{B}_i, \overline{C}_i, \overline{D}_i, \overline{E}_i$ and $\overline{G}_i$ (i=1,2,3) using the initial conditions (27) and (28).
(ii) We then evaluate the determinant in (31).
(iii) Growth rate $\overline{\beta}$ is then incremented by a given amount $\Delta\overline{\beta}$ and the determinant in (31) is again evaluated.
(iv) We continue incrementing $\Delta\overline{\beta}$ till the determinant changes sign. At this stage $\Delta\overline{\beta}$ is halved and the procedure is repeated until $|\Delta\overline{\beta}| < 10^{-5}$.

RESULTS AND DISCUSSION

The growth rate $\overline{\beta}$ is computed for R=0,1,2,...10 and M=0,1,2,...7 for various non-zero values of $\overline{\alpha}$. It is found that $\overline{\beta} < 0$ for all values of $\overline{\alpha}$ ($\neq 0$), R and , M which implies that the flow is stable in the linear theory. Further $\overline{\beta}$ increases with increasing R for given $\overline{\alpha}$ and M and this shows that suction exerts a stabilizing influence on the flow. Physically suction has a thinning effect on the boundary layer which becomes less prone to instability.

Figure 2 shows the plot of $-\overline{\beta}$ versus $\overline{\alpha}$ for M=1,3 and 5 with R=2. It is found that the decay rate of the disturbance increases with increase in the magnetic field which therefore has a stabilizing influence on the flow. While this is not unexpected, the interesting aspect of the analysis lies in the fact that the flow is stable (in the linear theory) even in the absence of the magnetic field. This may be explained physically as follows. In such a flow there are certain regions where the circulation (product of the local velocity and the local radius of curvature of the streamline of the basic flow) decreases as the local centre of curvature is approached normal to the streamlines. This stems from the fact that the fluid velocity increases as the sheet is approached due to the stretching of the sheet. Thus Rayleigh's criterion [6] suggests stability.

REFERENCES

[1] PAVLOV, K.B., 1974, 'Magnetohydrodynamic flow of an incompressible viscous fluid caused by the deformation of a plane surface', Magnit.Gidro., Vol.4, 146.

[2] CHAKRABARTI, A. and GUPTA, A.S., 1979, 'Hydromagnetic flow and heat transfer over a stretching sheet', Quart.Appl.Math., Vol.37, 73.

[3] GORTLER, H., 1955, 50 Jahre Grenzschichtforschung, Friedr. Vieweg and Sohn, Braunschweig, 304.

[4] HAMMERLEIN, G., 1955, 50 Jahre Grenzschichtforschung, Friedr. Vieweg and Sohn, Braunschweig, 315.

[5] HARRIS, D.L. and REID, W.H., 'On the stability of viscous flow between rotating cylinders. Part 2. Numerical analysis', J. Fluid Mech., Vol.20, 95.

[6] CHANDRASEKHAR, S., 1961, Hydrodynamic and Hydromagnetic Stability, Oxford University Press, 273.

STABILITY OF CLOSED AZIMUTHAL JET OF LIQUID METAL

K.A. SERGEEV and V.N. SHTERN
Institute of Thermophysics of the Siberian Branch
of the USSR Academy of Sciences
630090 Novosibirsk
pr. Academician Lavrentyev, 1
USSR

ABSTRACT. Stability of a closed azimuthal jet of liquid metal is investigated. Structures appearing at the loss of stability by this flow, are investigated on the basis of amplitude equations. The latter are obtained by applying the Galerkin procedure to equations of viscous fluid movement. In the above procedure, eigenfunction of two different types of the linear theory disturbances are used as basic functions.

I. Introduction

Free rotating layer is an important element in flowing liquid-metal paths of the fusion reactor blanket, in a centrifugal conductive MHD-pump and other MHD facilities. Modelling it by closed azimuthal jet of liquid metal allows to investigate many important magnetic hydrodynamic properties in the above technical applications in a number of cases. Investigation of hydrodynamic stability of these flows is an inportant problem since it leads to practical recommendations concerning their stabilization.

Moreover the problem also has in independent significance for describing turbulence and understanding its nature. The choice of a closed azimuthal jet of liquid metal as an object of investigating instability and dynamics of laminar-turbulent transition is attractive due to comparitive simplicity of the model. First, the transverse magnetic field used as a tool, allows to obtain a two-dimensional flow structure in the plane perpendicular to it. Secondly, both the averaged flow and the disturbances generated by the flow become two-dimensional at sufficiently strong magnetic field, i.e. a flow with a two-dimensional turbulence structure occurs. Thirdly, the presence of inflection points in the velocity profile, which is characteristic for such flows of the jet type, promotes decrease of critical parameter values. And finally, the next advantage is the absence of introduced disturbances due to the jet closure.

As for the two-dimensional turbulence, it should be noted that the study of its physical laws presents a topical problem both for the development of the general turbulence theory and for the meteorology due to its application significance. Since two-dimensional coherent vortex structures appearing after the loss of the azimuthal jet stability, are similar to the chain of vortices observed over the Antartic, the investigation of the supercritical regime of a liquid metal jet is directly related to the dynamics of large-scale atmospheric phenomena, with the difference that these phenomena are caused by mass forces of different nature.

2. Problem definition

Closed azimuthal jet of liquid metal results from interaction of the transverse magnetic field with the radial component of the electric current supplied by means of the annular electrodes (Fig. 1), where r, z,

J. Lielpeteris and R. Moreau (eds.), Liquid Metal Magnetohydrodynamics, 473–479.
© *1989 by Kluwer Academic Publishers.*

φ are cylindrical coordinates ; h is the distance between the isolated planes ; **B** is the outer magnetic field, I is the total electric current. The flow has Hartmann's boundary layers near the isolated planes and free shear layers, which are parallel to the magnetic field.

The magnetic Reynolds number is small in technical applications, i.e. the magnetic field role is essentially reduced to outer driving effect. Thus at analyzing stability of a liquid-metal jet, main attention will be paid to the hydrodynamic aspect of the problem.

Since the axial magnetic field suppresses Taylor's instability more strongly than Kelvin-Helmholtz instability, the latter is more dangerous for the flow considered in this paper [1].

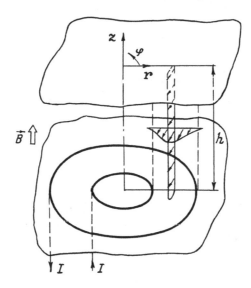

Figure 1 - Diagram of flow

A core, that is practically independent of the coordinate along the field, is formed in the central region of the flow as the magnetic field value increases. The potential difference induced by motion in this core, balances the one applied from the outside. In this case, the electric current is closed through the free shear layers parallel to the field and Hartmann boundary layers. The governing magnetic field influence on the flow proceeds just through the latter, and the main losses of mechanical energy due to Joule's and viscous dissipation occur just in them. I is suggested in [2] to take Hartmann boundary layers into account by averaging the stability equations with respect to the z-coordinate. The viscous term will only change in this case. Since it is linear, the original equation describing our model in terms of the stream function ψ, may be written in an analogous way without loss of generality. Its dimensionless form is :

$$\frac{\partial}{\partial t}\Delta_1\psi + \frac{1}{r}\left[\frac{\partial\psi}{\partial r}\frac{\partial}{\partial\varphi}(\Delta_1\psi) - \frac{\partial\psi}{\partial\varphi}\frac{\partial}{\partial r}(\Delta_1\psi)\right]$$

$$= \frac{1}{Re}\Delta_1\Delta_1\psi - \frac{M}{Re}\Delta_1\psi - \int_0^h Al\ rot_z\ [\mathbf{j}\times\mathbf{B}]dz,$$

$$\Delta_1\psi \equiv \Delta_2\frac{\partial\psi}{\partial r} + \frac{1}{r^2}\frac{\partial^2\psi}{\partial\varphi^2},\ \ \Delta_2(...) \equiv \frac{1}{r}\frac{\partial}{\partial r}(r...),\qquad(1)$$

here Re is the Reynolds number ; Al is the Alfen number ; j is the electric current density. The distance between the outer edge of the outer electrode and the inner edge of the inner electrode is chosen as scale of the length $b_0 = r_4 - r_1$. Complex b_0I [$\sqrt{\sigma\eta}\pi$ $(r_2^2 - r_1^2)$] is the velocity scale, where σ is electrical conductivity, η is dynamic visocisty, r_2 is the radius of the outer edge of the inner electrode. The stream function is determined from the following relations

$$V_r = -\frac{1}{r}\frac{\partial\psi}{\partial\varphi}, \quad V_\varphi = \frac{\partial\psi}{\partial r},$$

where V_r and V_φ are velocity components in the cylindrical coordinate system.

Taking the Hartmann boundary layers into account results in the appearance of the term - (M/Re) $\Delta_1\psi$ in the right part of Eq. (1), where $M = 2Ha/h$ is a modified Hartmann number. It may be considered as an effective friction force on the boundaries perpendicular to the magnetic field.

3. The linear theory

Linear stability of the azimuthal jet of conductive liquid in the axial magnetic field has been studied in the paper [3]. Solution for disturbances was searched for in the form of harmonic oscillations

$$\widetilde{\psi}\,(r)\exp\,[i(m\varphi - \omega t)],$$

where $\widetilde{\psi}$ is the eigenfunction, m is the azimuthal wave number, and $\omega = \omega_r + i\omega_i$ is the complex frequency. Since the calculated laminar profile of the flow considered is assymetric about the centre of the interelectrode interval, two critical points occur. This is expected to result in the appearance of two types of disturbances, namely : the disturbance with the maximal intensity near the electrode, that is close to the axis (the first type), and the disturbance with the maximal intensity near the distant electrode (the second type). The results of the neutral curve calculation for the flow at the Hartmann number Ha = 421 and the channel hight h = 7, are given in Fig. 2. The curve 1 corresponds to the

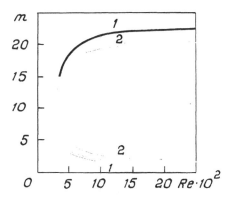

Figure 2 - Neutral curves

disturbances of the second type, and the curve 2 - the first type. The parameters of the "nose" of the neutral curve 1 have the values : $m = 9$, $Re = 252$, $\omega_r = = 1.028$, and those of the curve 2 have the values : $m = 11$, $Re = 304.1$, $\omega_r = 1.851$. The neutral curves have the form, that is characteristic of jet flows. Note that these curves possess the physical sense only in the points corresponding to the integer values of the wave number m.

Comparison of the numerical calculation results with the experimental ones has shown that the critical Reynolds numbers appear to be two times less than the experimental ones. This is related to the fact that according to [4], the dependance of disturbances on z differs from the Hartmann one accepted in [3]. Thus the effective friction force has appeared to be lowered.

4. The dynamic system

Having chosen two the most dangerous types of disturbances by means of the linear theory, and using them as basic functions in Galerkin's method, it is possible to construct an approximate nonlinear theory taking the interaction of the developing disturbances with the main flow into account, and to calculate the stationnary values of the disturbance amplitudes.

Let the stream function be represented in the following form

$$\psi(r, \varphi ; t) = \psi_0(r) + \psi_g(r ; t) + \psi'(r, \varphi ; t) , \qquad (2)$$

where ψ_0 is the stream function the laminar flow U_0, ψ_g is an axisymmetric part of the stream function of the pulsations appearing due to the Reynolds stresses, ψ' is the nonaxisymmetric component of stream function. Substituting (2) into Eq. (1) and subtracting the laminar flow equation from the obtained expression, we obtain

$$\frac{\partial}{\partial t}(\Delta_2 U_g) + \frac{\partial}{\partial t}(\Delta_1 \psi') + \frac{1}{r}[U_0 \frac{\partial}{\partial \varphi}(\Delta_1 \psi') + U_g \frac{\partial}{\partial \varphi}(\Delta_1 \psi') +$$

$$\frac{\partial \psi'}{\partial r} \frac{\partial}{\partial \varphi}(\Delta_1 \psi') - \frac{\partial \psi'}{\partial \varphi} \frac{\partial}{\partial r}(\Delta_2 U_0) - \frac{\partial \psi'}{\partial \varphi} \frac{\partial}{\partial r}(\Delta_2 U_g) - \frac{\partial \psi'}{\partial \varphi} \frac{\partial}{\partial r}(\Delta_1 \psi')]$$

$$= \frac{1}{Re}(\Delta_1 - M)\Delta_2 U_g + \frac{1}{Re}(\Delta_1 - M)\Delta_1 \psi' ; \qquad (3)$$

$$U_g = \frac{\partial \psi_g}{\partial r} .$$

Averaging Eq. (3) with respect to the homogeneous φ-coordinate, we obtain Reynolds equation

$$\Delta_2(\frac{\partial U_g}{\partial t}) + \frac{1}{r}[\int_0^{2\pi} \frac{\partial \psi'}{\partial r} \frac{\partial}{\partial \varphi}(\Delta_1 \psi') d\varphi - \int_0^{2\pi} \frac{\partial \psi'}{\partial \varphi} \frac{\partial}{\partial r}(\Delta_1 \psi') d\varphi]$$

$$= \frac{1}{Re} \Delta_2 [\frac{\partial}{\partial r}(\Delta_2 U_g)] - \frac{M}{Re} \Delta_2 U_g . \qquad (4)$$

Substracting Eq. (4) from Eq. (3), we obtain an equation for pulsations

$$\frac{\partial}{\partial t}(\Delta_1 \psi') + \frac{1}{r}\left[(U_0 + U_g)\frac{\partial}{\partial \varphi}(\Delta_1 \psi') + \frac{\partial \psi'}{\partial r}\frac{\partial}{\partial \varphi}(\Delta_1 \psi') - \right.$$

$$-\frac{\partial \psi'}{\partial \varphi}\frac{\partial}{\partial r}\Delta_2(U_0 + U_g) - \frac{\partial \psi'}{\partial \varphi}\frac{\partial}{\partial r}(\Delta_1 \psi') - \int_0^{2\pi}\frac{\partial \psi'}{\partial r}\frac{\partial}{\partial \varphi}(\Delta_1 \psi')d\varphi + \tag{5}$$

$$\left.+\int_0^{2\pi}\frac{\partial \psi'}{\partial \varphi}\frac{\partial}{\partial r}(\Delta_1 \psi')\, d\varphi\right] = \frac{1}{Re}(\Delta_1 - M)\,\Delta_1 \psi'$$

Let us consider the supercriticalilty being small and the disturbance form coinciding with that calculated by the linear theory. Then the following approximation of the current function pulsations

$$\psi' = A_1(t)\,\psi_1(r)\,e^{im_1\varphi} + A_2(t)\,\psi_2(r)\,e^{im_2\varphi} +$$

$$A_1^*\,\psi_1^*\,e^{-im_1\varphi} + A_2^*\,\psi_2^*\,e^{-im_2\varphi} \tag{6}$$

can be written, where the index 1 denotes the disturbances of the first type, and the index 2 denotes the disturbances of the second type. Substituting the latter into Eq. (4), we find corrections to the laminar velocity profile at a given moment of time. Then the average profile is

$$U = U_0(r) + A_3(t)\,U_{g1}(r) + A_4(t)\,U_{g2}(r) + A_5(t)\,U_{g3}(r) + A_5^*U_{g3}^* \ . \tag{7}$$

The most interesting cases are : when (a) the wave member used in (6) coincide ; (b) $m_1 = 2\,m_2$ or $2m_1 = m_2$. Here the case $m_1 = m_2$ is considered. Strong interaction both of the disturbance with the main flow and between them, occurs in this case.

Application of the Galerkin procedure to Eqs. (4-5) using (6) as basic functions and taking (7) into account, results in the following dynamic equation system

$$A'_i = -L_{ij}A_j - B_{ijk}\,A_j\,A_k \qquad i, j, k = 1,5$$

which is a system of the hydrodynamic type.

5. Calculation of self-oscillations

Figure 3 shows the dependence of the pulsation intensity ε (the ratio of the maximal velocity pulsation value to the maximal flow velocity) on the Reynolds number at Ha = 421 and h = 7. The curve 1 corresponds to the radial velocity pulsation and the curve 2 corresponds to the azimuthal one. It is seen that the azimuthal pulsation intensity is higher. The dashed line designates the dependences in the region, where the primary self-oscillations lose stability. A two-periodical regime is generated in the point of self-oscillation stability loss. The values of the parameters and amplitudes in the bifurcartion point are the following : $A_3 = 0.14$, $A_4 = 0.88$, $A_5 = (-0.07, 0.35)$, $A_1 = (-0.004, -0.018)$, $A_2 = (0.045, 0.)$, $\omega_r = 1.04$, Re = 442. The isobars and the average flow velocity profile corresponding to them, are given in Fig. 4.

478

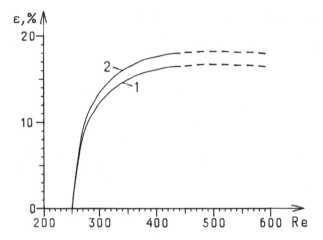

Figure 3 - Pulsation intensity dependance on the Reynolds number.

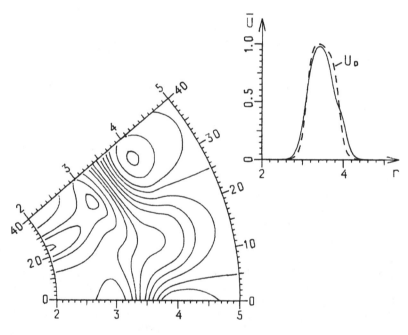

Figure 4. The isobars and the average flow velocity profile in the bifucartion point.

6. Conclusion

Stationary self-oscillations have been calculated in the paper. It has been also found out that bifucartion occurs at some supercritical Reynolds number. Further investigations are needed to demonstrate chaotic regime developing.

REFERENCES

[1] LEVIN V.B., 1980, 'Free rotating layer of conducting fluid in axial magnetic field', Magnetohydrodyn., vol. 1, 86.

[2] LEVIN V.B. & SHTERN V.N., 1985, 'Stability of jet MHD flow between insulating plates in transversal magnetic field', Magnetohydrodyn., vol. 2, 23.

[3] SERGEEV K.A. & SHTERN V.N., 1987, 'Stability of azimuthal jet flow of a conducting fluid in an axial magnetic field', Magnetohydrodyn., vol. 3, 39.

[4] KLYUKIN A. A., KOLESNIKOV YU. B., LEVIN V. B., 1980, 'Experimental study of free rotating layer in axial magnetic field. Part 2. Stability limits and perturbation structure', Magnetohydrodyn., vol. 1, 140.

MECHANICS OF FLUIDS AND TRANSPORT PROCESSES

Editors: R. J. Moreau and G. Æ. Oravas

1. J. Happel and H. Brenner, Low Reynolds Number Hydrodynamics. 1983.
   ISBN 90–247–2877–0
2. S. Zahorski, Mechanics of Viscoelastic Fluids. 1982.
   ISBN 90–247–2687–5
3. J. A. Sparenberg, Elements of Hydrodynamic Propulsion. 1984.
   ISBN 90–247–2871–1
4. B. K. Shivamoggi, Theoretical Fluid Dynamics. 1984.
   ISBN 90–247–2999–8
5. R. Timman, A. J. Hermans and G. C. Hsiao, Water Waves and Ship
   Hydrodynamics: An Introduction. 1985.
   ISBN 90–247–3218–2
6. M. Lesieur, Turbulence in Fluids. 1987.
   ISBN 90–247–3470–3
7. L. A. Lliboutry, Very Slow Flows of Solids. 1987.
   ISBN 90–247–3482–7
8. B. K. Shivamoggi, Introduction to Nonlinear Fluid-Plasma Waves. 1988.
   ISBN 90–247–3662–5
9. V. Bojarevičs, Ya. Freibergs, E. I. Shilova and E. V. Shcherbinin, Electrically
   Induced Vortical Flows. 1989.
   ISBN 90–247–3712–5